EARTH SYSTEM EVOLUTION

EarthComm®

EARTH SYSTEM SCIENCE IN THE COMMUNITY

Michael J. Smith Ph.D.
Principal Investigator

John B. Southard Ph.D.
Senior Writer

Ruta Demery
Editor

Emily Crum
Contributing Writer

Developed by the American Geological Institute
Supported by the National Science Foundation and
the American Geological Institute-Foundation

Published by
It's About Time, Inc., Armonk, NY

It's About Time, Inc.
84 Business Park Drive, Armonk, NY 10504
Phone (914) 273-2233 Fax (914) 273-2227
Toll Free (888) 698-TIME
www.Its-About-Time.com

Publisher
Laurie Kreindler

Project Manager	**Project Coordinator**	**Design**
Ruta Demery	Matthew Smith	John Nordland
Editor	**Contributing Writer**	**Production Manager**
Ena de Jong	Emily Crum	Joan Lee

Studio Manager
Jon Voss

All student activities in this textbook have been designed to be as safe as possible, and have been reviewed by professionals specifically for that purpose. As well, appropriate warnings concerning potential safety hazards are included where applicable to particular activities. However, responsibility for safety remains with the student, the classroom teacher, the school principals, and the school board.

EarthComm® is a registered trademark of the American Geological Institute. Registered names and trademarks, etc., used in this publication, even without specific indication thereof, are not to be considered unprotected by law.

It's About Time® is a registered trademark of It's About Time, Inc. Registered names and trademarks, etc., used in this publication, even without specific indication thereof, are not to be considered unprotected by law.

Printed and bound in the United States of America

ISBN #1-58591-069-4

2 3 4 5 QC 06 05 04 03

This project was supported, in part, by the
National Science Foundation (grant no. ESI-9452789)

Opinions expressed are those of the authors and not necessarily those of the National Science
Foundation or the donors of the American Geological Institute Foundation.

Student's Edition Illustrations and Photos

E175, photo by Cody Arenz, Nebraska Wesleyan University

E10, E11, E20, E22, E24, E29, E30, E31, E33, E34 Fig. 4, E50, E52, E59, E64, E71, E75 Fig. 2, E100, E103, E106, E107, E108, E109, E111, E113, E114, E118, E119, E121, E129, E130, E131, E140 Fig. 2, E149, E150, E151 Fig. 1, E184, illustrations by Stuart Armstrong

E85, illustration by Stuart Armstrong, source: Köppen Climate Map

E97 (top and bottom right), illustration by Stuart Armstrong; source: R. S. Bradley and J. A. Eddy based on J. T. Houghton et al., Climate Change: The IPCC Assessment, Cambridge University Press, Cambridge, 1990 and published in EarthQuest, vol. 5, no. 1, 1991.

E100 Fig. 1, by Stuart Armstrong, source: redrafted from Muller and MacDonald, Ice Ages and Astronomical Causes, Springer-Praxis, 2000

E101 Fig. 3, E132, E169, illustrations by Stuart Armstrong, source: Jouzel, J.C., et al, 1987, Nature 329:403-8; Jouzel, J.C., et al, 1993, Nature 364:407-12; Jouzel, J.C., et al, 1996, Climate Dynamics 12:513-521; and Jouzel, J.C., et al, 1999, Nature 399: 429-436

E114, illustration by Stuart Armstrong, source: Skinner and Porter, The Dynamic Planet, John Wiley & Sons, 2nd Edition, February 2000

E126 Part A Table, illustration by Stuart Armstrong, source: Global Historical Climatology Network 1701-07/2000 (meteorological stations only) and Hansen, J., Sato, M., Lacis, A., Ruedy, R., Tegen, I. & Mathews, E. (1998) Climate forcings in the Industrial era, Proc. Natl. Acad. Sci. USA, 95, 12753-12758.

E126 Part B graph, illustration by Stuart Armstrong, source: Methane data: Chappellaz et al, Nature, v. 345, p. 127-131, 1990; CO_2 data: Barnola et al., Nature, v. 329, p. 408-414, 1987; Change in temperature data: Jouzel et al, Climate Dynamics, 12:8, p. 513-521, 1996.

E139, redrafted by Stuart Armstrong from a diagram by John Woolsey of Woolsey Studio, Boston, source: Stanley Chernicoff

E159, illustration by Stuart Armstrong, source: Tom Crowley and Consequences Magazine E161

E166, illustration by Stuart Armstrong, source: Gastaldo, Savrda, Lewis. Deciphering Earth History. Page 13-11.

E66, Figs. 4, 5, Dana Berry, Space Telescope Science Institute

E4, E14, E28, E37, E47, E58, E69, E105, E117, E125, E136, E148, E156, E165, E173, E182, illustrations by Tomas Bunk

E176 Tyrannosaurs Rex, University of California Museum of Paleontology

E51, photo by Digital Stock Corporation Royalty Free Images: Backgrounds from Nature

E18, E43, E54, photos by Digital Stock Corporation Royalty Free Images: Space Exploration

E9, E19, photos by Digital Vision Royalty Free Images: Astronomy and Space

E157 desert, photo by Digital Vision Royalty Free Images: North American Landscapes

E161 Fig. 2, photo by Digital Vision Royalty Free Images: American Highlights

E75 Fig. 4, E76 Figs. 5A, 5C, photo by European Southern Observatory

E176, Archaeohippus, Florida Museum of Natural History

E176 Wolf skull, Bill Forbes, Biology Department, Indiana University of Pennsylvania

E180, image courtesy of Geological Survey of Canada

E76 Fig. 5B, photo by Pat Harrington at University of Maryland

E77, photo by Jeff Hester and Paul Scowen, Arizona State University, NASA, and National Space Science Data Center

E34 Fig. 5, International Astronomical Union

E97 (top left), photo courtesy of Laboratory of Tree Ring Research, University of Tennessee

E92, E102 Fig. 4, E140 Fig. 3, E152 Fig. 3, E157 grassland, taiga, photos by Bruce Molnia

E187 Fig. 5, photo courtesy of the Morton Arboretum, Lisle, Illinois.

E7, photo by NASA and European Space Agency, A. Dupree and Ronald Gilliand

E8, E13, photos by NASA and the Hubble Heritage Team

E41, Photo by NASA and National Space Science Data Center

E42 Table 1, E44 Table 2, data from NASA

E45, E65 Fig. 3, photos by NASA

E53, photo by NASA Goddard Space Flight Center

E83 (top right), image by NASA

E2, E3, E25-27, E32, E40, E55, E68, E84, E96, E82 (left, bottom), E110, E122 Fig. 4, E135, E146, E147, E157 tundra, chaparral, mountain zones, E160, E161 Fig. 1, E163, E171, by PhotoDisc

E187 Fig. 4, from a painting by Alice Prickett and published in black and white in Phillips and Cross (1991, pl. 4) Phillips, T.L., and Cross, A.T., 1991, Paleobotany and paleoecology of coal, in Gluskoter, H.J., et al., eds., Economic Geology: U.S.: Boulder, Colorado, Geological Society of America, Geology of North America, v. P-2, p. 483-502.

E168, photos courtesy of F.W. Potter and D.L. Dilcher, Florida Museum of Natural History

E116, photo courtesy of T. Rimmele, M. Hanna/NOAO/AURA/NSF

E93, E133, E151 Fig. 2, E152 Fig. 4, E154, E158 tropical rainforest, temperate evergreen forest, temperate deciduous forest, E162, E178, photos by Doug Sherman, Geo File Photography

E141, photo by Mike Smith

E120, photo courtesy of Walter Smith, NASA and David Sandwell, Scripps Institution of Oceanography

E179, E185, image courtesy of Smithsonian Institution

E186, photo courtesy of Smithsonian Institution

E176 Smilodon Californicus and Diplodocus, photos courtesy of Smithsonian Institution

E73, photo by Susan Tereby, Extrasolar Research Group, NASA, and National Space Science Data Center

E100 Fig. 2, E122 Fig. 3, E158 polar ice, photos by Mark Twickler, Institute for the Study of Earth, Oceans and Space, University of New Hampshire

E65 Fig. 2, photo by Dave Westpfahl, New Mexico Tech University, and Dave Finely, National Radio Astronomy Observatory

E23, photo by Jerome Wycoff

E75 Fig. 3, photo by Hui Yang, University of Illinois and NASA

E60, table modified with permission from "Scaling the Spectrum," courtesy of Donna Young, Tufts University

E123, photo by Barbara Zahm

Taking Full Advantage of *EarthComm*
Through Professional Development

Implementing a new curriculum is challenging. That is why It's About Time Publishing has partnered with the American Geological Institute, developers of *EarthComm*, to provide a full range of professional development services. The sessions described below were designed to help you deepen your understanding of the content, pedagogy, and assessment strategies outlined in this Teacher's Edition, and adapt the program to suit the needs of your students and your local and state standards and curriculum frameworks.

Professional Development Services Available

Implementation Workshops

Two- to five-day sessions held at your site that prepare you to implement the inquiry, systems, and community-based approach to learning Earth Science featured in *EarthComm*. These workshops can be tailored to serve the needs of your school district, with chapters selected from the modules based on local or state curricula and framework criteria.

Program Overviews

One- to three-day introductory sessions that provide a complete overview of the content and pedagogy of the *EarthComm* program, as well as hands-on experience with activities from specific chapters. Program overviews are designed in consultation with school districts, counties, and SSI organizations.

Regional New-User Summer Institutes

Two- to five-day sessions that are designed to deepen your Earth Science content knowledge, and to prepare you to teach through inquiry. Guidance is provided in the gathering and use of appropriate materials and

resources and specific attention is directed to the assessment of student learning.

Leadership Institutes

Six-day summer sessions supported by the American Geological Institute that are designed to prepare current users for professional development leadership and mentoring within their districts.

Follow-up Workshops

One- to two-day sessions that provide additional Earth Science content and pedagogy support to teachers using the program. These workshops focus on identifying and solving practical issues and challenges to implementing an inquiry-based program.

Mentoring Visits

One-day visits that can be tailored to your specific needs that include class visits, mentoring users of the program, and in-service sessions.

Please fill in the form below to receive more information about participating in one of these Professional Development Services. The form can be directly faxed to our Professional Development at 914-273-2227. Our department will contact you to discuss further details and fees.

District/School: _____ Phone: _____

Address: _____

Contact Name: _____ Title: _____

E-mail: _____ Fax: _____

School Enrollment: _____ Number of Students Impacted: _____ Grade Level: _____

Have you purchased the following: ❏ Student Editions ❏ Teacher Editions ❏ Kits

Briefly explain how you plan to implement or how you are implementing the program in your school.

_____

_____

_____

Table of Contents

EarthComm Team

EarthComm Project Staff

Michael J. Smith, Principal Investigator
Director of Education, American Geological Institute
John B. Southard, Senior Writer
Professor of Geology Emeritus, Massachusetts Institute of Technology
Emily J. Crum, Contributing Writer
American Geological Institute
Matthew Smith, Project Coordinator
American Geological Institute
Caitlin N. Callahan, Project Assistant
American Geological Institute
William S. Houston, Field Test Coordinator
American Geological Institute
Robert A. Bernoff, Field Test Evaluator
Professor Emeritus, Penn State University
Do Yong Park, Field Test Evaluator
University of Iowa
Larry G. Enochs, Pilot Test Evaluator
Professor of Science Education, Oregon State University

Original *EarthComm* Project Personnel

Charles Groat, United States Geological Survey
Marilyn Suiter, American Geological Institute
Bonnie Brunkhorst, UC San Bernardino
Richard M. Busch, West Chester University
Steven C. Good, West Chester University
John Carpenter, University of South Carolina
Linda Knight, Houston, Texas
Bob Ridky, University of Maryland

National Advisory Board

Harold Pratt, Chair
Jane Crowder, Bellevue, Washington
Don Lewis, Lafayette, California
Arthur Eisenkraft, Bedford Public Schools, New York
Tom Ervin, LeClaire, Iowa
Mary Kay Hemenway, University of Texas at Austin
William Leonard, Clemson University
Wendell Mohling, National Science Teachers Association
Barb Tewksbury, Hamilton College
Laure Wallace, United States Geological Survey

National Science Foundation Program Officers

Gerhard Salinger
Patricia Morse

Acknowledgements

Principal Investigator

Michael Smith is Director of Education at the American Geological Institute in Alexandria, Virginia. Dr. Smith worked as an exploration geologist and hydrogeologist. He began his Earth Science teaching career with Shady Side Academy in Pittsburgh, PA in 1988 and most recently taught Earth Science at the Charter School of Wilmington, DE. He earned a doctorate from the University of Pittsburgh's Cognitive Studies in Education Program and joined the faculty of the University of Delaware School of Education in 1995. Dr. Smith received the Outstanding Earth Science Teacher Award for Pennsylvania from the National Association of Geoscience Teachers in 1991, served as Secretary of the National Earth Science Teachers Association, and is a reviewer for Science Education and The Journal of Research in Science Teaching. He worked on the Delaware Teacher Standards, Delaware Science Assessment, National Board of Teacher Certification, and AAAS Project 2061 Curriculum Evaluation programs.

Senior Writer

Dr. Southard received his undergraduate degree from the Massachusetts Institute of Technology in 1960 and his doctorate in geology from Harvard University in 1966. After a National Science Foundation postdoctoral fellowship at the California Institute of Technology, he joined the faculty at the Massachusetts Institute of Technology, where he is currently Professor of Geology Emeritus. He was awarded the MIT School of Science teaching prize in 1989 and was one of the first cohorts of the MacVicar Fellows at MIT, in recognition of excellence in undergraduate teaching. He has taught numerous undergraduate courses in introductory geology, sedimentary geology, field geology, and environmental Earth Science both at MIT and in Harvard's adult education program. He was editor of the Journal of Sedimentary Petrology from 1992 to 1996, and he continues to do technical editing of scientific books and papers for SEPM, a professional society for sedimentary geology. Dr. Southard received the 2001 Neil Miner Award from the National Association of Geoscience Teachers.

Safety Reviewer Dr. Ed Robeck, Salisbury University, MD.

PRIMARY AND CONTRIBUTING AUTHORS

Earth's Dynamic Geosphere

Daniel J. Bisaccio
Souhegan High School
Amherst, NH

Steve Carlson
Middle School, OR

Warren Fish
Paul Revere School
Los Angeles, CA

Miriam Fuhrman
Carlsbad, CA

Steve Mattox
Grand Valley State University

Keith McKain
Milford Senior High School
Milford, DE

Mary McMillan
Niwot High School
Niwot, CO

Bill Romey
Orleans, MA

Michael Smith
American Geological Institute

Tom Vandewater
Colton, NY

Understanding Your Environment

Geoffrey A. Briggs
Batavia Senior High School
Batavia, NY

Cathey Donald
Auburn High School
Auburn, AL

Richard Duschl
Kings College
London, UK

Fran Hess
Cooperstown High School
Cooperstown, NY

Laurie Martin-Vermilyea
American Geological Institute

Molly Miller
Vanderbilt University

Mary-Russell Roberson
Durham, NC

Charles Savrda
Auburn University

Michael Smith
American Geological Institute

Earth's Fluid Spheres

Chet Bolay
Cape Coral High School
Cape Coral, FL

Steven Dutch
University of Wisconsin

Virginia Jones
Bonneville High School
Idaho Falls, ID

Acknowledgements (continued)

Laurie Martin-Vermilyea
American Geological Institute
Joseph Moran
University of Wisconsin
Mary-Russell Roberson
Durham, NC
Bruce G. Smith
Appleton North High School
Appleton, WI
Michael Smith
American Geological Institute

Earth's Natural Resources
Chuck Bell
Deer Valley High School
Glendale, AZ
Jay Hackett
Colorado Springs, CO
John Kemeny
University of Arizona
John Kounas
Westwood High School
Sloan, IA
Laurie Martin-Vermilyea
American Geological Institute
Mary Poulton
University of Arizona
David Shah
Deer Valley High School
Glendale, AZ
Janine Shigihara
Shelley Junior High School
Shelley, ID
Michael Smith
American Geological Institute

Earth System Evolution
Julie Bartley
University of West Georgia
Lori Borroni-Engle
Taft High School
San Antonio, TX
Richard M. Busch
West Chester University
West Chester, PA
Kathleen Cochrane
Our Lady of Ransom School
Niles, IL
Cathey Donald
Auburn High School, AL
Robert A. Gastaldo
Colby College
William Leonard
Clemson University
Tim Lutz
West Chester University
Carolyn Collins Petersen
C. Collins Petersen Productions
Groton, MA
Michael Smith
American Geological Institute
Matthew Smith
American Geological Institute

Content Reviewers
Gary Beck
BP Exploration
Phil Bennett
University of Texas, Austin

Steve Bergman
Southern Methodist University
Samuel Berkheiser
Pennsylvania Geologic Survey
Arthur Bloom
Cornell University
Craig Bohren
Penn State University
Bruce Bolt
University of California, Berkeley
John Callahan
Appalachian State University
Sandip Chattopadhyay
R.S. Kerr Environmental Research
Center
Beth Ellen Clark
Cornell University
Jimmy Diehl
Michigan Technological University
Sue Beske-Diehl
Michigan Technological University
Neil M. Dubrovsky
United States Geological Survey
Frank Ethridge
Colorado State University
Catherine Finley
University of Northern Colorado
Ronald Greeley
Arizona State University
Michelle Hall-Wallace
University of Arizona
Judy Hannah
Colorado State University
Blaine Hanson
Dept. of Land, Air, and Water
Resources
James W. Head III
Brown University
Patricia Heiser
Ohio University
John R. Hill
Indiana Geological Survey
Travis Hudson
American Geological Institute
Jackie Huntoon
Michigan Tech. University
Teresa Jordan
Cornell University
Allan Juhas
Lakewood, Colorado
Robert Kay
Cornell University
Chris Keane
American Geological Institute
Bill Kirby
United States Geological Survey
Mark Kirschbaum
United States Geological Survey
Dave Kirtland
United States Geological Survey
Jessica Elzea Kogel
Thiele Kaolin Company
Melinda Laituri
Colorado State University
Martha Leake
Valdosta State University
Donald Lewis
Happy Valley, CA

Steven Losh
Cornell University
Jerry McManus
Woods Hole Oceanographic
Institution
Marcus Milling
American Geological Institute
Alexandra Moore
Cornell University
Jack Oliver
Cornell University
Don Pair
University of Dayton
Mauri Pelto
Nicolas College
Bruce Pivetz
ManTech Environmental Research
Services Corp.
Stephen Pompea
Pompea & Associates
Peter Ray
Florida State University
William Rose
Michigan Technological Univ.
Lou Solebello
Macon, Gerogia
Robert Stewart
Texas A&M University
Ellen Stofan
NASA
Barbara Sullivan
University of Rhode Island
Carol Tang
Arizona State University
Bob Tilling
United States Geological Survey
Stanley Totten
Hanover College
Scott Tyler
University of Nevada, Reno
Michael Velbel
Michigan State University
Ellen Wohl
Colorado State University
David Wunsch
State Geologist of New Hampshire

Pilot Test Evaluator
Larry Enochs
Oregon State University

Pilot Test Teachers
Rhonda Artho
Dumas High School
Dumas, TX
Mary Jane Bell
Lyons-Decatur Northeast
Lyons, NE
Rebecca Brewster
Plant City High School
Plant City, FL
Terry Clifton
Jackson High School
Jackson, MI
Virginia Cooter
North Greene High School
Greeneville, TN

Acknowledgements (continued)

Monica Davis
North Little Rock High School
North Little Rock, AR

Joseph Drahuschak
Troxell Jr. High School
Allentown, PA

Ron Fabick
Brunswick High School
Brunswick, OH

Virginia Jones
Bonneville High School
Idaho Falls, ID

Troy Lilly
Snyder High School
Snyder, TX

Sherman Lundy
Burlington High School
Burlington, IA

Norma Martof
Fairmont Heights High School
Capitol Heights, MD

Keith McKain
Milford Senior High School
Milford, DE

Mary McMillan
Niwot High School
Niwot, CO

Kristin Michalski
Mukwonago High School
Mukwonago, WI

Dianne Mollica
Bishop Denis J. O'Connell
High School
Arlington, VA

Arden Rauch
Schenectady High School
Schenectady, NY

Laura Reysz
Lawrence Central High School
Indianapolis, IN

Floyd Rogers
Palatine High School
Palatine, IL

Ed Ruszczyk
New Canaan High School
New Canaan, CT

Jane Skinner
Farragut High School
Knoxville, TN

Shelley Snyder
Mount Abraham High School
Bristol, VT

Joy Tanigawa
El Rancho High School
Pico Rivera, CA

Dennis Wilcox
Milwaukee School of Languages
Milwaukee, WI

Kim Willoughby
SE Raleigh High School
Raleigh, NC

Field Test Workshop Staff

Don W. Byerly
University of Tennessee

Derek Geise
University of Nebraska

Michael A. Gibson
University of Tennessee

David C. Gosselin
University of Nebraska

Robert Hartshorn
University of Tennessee

William Kean
University of Wisconsin

Ellen Metzger
San Jose State University

Tracy Posnanski
University of Wisconsin

J. Preston Prather
University of Tennessee

Ed Robeck
Salisbury University

Richard Sedlock
San Jose State University

Bridget Wyatt
San Jose State University

Field Test Evaluators

Bob Bernoff
Dresher, PA

Do Yong Park
University of Iowa

Field Test Teachers

Kerry Adams
Alamosa High School
Alamosa, CO

Jason Ahlberg
Lincoln High
Lincoln, NE

Gregory Bailey
Fulton High School
Knoxville, TN

Mary Jane Bell
Lyons-Decatur Northeast
Lyons, NE

Rod Benson
Helena High
Helena, MT

Sandra Bethel
Greenfield High School
Greenfield, TN

John Cary
Malibu High School
Malibu, CA

Elke Christoffersen
Poland Regional High School
Poland, ME

Tom Clark
Benicia High School
Benicia, CA

Julie Cook
Jefferson City High School
Jefferson City, MO

Virginia Cooter
North Greene High School
Greeneville, TN

Mary Cummane
Perspectives Charter
Chicago, IL

Sharon D'Agosta
Creighton Preparatory
Omaha, NE

Mark Daniels
Kettle Morraine High School
Milwaukee, WI

Beth Droughton
Bloomfield High School
Bloomfield, NJ

Steve Ferris
Lincoln High
Lincoln, NE

Bob Feurer
North Bend Central Public
North Bend, NE

Sue Frack
Lincoln Northeast High
Lincoln, NE

Rebecca Fredrickson
Greendale High School
Greendale, WI

Sally Ghilarducci
Hamilton High School
Milwaukee, WI

Kerin Goedert
Lincoln High School
Ypsilanti, MI

Martin Goldsmith
Menominee Falls High School
Menominee Falls, WI

Randall Hall
Arlington High School
St. Paul, MN

Theresa Harrison
Wichita West High
Wichita, KS

Gilbert Highlander
Red Bank High School
Chattanooga, TN

Jim Hunt
Chattanooga School of Arts
& Sciences
Chattanooga, TN

Patricia Jarzynski
Watertown High School
Watertown, WI

Pam Kasprowicz
Bartlett High School
Bartlett, IL

Caren Kershner
Moffat Consolidated
Moffat, CO

Mary Jane Kirkham
Fulton High School

Ted Koehn
Lincoln East High
Lincoln, NE

Philip Lacey
East Liverpool High School
East Liverpool, OH

Joan Lahm
Scotus Central Catholic
Columbus, NE

Erica Larson
Tipton Community

Michael Laura
Banning High School
Wilmington, CAFawn LeMay

Plattsmouth High
Plattsmouth, NE

Acknowledgements (continued)

Christine Lightner
Smethport Area High School
Smethport, PA

Nick Mason
Normandy High School
St. Louis, MO

James Matson
Wichita West High
Wichita, KS

Jeffrey Messer
Western High School
Parma, MI

Dave Miller
Parkview High
Springfield, MO

Rick Nettesheim
Waukesha South
Waukesha, WI

John Niemoth
Niobrara Public
Niobrara, NE

Margaret Olsen
Woodward Academy
College Park, GA

Ronald Ozuna
Roosevelt High School
Los Angeles, CA

Paul Parra
Omaha North High
Omaha, NE

D. Keith Patton
West High
Denver, CO

Phyllis Peck
Fairfield High School
Fairfield, CA

Randy Pelton
Jackson High School
Massillon, OH

Reggie Pettitt
Holderness High School
Holderness, NH

June Rasmussen
Brighton High School
South Brighton, TN

Russ Reese
Kalama High School
Kalama, WA

Janet Ricker
South Greene High School
Greeneville, TN

Wendy Saber
Washington Park High School
Racine, WI

Garry Sampson
Wauwatosa West High School
Tosa, WI

Daniel Sauls
Chuckey-Doak High School
Afton, TN

Todd Shattuck
L.A. Center for Enriched Studies
Los Angeles, CA

Heather Shedd
Tennyson High School
Hayward, CA

Lynn Sironen
North Kingstown High School
North Kingstown, RI

Jane Skinner
Farragut High School
Knoxville, TN

Sarah Smith
Garringer High School
Charlotte, NC

Aaron Spurr
Malcolm Price Laboratory
Cedar Falls, IA

Karen Tiffany
Watertown High School
Watertown, WI

Tom Tyler
Bishop O'Dowd High School
Oakland, CA

Valerie Walter
Freedom High School
Bethlehem, PA

Christopher J. Akin Williams
Milford Mill Academy
Baltimore, MD

Roseanne Williby
Skutt Catholic High School
Omaha, NE

Carmen Woodhall
Canton South High School
Canton, OH

Field Test Coordinator
William Houston
American Geological Institute

Advisory Board
Jane Crowder
Bellevue, WA

Arthur Eisenkraft
Bedford (NY) Public Schools

Tom Ervin
LeClaire, IA

Mary Kay Hemenway
University of Texas at Austin

Bill Leonard
Clemson University

Don Lewis
Lafayette, CA

Wendell Mohling
National Science Teachers Association

Harold Pratt
Littleton, CO

Barb Tewksbury
Hamilton College

Laure Wallace
USGS

AGI Foundation
Jan van Sant
Executive Director

The American Geological Institute and EarthComm

Imagine more than 500,000 Earth scientists worldwide sharing a common voice, and you've just imagined the mission of the American Geological Institute. Our mission is to raise public awareness of the Earth sciences and the role that they play in mankind's use of natural resources, mitigation of natural hazards, and stewardship of the environment. For more than 50 years, AGI has served the scientists and teachers of its Member Societies and hundreds of associated colleges, universities, and corporations by producing Earth science educational materials, *Geotimes*–a geoscience news magazine, GeoRef–a reference database, and government affairs and public awareness programs.

So many important decisions made every day that affect our lives depend upon an understanding of how our Earth works. That's why AGI created *EarthComm*. In your *EarthComm* classroom, you'll discover the wonder and importance of Earth science by studying it where it counts—in your community. As you use the rock record to investigate climate change, do field work in nearby beaches, parks, or streams, explore the evolution and extinction of life, understand where your energy resources come from, or find out how to forecast severe weather, you'll gain a better understanding of how to use your knowledge of Earth science to make wise personal decisions.

We would like to thank the AGI Foundation Members that have been supportive in bringing Earth science to students. These AGI Foundation Members include: Anadarko Petroleum Corp., The Anschutz Foundation, Baker Hughes Foundation, Barrett Resources Corp., Elizabeth and Stephen Bechtel, Jr. Foundation, BPAmoco Foundation, Burlington Resources Foundation, CGG Americas, Inc., ChevronTexaco Corp., Conoco Inc., Consolidated Natural Gas Foundation, Diamond Offshore Co., Dominion Exploration & Production, Inc., EEX Corp., ExxonMobil Foundation, Global Marine Drilling Co., Halliburton Foundation, Inc., Kerr McGee Foundation, Maxus Energy Corp., Noble Drilling Corp., Occidental Petroleum Charitable Foundation, Parker Drilling Co., Phillips Petroleum Co., Santa Fe Snyder Corp., Schlumberger Foundation, Shell Oil Company Foundation, Southwestern Energy Co., Texaco, Inc., Texas Crude Energy, Inc., Unocal Corp. USX Foundation (Marathon Oil Co.).

We at AGI wish you success in your exploration of the Earth System and your Community.

Michael J. Smith
Director of Education, AGI

Marcus E. Milling
Executive Director, AGI

EarthComm: Earth System Science in the Community

Goals of *EarthComm*

Earth System Science in the Community (*EarthComm*) is an NSF-funded curriculum project guided in design and approach by the National Science Education Standards (1996), AGI's *Earth Science Content Guidelines for Grades K-12*, and other major science education curriculum and reform programs. This program builds on the strength of other successful AGI Earth Science education projects such as the *Earth Science Curriculum Project* (known to many as *Investigating the Earth*). *EarthComm* provides a comprehensive secondary-level educational program in the Earth Sciences that includes student learning materials, teacher resources (both materials and teacher-support networks), and assessment tools for a hands-on, inquiry-driven, instructional program and an *EarthComm* web site.

EarthComm covers fewer topics than the traditional Earth Science textbook. It emphasizes important concepts, understandings, and abilities that all students can use to make wise decisions, think critically, and understand and appreciate the Earth system. The goals of the *EarthComm* program are:

- To teach students the principles and practices of Earth Science and to demonstrate the relevance of Earth Science to their life and environment.
- To approach Earth Science through the problem-solving, community-based model in which the teacher plays the role of facilitator.
- To establish an expanded learning environment which incorporates field work, technological access to data, and traditional classroom and laboratory activities.
- To support the development of communities of learners by establishing student teams and by building a greater regional and national community through telecommunication access.
- To utilize local and regional issues and concerns to stimulate problem-solving activities and to foster a sense of Earth stewardship by students in their communities.

Developing *EarthComm*

Hundreds of teachers, scientists, and students helped develop *EarthComm*. In the summer of 1998, six teams of Earth Science educators wrote 122 inquiry-based investigations. Teachers and scientists reviewed draft chapters, which were then revised for pilot testing by 26 teachers in the spring of 1999. Seventeen teachers from the National Earth Science Teachers Association collaborated with project staff to revise *EarthComm* in the summer of 1999. In the 1999-2000 school year, *EarthComm* underwent a national field test with 77 teachers in 27 states. Results of field testing and further content review by more than 40 professional scientists were used to produce the commercial edition of *EarthComm*.

EarthComm Modules and Chapters

I. Earth's Dynamic Geosphere

Volcanoes and Your Community
Plate Tectonics and Your Community
Earthquakes and Your Community

II. Understanding Your Environment

Bedrock Geology and Your Community
River Systems and Your Community
Land Use Planning and Your Community

III. Earth's Fluid Spheres

Oceans and Your Community
Severe Weather and Your Community
Cryosphere and Your Community

IV. Earth's Natural Resources

Energy Resources and Your Community
Mineral Resources and Your Community
Water Resources and Your Community

V. Earth System Evolution

Astronomy and Your Community
Climate Change and Your Community
Changing Life and Your Community

EarthComm: Correlation to the National Science Education Standards

National Science Education Content Standards	Earth's Dynamic Geosphere			Understanding Your Environment			Earth's Fluid Spheres			Earth's Natural Resources			Earth System Evolution		
	1	2	3	1	2	3	1	2	3	1	2	3	1	2	3
UNIFYING CONCEPTS AND PROCESSES															
System, order and organization	•	•	•	•	•	•	•	•	•	•	•	•	•	•	•
Evidence, models, and explanation	•	•	•	•	•	•	•	•	•	•	•	•	•	•	•
Constancy, change, and measurement	•	•	•	•	•	•	•	•	•	•	•	•	•	•	•
Evolution and equilibrium		•		•	•	•	•	•	•	•	•	•	•	•	•
Form and function						•	•								•
SCIENCE AS INQUIRY															
Identify questions and concepts that guide scientific investigations	•	•	•	•	•	•	•	•	•	•	•	•	•	•	•
Design and conduct scientific investigations	•	•	•	•	•	•	•	•	•	•	•	•	•	•	•
Use technology and mathematics to improve investigations	•	•	•	•	•	•	•	•	•	•	•	•	•	•	•
Formulate and revise scientific explanations and models using logic and evidence	•	•	•	•	•	•	•	•	•	•	•	•	•	•	•
Communicate and defend a scientific argument	•	•	•	•	•	•	•	•	•	•	•	•	•	•	•
Understand scientific inquiry	•	•	•	•	•	•	•	•	•	•	•	•	•	•	•
EARTH AND SPACE SCIENCE															
Energy in the Earth system	•	•			•	•	•	•	•	•	•	•	•	•	•
Geochemical cycles	•	•		•	•	•	•	•	•	•	•	•	•	•	
Origin and evolution of the Earth system	•	•	•	•	•	•	•	•	•	•	•	•	•	•	•
Origin and evolution of the universe												•	•		
SCIENCE AND TECHNOLOGY															
Identify a problem or design an opportunity	•		•			•				•		•			
Propose designs and choose between alternative solutions	•		•			•				•		•			
Implement a proposed solution	•		•			•				•		•			
Evaluate the solution and its consequences	•		•			•				•		•			
Communicate the problem, process, and solution	•	•	•	•	•	•	•	•	•	•	•	•	•	•	•
Understand science and technology	•	•	•				•	•	•	•	•	•	•	•	•
SCIENCE IN PERSONAL AND SOCIAL PERSPECTIVES															
Personal and community health	•		•		•	•		•		•	•	•	•	•	
Population growth						•					•			•	•
Natural Resources	•					•	•			•	•	•			
Environmental quality	•			•	•					•	•			•	
Natural and human-induced hazards	•	•	•		•	•			•	•	•	•	•	•	
Science and technology in local, national, and global challenges	•	•	•			•	•	•	•	•	•	•		•	
HISTORY AND NATURE OF SCIENCE															
Science as a Human Endeavor	•	•	•	•	•	•	•	•	•	•	•	•	•	•	•
Nature of Scientific Knowledge	•	•	•	•	•	•	•	•	•	•	•	•	•	•	•
Historical Perspectives		•	•	•					•		•	•	•	•	•

EarthComm "Big Ideas"

EarthComm curriculum development was guided by 10 fundamental ideas that are emphasized in the five modules and are the primary goals for student learning:

- Earth science literacy empowers us to understand our environment, make wise decisions that affect quality of life, and manage resources, environments, and hazards.

- Earth's dynamic equilibrium system contains subsystems from atoms to planetary spheres. Materials interact among these subsystems due to natural forces and energy that flows from sources inside and outside of the planet. These interactions, changes, forces, and flows tend to occur in offsetting directions and amounts. Materials tend to flow in chains, cycles, and webs that tend toward equilibrium states in which energy is distributed as uniformly as possible. The net result is a state of balanced change or dynamic equilibrium, a condition that appears to have existed for billions of years.

- Change through time produced Earth, the net result of constancy, gradual changes, and episodic changes over human, geological, and astronomical scales of time and space.

- Extraterrestrial influences upon Earth include extraterrestrial energy and materials, and influences due to Earth's position and motion as a subsystem of an evolving solar system, galaxy, and universe.

- The dynamic geosphere includes a rocky exterior upon which ecosystems and human communities developed and a partially molten interior with convection circulation that generates the magnetosphere and drives plate tectonics. It contains resources that sustain life, causes natural hazards that may threaten life, and affects all of Earth's other geospheres.

- Fluid spheres within the Earth system include the hydrosphere, atmosphere, and cryosphere, which interact and flow to produce ever-changing weather, climate, glaciers, seascapes, and water resources that affect human communities, and which shape the land, transfer Earth materials and energy, and change surface environments and ecosystems.

- Dynamic environments and ecosystems are produced by the interaction of all the geospheres at the Earth's surface and include many different environments, ecosystems, and communities that affect one another and change through time.

- Earth resources include the nonrenewable and renewable supplies of energy, mineral, and water resources upon which individuals and communities depend in order to maintain quality of human life, economic prosperity, and requirements for industrialization.

- Natural hazards associated with Earth processes and events include drought, floods, storms, volcanic activity, earthquakes, and climate change and can pose risks to humans, their property, and communities. Earth science is used to study, predict, and mitigate natural hazards so that we can assess risks, plan wisely, and adapt to the effects of natural hazards.

- In order to sustain the presence and quality of human life, humans and communities must understand their dependence on Earth resources and environments, realize how they influence Earth systems, appreciate Earth's carrying capacity, manage and conserve nonrenewable resources and environments, develop alternate sources of energy and materials needed for human sustenance, and invent new technologies.

EarthComm Goals and Expectations for Teachers

- Use motivational teaching methods, interactive technologies, and manipulatives to pique student interest and help all students to understand the practical effects of Earth science and essential concepts and principles that underlie energy within the Earth system, geochemical cycles, and the origin and evolution of the Earth system.

- Facilitate students' understanding of inquiry and ability to inquire scientifically by having students answer questions about local problems and issues, design and conduct investigations, use technology and mathematics, form scientific explanations using logic and evidence, analyze alternative explanations, and communicate and defend scientific arguments.

- Emphasize the connections and relationships between Earth science and other academic disciplines.

- Establish an expanded learning environment for students through fieldwork, technological access to data, laboratory and other classroom activities.

- Nurture communities of science learners by establishing student teams, orchestrating discourse about scientific ideas, building networks of local, regional and national information exchange, and using the services of Earth and space organizations.

- Raise students' awareness of environmental and resource issues and problems in their communities.

EarthComm Goals and Expectations for Students

- Develop knowledge and understanding of practical and essential Earth science concepts and the principles Earth science shares with other disciplines.

- Understand basic principles of Earth system science and think from an Earth system science perspective.

- Develop an understanding of scientific inquiry and abilities needed to conduct scientific inquiry.

- Develop technology-oriented abilities for human enterprises in Earth and space.

- Understand the nature, origin, and distribution of Earth's energy, mineral, and water resources; technologies used to locate, extract, and process these resources; and dependence on these resources to satisfy our wants, needs, and expectations.

- Understand how terrestrial and extraterrestrial processes affect Earth's materials, environments, and organisms, how scientists study these processes on Earth and from space, and how some processes benefit humans while others pose risks.

- Understand how human activities influence Earth's spheres, processes, resources, and environments — factors that affect the size and distribution of human population and Earth's capacity to support life.

- Become aware of career opportunities in the Earth and space sciences, how professions and businesses benefit from technologies used by Earth and space scientists, and how these combined professions and businesses are related to regional economies.

EarthComm Curriculum Design

EarthComm modules are each three chapters connected to a common theme. Every chapter begins with a community-based problem or issue that can only be solved by developing key ideas and understandings in the chapter activities. Activities follow a learning cycle model.

Component of *EarthComm*	What Happens in the Classroom	Stages of 5-E Learning Cycle Model
Chapter Challenge	Students read and discuss a scenario that presents a community-based issue to solve through Earth Science and inquiry. They also explore the criteria and expectations for solving the challenge. Teachers allow students to share their current thinking openly and without closure.	Engage
1. Think about It	Students answer an open-ended question (or two) that sets the context for an activity and provides the teacher with a pre-assessment of their ideas. They briefly discuss their ideas in groups and/or as a class. Teachers allow students to share their ideas openly. They avoid assigning formal labels to concepts or seeking closure.	Engage
2. Investigate	Students collaborate on an inquiry activity that requires hands-on work, literature or web research, or fieldwork. Teachers facilitate and guide student-driven inquiry.	Explore
3. Reflecting upon the Activity and the Challenge	Students read a brief summary of the main ideas explored in the investigation and their relationship to the challenge. Teachers review the main ideas with students and affirm the relevance of the activity to the challenge.	Explain
4. Digging Deeper	Students read text, illustrations, and photographs that explain concepts explored in the investigation. Terms are defined and clarified. Teachers provide further information and clarification of concepts through lecture, slides, videos, or laser disk presentations.	Explain

Component of *EarthComm*	What Happens in the Classroom	Stages of 5-E Learning Cycle Model
5. Understanding and Applying What You Have Learned	Students respond to questions that check their understanding of key principles and concepts (learning goals) for the activity. New, yet familiar situations and scenarios provide contexts for students to apply their developing understandings. Teachers review student responses and use the questions to further probe and hone understanding of key learning goals.	Elaborate
6. Preparing for the Chapter Challenge	Students put their investigative results into the context of the challenge by preparing or organizing their work as it relates to their final product. Teachers review student performance in terms of its consistency with criteria set forth in the expectations for the activity and the challenge.	Elaborate/ Evaluate
7. Inquiring Further	Students are presented with options for deepening their understanding of concepts and skills developed within the activity. Teachers promote and encourage further inquiry.	Elaborate/ Evaluate
Chapter Assessment	Students present their solution to the Chapter Challenge in a variety of formats and consider ways to share their findings beyond the classroom. Teachers use the Assessment Criteria to assess the extent to which student work demonstrates mastery of concepts and skills. They also explore creative ways to share student solutions with the community.	Evaluate
Alternative Assessment	Students respond to a chapter test of essential knowledge and skills targeted throughout the chapter. Teachers score and review the test with students. They help students to understand how to use the results to guide future efforts.	Evaluate

Using *EarthComm* Features in Your Classroom

I. Getting Started

Each *EarthComm* chapter begins with one or more open-ended questions that give teachers the opportunity to explore what their students know about the central concepts of the chapter. Uncovering students' thinking (their prior knowledge) and exposing the diversity of ideas in the classroom are the first steps in the learning cycle. Some teachers prefer to have students record their responses to these questions. They then call for volunteers to offer ideas up for discussion. Other teachers prefer to start with discussion by asking students to volunteer their ideas. In either situation, it is important that teachers encourage the sharing of ideas by not judging responses as "right" or "wrong." It is also important that teachers keep a record of the variety of ideas, which can be displayed in the classroom (on a sheet of easel pad paper or on an overhead transparency) and referred to as students explore the concepts in the chapter. Teachers often find that they can group responses into a few categories and record the number of students who hold each idea.

2. Scenario

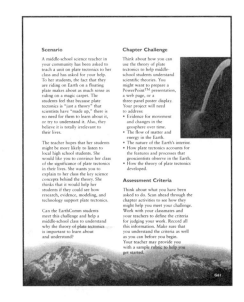

Each *EarthComm* chapter begins with an engaging description of an event or situation in the Earth system that has happened or could actually take place. The scenario (only a paragraph or two in length) sets the stage for the **Chapter Challenge**, which comes next. Many teachers read the scenario aloud to the class as a way to introduce the new chapter. Some teachers expand on the scenario by using videos of actual events, or by inviting persons from the field to present the scenario.

3. Chapter Challenge

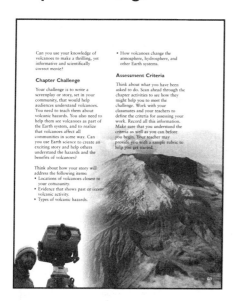

The **Chapter Challenge** is the central core of *EarthComm*. The challenge provides the context for all activities within the chapter. The **Chapter Challenge** provides a ready answer to the question asked all too often by students, "Why am I doing this?" because every activity contributes to solving the central problem set forth in the challenge. It also makes learning relevant to high school students. Each challenge is grounded in the community and designed to make the learning of Earth Science more relevant to the lives of students.

For example, in *Earth's Dynamic Geosphere*, Chapter 1: Volcanoes and Your Community, students are asked to create a screenplay for a thrilling yet informative and scientifically correct movie about volcanoes. Students are naturally intrigued by the dramatic effects and forces of natural hazards, and can easily relate to films they have seen that focus on natural disasters (*Volcano, Dante's Peak, Deep Impact*, and so on).

But unless they live in a volcanic region, it is unlikely that they have contemplated how volcanoes might affect their lives. Writing a story that is set in their community makes students think more deeply about the causes and effects of volcanism, and how volcanoes impact all communities because we live within a set of interconnected systems on Earth. All challenges require that students demonstrate solid understanding of Earth Science concepts and principles.

Another important element of the **Chapter Challenge** is that it provides opportunities for students with diverse interests and abilities to express their understanding in different ways. Challenges are completed in various ways. All involve writing to one extent or another, but some feature oral presentations, teaching, designing brochures, constructing models, creating web sites, or preparing formal presentations. Students who express themselves artistically will shine in some challenges, while those who enjoy designing and constructing will take a leading role in others.

Challenges are flexible enough to engage students at all levels of high school. Classes ranging from 9th grade integrated science to grades 11-12 honors, studying Earth Science, tested the challenges. Teachers establish different expectations for the students they teach, but the challenge is consistent.

4. Assessment Criteria

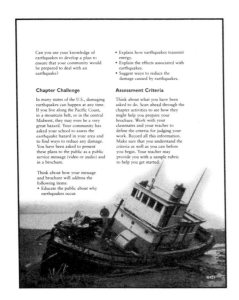

The completion of the challenge (the final report or project) serves as the primary source of summative assessment information. Traditional assessment strategies often give too much attention to the memorization of terms or the recall of information. As a result, they often fall short of providing information about students' ability to think and reason critically and apply information that they have learned. In *EarthComm*, the solutions students provide to **Chapter Challenges** provide information used to assess thinking, reasoning, and problem-solving skills that are essential to lifelong learning.

Assessment is one of the key areas that teachers need to be familiar with and understand when trying to envision implementing *EarthComm*. In any curriculum model, the mode of instruction and the mode of assessment are connected. In the best scheme, instruction and assessment are aligned in both content and process. However, to the extent that one becomes an

impediment to reform the other, they can also be uncoupled. *EarthComm* uses multiple assessment formats. Some are non-traditional and are consistent with reform movements in science education that *EarthComm* is designed to promote. **Project-based assessment,** for example, is built into every *EarthComm* **Chapter Challenge**. At the same time, the developers acknowledge the need to support teachers whose classroom context does not allow them to depart completely from traditional assessment formats, such as paper and pencil tests.

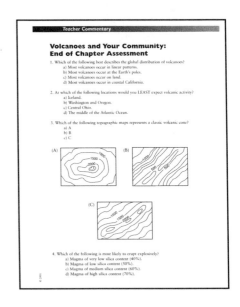

An assessment instrument can imply but not determine its own best use. This means that *EarthComm* teachers can inadvertently assess chapter reports in ways that work against integrative thinking, a focus on important ideas, flexibility in approach, and consistency between assessment and the inferences made from that assessment.

All expectations should be communicated to students. Discussing the grading criteria and creating a general rubric are critical to student success. Better still, teachers can engage students in modifying and/or creating the criteria that will be used to assess their performance. Start by sharing the sample rubric with students and holding a class discussion. Questions that can be used to focus the discussion include: Why are these criteria included? Which activities will help you to meet these expectations? How much is required? What does an "A" presentation or report look like? The criteria should be revisited throughout the completion of the chapter, but for now students will have a clearer understanding of the challenge and the expectations they should set for themselves.

Teacher Commentary

Assessment Rubric for Chapter Report on Volcanoes

Meets the standard of excellence. **5**	_Significant_ information is presented about all of the following: • Locations of volcanoes closest to your community • Evidence of past or recent volcanic activity • Volcanic hazards • How volcanoes change Earth systems All the information is accurate and appropriate The writing is clear and interesting
Approaches the standard of excellence. **4**	_Significant_ information is presented about most of the following: • Locations of volcanoes closest to your community • Evidence of past or recent volcanic activity • Volcanic hazards • How volcanoes change Earth systems All the information is accurate and appropriate The writing is clear and interesting
Meets an acceptable standard. **3**	_Significant_ information is presented about most of the following: • Locations of volcanoes closest to your community • Evidence of past or recent volcanic activity • Volcanic hazards • How volcanoes change Earth systems Most of the information is accurate and appropriate The writing is clear and interesting
Below acceptable standard and requires remedial help. **2**	_Limited_ information is presented about the following: • Locations of volcanoes closest to your community • Evidence of past or recent volcanic activity • Volcanic hazards • How volcanoes change Earth systems Most of the information is accurate and appropriate Generally, the writing does not hold the reader's attention
Basic level that requires remedial help or demonstrates a lack of effort. **1**	_Limited_ information is presented about the following: • Locations of volcanoes closest to your community • Evidence of past or recent volcanic activity • Volcanic hazards • How volcanoes change Earth systems Little of the information is accurate and appropriate The writing is difficult to follow

10

By the conclusion of the discussion of Assessment Criteria, students should have a clear "road map" of the structure of the chapter. They should have a sense as to why it is important to complete each activity in order to successfully meet the **Chapter Challenge.** They should be able to describe how each activity contributes toward the long-term goal.

5. Goals

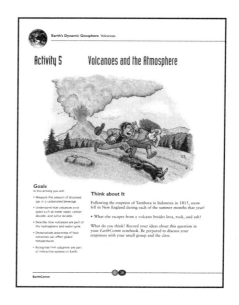

Earth's Dynamic Geosphere Volcanoes

Activity 5   Volcanoes and the Atmosphere

Goals
In this activity you will:
• Measure the amount of dissolved gas in a carbonated beverage.
• Understand that volcanoes emit gases such as water vapor, carbon dioxide, and sulfur dioxide.
• Describe how volcanoes are part of the hydrosphere and water cycle.
• Demonstrate awareness of how volcanoes can affect global temperatures.
• Recognize that volcanoes are part of interactive systems on Earth.

Think about It
Following the eruption of Tambora in Indonesia in 1815, snow fell in New England during each of the summer months that year!

• What else escapes from a volcano besides lava, rock, and ash?

What do you think? Record your ideas about this question in your _EarthComm_ notebook. Be prepared to discuss your responses with your small group and the class.

EarthComm

At the beginning of each activity students are provided with a list of goals that they should be able to achieve by completing the activity. Throughout this Teacher's Edition, we point out where each goal is addressed within each activity and provide some suggestions for assessing the goal. In most cases the goals are addressed directly by the hands-on investigation, as well as through reading the text or working on the **Chapter Challenge.** Pointing out the goals at the start of the activity reminds students about the expectations for learning. It is often helpful to point out how specific goals relate to the **Chapter Challenge.** For example, one element of the **Chapter Challenge** in _Earth's Dynamic Geosphere_, Chapter 1, Volcanoes and Your Community, is for students to address how volcanoes change the atmosphere, hydrosphere, and other Earth systems. One of the goals of Activity 5 in that chapter (Volcanoes and the Atmosphere) is to "describe how volcanoes are part of the hydrosphere and water cycle." When introducing the activity to students, it helps to point out how this particular goal contributes to the **Chapter Challenge**. It also serves to remind students "why we are doing this."

6. Think about It

One of the most fundamental principles derived from many years of research on student learning is that:

"Students come to the classroom with preconceptions about how the world works. If their initial understanding is not engaged, they may fail to grasp the new concepts and information that are taught, or they may learn them for the purposes of a test but revert to their preconceptions outside the classroom." (*How People Learn: Bridging Research and Practice*, National Research Council, 1999, P. 10.)

This principle has been illustrated through the Private Universe series of videotapes that show Harvard graduates responding to basic science questions in much the same way that fourth grade students do. Although the videotapes revealed that the Harvard graduates used a more sophisticated vocabulary, the majority held onto the same naïve incorrect conceptions of elementary school students. Research on learning suggests that the belief systems of students who are not confronted with what they believe and adequately shown why they should give up that belief system remain intact. Real learning requires confronting one's beliefs and testing them in light of competing explanations.

Drawing out and working with students' preconceptions is important for learners. In *EarthComm*, **Think about It** is used to ascertain students' prior knowledge about the key concept or Earth Science processes or events explored in each activity. Students verbalize what they think about the age of the Earth, the causes of volcanoes, or the way that the landscape changes over time before they embark on an activity designed to challenge and test these beliefs. A brief discussion about the diversity of beliefs in the classroom makes students consider how their ideas compare to others and the evidence that supports their view of volcanoes, earthquakes, or seasons.

The **Think about It** question is not a conclusion, but a lead into inquiry. It is not designed to produce the correct answer or a debate about the features of the question, or to bring closure. The activity that follows will provide that discussion as the results of inquiry are analyzed. Students are encouraged to record their ideas in words and/or drawings to ensure that they have considered their prior knowledge. After students discuss their ideas in pairs or in small groups, teachers activate a class discussion. A discussion with fellow students prior to class discussion may encourage students to exchange ideas without the fear of personally giving a "wrong answer." Teachers sometimes have students exchange papers and volunteer responses that they find interesting.

The "humorous illustration" above each **Think about It** section was designed to stimulate student thinking. In our field test edition of *EarthComm*, we used photographs of events and processes to stimulate thinking. However, we came to realize that illustrations would provide greater flexibility to stimulate students to begin to make the specific kinds of connections emphasized in each activity. For example, the first activity in *Earth's Dynamic Geosphere* Chapter 1 is titled "Where are the Volcanoes?" The **Think about It** question asks students "Can volcanoes form anywhere on Earth? Why or why not?" While some might argue that a photograph of a volcanic eruption would be most appropriate here, the drawing of a volcanic eruption occurring in a backyard barbeque used in the activity stimulates further thinking: "Could a volcano occur in my backyard?" Most students have experienced a summer cookout, but few have experienced a volcanic eruption. The context of the drawing is more relevant to students and makes it easier to stimulate student thinking about phenomena that they have never experienced.

7. Investigate

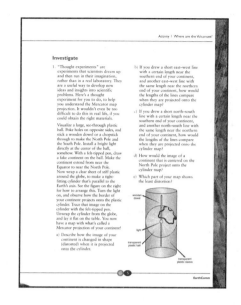

EarthComm is a hands-on, minds-on curriculum. In designing *EarthComm*, we were guided by the belief that doing Earth Science is essential to learning Earth Science. Testing of *EarthComm* activities by teachers across America provided critical testimonial about the importance of the activities to student learning. In small groups and as a class, students take part in doing hands-on experiments, participating in field work, or searching for answers using the Internet and reference materials. Blackline Masters are included in the Teacher's Edition for any maps or illustrations that are essential for students to complete the activity.

Each part of an *EarthComm* activity, as well as the sequence of activities within a chapter, moves from concrete to abstract. Hands-on activities provide the basis for exploring student beliefs about how the world works and to manipulate variables that affect the outcomes of experiments, models, or simulations. Later in each activity, formal labels are applied to concepts by introducing terminology used to describe the processes that students have explored through hands-on activity. This flow from concrete (hands-on) to abstract (formal explanations) is progressive – students begin to develop their own explanations for phenomena by responding to questions within the **Investigate** section.

Each activity has instructions for each part of the investigation. The community focus of *EarthComm* makes investigating the world more relevant to students. Have any volcanoes occurred in my state in the past? Have we ever experienced a major earthquake? What mineral resources exist in my community? Activities were designed with regard to the cost of materials and equipment needed. Many resources can be readily obtained in the community (local rock or soil samples, for example) or brought in to school by students (plastic two-liter soda bottles). Materials kits are available for purchase, but you will also need to obtain some resources from outside suppliers, such as topographic and geologic maps of your community, state, or region. The *EarthComm* web site will direct you to sources where you can gather such materials.

Most **Investigate** activities will require between one and two class periods. The variety of school schedules and student needs makes it difficult to predict exactly how much time your class will need. For example, if students need to construct a graph for part of an investigation, and the students have never been exposed to graphing, then this may require additional introductory lessons on the construction and interpretation of graphs. The most challenging aspect of

EarthComm for teachers to "master" is that the **Investigate** section of each activity has been designed to be student-driven. Students learn more when they have to struggle to "figure things out" and work in collaborative groups to solve problems as a team. Teachers will have to resist the temptation to provide the answers to students when they get "stuck" or hung up on part of a problem. Eventually, students learn that while they can call upon their teacher for assistance, the teacher is not going to "show them the answer." Field testing of *EarthComm* revealed that teachers who were most successful in getting their students to solve problems as a team were patient with this process and steadfast in their determination to act as facilitators of learning during the **Investigate** portion of activities. As one teacher noted, "My response to questions during the investigation was like a mantra, 'What do you think you need to do to solve this?' My students eventually realized that although I was there to provide guidance, they weren't going to get the solution out of me."

Another concern that many teachers have when examining *EarthComm* for the first time is that their students do not have the background knowledge to do the investigations. They want to deliver a lecture about the phenomena before allowing students to do the investigation. Such an approach is common to many traditional programs and is inconsistent with the pedagogical theory used to design *EarthComm*. The appropriate place for delivering a lecture or reading text in *EarthComm* is following the investigation, not preceding it. For example, suppose a group of students has been asked to interpret a map. The traditional approach to science education is for the teacher to give a lecture or assign a reading, "How to Interpret Maps," then give students practice reading maps. *EarthComm* teachers recognize that while students may lack some specific skills (reading latitude and longitude, for example), within a group of four students, it is not uncommon for at least one of the students to have a vital skill or piece of knowledge that is required to solve a problem. The one or two students who have been exposed to (or better yet, understood) latitude and longitude have the opportunity to shine within the group by contributing that vital piece of information or demonstrating a skill. That's how scientific research teams work – specialists bring expertise to the group, and by working together, the group achieves something that no one could achieve working alone. The **Investigate** section of *EarthComm* is modeled in the spirit of the scientific research team.

8. Reflecting on the Activity and the Challenge

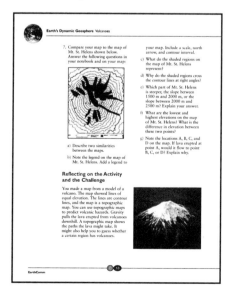

Each activity contributes to the solution of the **Chapter Challenge**. This feature gives students a brief summary of the activity. It helps students to relate the activity that they just completed to the "big picture." Teachers also find this section useful for students who were absent for an investigation. In situations where students cannot make up the investigation (after school or during off-hours), teachers can use this section to provide an overview of what was missed. Although reading about the main point of an activity is a poor substitute to actually doing it, teachers find that the overview helps them deal with the reality of student absences and the hectic pace of school schedules.

9. Digging Deeper

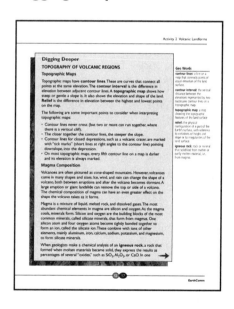

This section provides text, illustrations, data tables, and photographs that give students greater insight into the concepts explored in the activity. Words that may be new or unfamiliar to students are defined and explained (so-called **Geo Words**). These are words that geologists use when discussing the concepts presented. This is not the same thing as stating that **Geo Words** are "important words," or "words to be memorized." Teachers use their own judgment about selecting the **Geo Words** that are most important for their students to learn. Teachers typically use discretion and consider their state and local guidelines for science content understanding when assigning importance to particular vocabulary, which in most cases is very likely to be a small subset of all the **Geo Words** introduced in each chapter.

Teachers often assign **Check Your Understanding** questions as homework to guide students to think about the major ideas in the text. Teachers can also select questions to use as quizzes, rephrasing the questions into multiple choice or "true/false" formats. This provides assessment information about student understanding and as a "motivational tool" to ensure that students complete the reading assignment and comprehend the main ideas.

This is the stage of the activity that is most appropriate for teachers to explain concepts to students in whole-class lectures or discussions. References to Blackline Masters are available throughout the Teacher's Edition. They refer to illustrations from the textbook that teachers may photocopy and distribute to students or make overhead transparencies for lectures or presentations.

10. Understanding and Applying What You Have Learned

Questions in this feature ask students to use the key principles and concepts introduced in the activity. Students are sometimes presented with new situations in which they are asked to apply what they have learned. The questions in this section typically require higher-order thinking and reasoning skills than **Check Your Understanding**. Teachers can assign these questions as homework, or have students complete them in groups during class. Assigning them as homework economizes time available in class, but has the drawback of making it difficult for students to collectively revisit the understanding that they developed as they worked through the concepts as a group during the investigation. A third alternative is of course to assign the work individually in class. When students work through application problems in class, teachers have the opportunity to interact with students at a critical juncture in their learning – when they may be just on the verge of "getting it."

11. Preparing for the Chapter Challenge

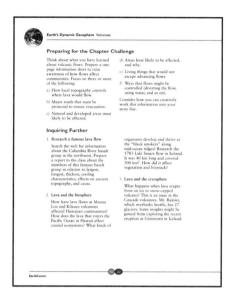

This feature suggests ways in which students can organize their work and get ready for the challenge. It prompts students to combine the results of their inquiry as they work through the chapter. Another one of the important principles of learning used to guide the selection of content in *EarthComm* is that:

"To develop competence in an area of inquiry, students must (a) have a deep foundation of factual knowledge, (b) understand facts and ideas in the context of a conceptual framework, and (c) organize knowledge in ways that facilitate retrieval and application." (*How People Learn: Bridging Research and Practice*, National Research Council, 1999, P. 12.)

This phase of an activity (**Preparing for the Chapter Challenge**) is an important metacognitive tool that makes students examine what they have learned in the activity and then think critically about

the usefulness of the results of their inquiry. The process of synthesizing what they have learned in order to solve the **Chapter Challenge** forces students to take stock of their learning and evaluate whether or not they really understand "how it fits into the big picture." It is important for teachers to guide students through this process with questions such as: "What part of your work best helps you to solve the challenge? How does what you learned help you to solve the challenge? How does this assignment relate to the criteria that we established for your chapter report? Are you making the best possible use of the evidence you have gathered?"

12. Inquiring Further

This feature provides lots of suggestions for helping students to deepen their understanding of the concepts and skills developed in the activity. It also gives students the opportunity to relate what they have learned to the Earth system. Teachers should review the suggestions and consider how the time available in class, the specific resources available to their students and/or in the school, and the needs and abilities of their students. Some of these suggestions make for excellent "do at home" investigations or Internet and library-based research projects. Some teachers assign **Inquiring Further** as "extra credit" projects. Some of the suggested activities in **Inquiring Further** may have particular relevance to your community. In such cases, make every attempt to integrate the activity into your instruction.

The most common complaint teachers make about Internet-based research stems from a concern about the limited amount of time students have available at school computers. The *EarthComm* web site has been designed to help students focus their research. By providing specific links helpful to each **Inquiring Further** activity on the site, students will gain access to useful information from stable web sites without spending time searching for information.

(Reference: *How People Learn: Bridging Research and Practice* (1999) Suzanne Donovan, John Bransford, and James Pellegrino, editors. National Academy Press, Washington, DC. 78 pages. The report is also available online at http://www.nap.edu.

EarthComm Assessment Opportunities

In keeping with the discussion of assessment outlined in the *National Science Education Standards* (NSES), teachers must be careful while developing the specific expectations for each chapter. Four issues are of particular importance in that they may present somewhat new considerations for teachers and students:

1. Integrative Thinking

The *National Science Education Standards* (NSES) state: "Assessments must be consistent with the decisions they are designed to inform." This means that as a prerequisite to establishing expectations, teachers should consider the use of assessment information. In *EarthComm*, students must be able to articulate the connection between Earth Science concepts and their own community. This means that they have to integrate traditional Earth Science content with knowledge of their surroundings. It is likely that this kind of integration will be new to students, and that they will require some practice at accomplishing it. Assessment in one chapter can inform how the next chapter is approached so that the ability to apply Earth Science concepts to local situations is enhanced on an ongoing basis.

2. Importance

An explicit focus of NSES is to promote a shift to deeper instruction on a smaller set of core science concepts and principles. Assessment can support or undermine that intent. It can support it by raising the priority of in-depth treatment of concepts, such as students evaluating the relevance of core concepts to their communities. Assessment can undermine a deep treatment of concepts by encouraging students to parrot back large bodies of knowledge-level facts that are not related to any specific context in particular. In short, by focusing on a few concepts and principles, deemed to be of particularly fundamental importance, assessment can help to overcome a bias toward superficial learning. For example, assessment of terminology that emphasizes deeper understanding of science is that which focuses on the use of terminology as a tool for communicating important ideas. Knowledge of terminology is not an end in itself. Teachers must be watchful that the focus remains on terminology in use, rather than on rote recall of definitions. This is an area that some students will find unusual if their prior science instruction has led them to rely largely on memorization skills for success.

3. Flexibility

Students differ in many ways. Assessment that calls on students to give thoughtful responses must allow for those differences. Some students will find the open-ended character of the *EarthComm* chapter reports disquieting. They may ask many questions to try to find out exactly what the finished product should look like. Teachers will have to give a consistent and repeated message to those students, expressed in many different ways, that the ambiguity inherent in the open-ended character of the assessments is an opportunity for students to show what they know in a way that makes sense to them. This also allows for the assessments to be adapted to students with differing abilities and proficiencies.

4. Consistency

While the chapter reports are intended to be flexible, they are also intended to be consistent with the manner in which instruction takes place, and the kinds of inferences that are going to be made about students' learning on the basis of them. The *EarthComm* design is such that students have the opportunity to learn new material in a way that places it in context. Consistent with that, the chapter reports also call for the new material to be expressed in context. Traditional tests are less likely to allow this kind of expression, and are more likely to be inconsistent with the manner of teaching that *EarthComm* is designed to promote. Likewise, in that *EarthComm* is meant to help students relate Earth Science to their community, teachers will be using the chapter reports as the basis for inferences regarding the students' abilities to do that. The design of the chapter reports is intended to facilitate such inferences.

EarthComm Assessment Tools

The series of evaluation sheets and scoring rubrics provided in the back of this Teacher's Edition should be available to students before they begin their first investigation. Consider photocopying a set of the sheets for each student to include in his or her *EarthComm* notebook. The purpose of distributing the evaluation sheets is to help students become familiar with the criteria and expectations for their work. If students have a complete set of the evaluation sheets, you can refer to the relevant evaluation sheet at the appropriate point within an *EarthComm* lesson.

Think about It Evaluation Sheet

This sheet will help students to learn the basic expectations for the warm-up activity. **Think about It** is intended to reveal student conceptions about the phenomena or processes explored in the activity. It is not intended to produce closure and so your assessment of student responses should not be driven by a concern for correctness. Instead, the evaluation sheet emphasizes that you want to see evidence of prior knowledge and that students should communicate their thinking clearly. It is unlikely that you will be able to apply this assessment every time students complete a warm-up activity (there are only so many hours in a teacher's day), yet in order to ensure that students value the importance of committing their initial conceptions to paper and taking the warm-up seriously, you can use this evaluation sheet as a spot check on the quality of their work.

Investigate Notebook Entry–Evaluation Sheet

This evaluation sheet is designed to allow the students to get a sense of the expectations for *EarthComm* notebook entries. When assessing student investigations, keep in mind that the **Investigate** section of an *EarthComm* activity equates to the exploration phase of the 5E learning-cycle model where students explore their conceptions of phenomena through hands-on activity. This evaluation sheet provides a variety of criteria for you to select from when students will be in a better position to ensure that the quality of their work meets the highest possible standards and expectations. Encourage students to internalize the criteria by making the criteria part of your "assessment conversations" with them as you circulate around the classroom.

For example, while students are working, you can ask them criteria-driven questions such as: "Is your work thorough and complete? Are all of you participating in the activity? Do you each have a role to play in solving this problem?" and so on.

EarthComm Notebook Entry–Checklist

The *EarthComm* Notebook–Entry Checklist provides a quick summary of important processes, concepts, and skills that you might wish to assess during and after an investigation. You can add further criteria specific to your classroom needs or a particular investigation. The checklist provides a quick guide for student self-assessment and also provides you with an opportunity to quickly score student work.

Check Your Understanding Notebook Entry–Evaluation Sheet

This evaluation sheet is used to help you evaluate the extent to which students understand the key concepts explored in the activity and explained in the **Digging Deeper** reading section. The two criteria used in the sheet include "Reflects an Understanding of Key Concepts" and "Clarity of Expression."

Student Presentation Evaluation Form

This evaluation form provides three simple yet powerful criteria to help your students prepare their presentations. In order to prepare properly, students must know that you have expectations for the quality of the ideas they present, their ability to answer questions during the presentation, and their overall comprehension of the material. When students work in groups and present the results of their inquiry, they often divide up the work, with some members preparing the presentation and others delivering the presentation. Students need to know that any member of the group can be called upon to demonstrate their understanding of the material during a presentation. This evaluation form will help you to make this clear to students.

Student Evaluation of Group Participation

One of the challenges to assessing students who work in collaborative teams is assessing group participation. Students need to know that each group member must pull his or her weight. As a component of a complete assessment system, especially in a collaborative learning environment, it is often helpful to engage students in conducting a self-assessment of their participation in a group. Knowing that their contributions to the group will be evaluated provides an additional motivational tool to keep students constructively engaged. This evaluation form provides students with an opportunity to assess group participation. In no case should the results of this evaluation be used as the sole source of assessment data. Rather, it is better to assign a weight to the results of this evaluation and factor it in with other sources of assessment data.

Student Ratings and Self Evaluation

This form provides an alternative to the student evaluations of group participation. You might alternate the use of these forms between chapters. The two-page form is flexible in that you can assign values as you see fit for your classroom and it allows students to provide extra comments to explain the ratings.

Reviewing and Reflecting upon Your Teaching

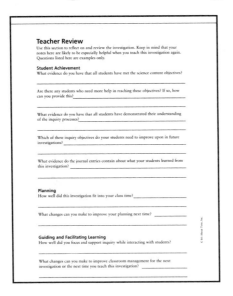
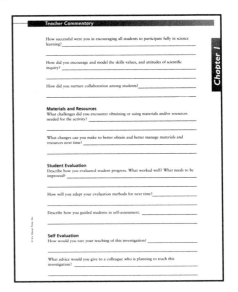

Reviewing and Reflecting upon Your Teaching provides an important opportunity for professional growth. A master copy of a two-page Teacher Review form is included at the back of each chapter. At the back of this Teacher's Edition are various Evaluation sheets for both students and teachers. They will help you to reflect upon your teaching for each investigation. We suggest that you try to answer each question at the completion of each investigation, then go back to the relevant section of this Teacher's Edition and write specific comments in the margins that will help you the next time you teach the investigation. For example, if you found that you were able to make substitutions to the list of materials needed, write a note about those changes in the margin of that page of this guide.

Using the *EarthComm* Web Site

http://www.agiweb.org/earthcomm
The *EarthComm* web site has been designed for teachers and students.

- Each *EarthComm* chapter has its own web page that has been designed specifically for the content addressed within that chapter.

- Chapter web sites are broken down by activity and also contain a section with links to relevant resources that are useful for the chapter.

- Each activity is divided into background information, materials and supplies needed, completing the investigation, and suggested links for completing the **Inquiring Further** portion of each activity.

Enhancing Teacher Content Knowledge

Each *EarthComm* chapter has a specific web page that will help teachers to gather further background information about the major topics covered in each activity. Example from Volcanoes and Your Community—Activity 1: Where are the Volcanoes?

To learn more about this topic, visit the following web sites:
1. **Volcanoes Beneath the Sea:**
Volcano World — "Submarine Volcanoes"
Reviews the basics of plate tectonics and examines closely submarine volcanoes at divergent and convergent boundaries and hot spots. The site has good images of underwater lava flows as well as images of the organisms that live near these submarine volcanoes. http://volcano.und.nodak.edu/vwdocs/Submarine/

Obtaining Resources

The community focus of *EarthComm* will require teachers to obtain local or regional maps, rocks, and data. The *EarthComm* web site helps teachers to find these materials.

Completing the Investigation

This portion of each activity provides suggestions for using Internet data in the classroom.

Example from Volcanoes and Your Community — Activity 4: Volcanic Hazards: Airborne Debris.

> Using the Internet during the investigation:
> To simulate the eruption of one of the three volcanoes, go to the Volcanic Ash Forecast Transport and Dispersion (http://www.arl.noaa.gov/ready/vaftadmenu.html) web site and click on "Run VAFTAD Model."

Inquiring Further

This section of each activity provides suggested web sites that will helps students to complete the **Inquiring Further** portion of *EarthComm* activities. This will help students to make the most of what is often limited time available to conduct Internet-based research during the school day.

Example from Volcanoes and Your Community — Activity 3: Volcanic Hazards: Flows.

> To complete the **Inquiring Further** section of this activity:
>
> Investigate how lava flows affect the biosphere:
> • USGS Hawaiian Volcano Observatory (http://hvo.wr.usgs.gov/) has information about Hawaiian volcanoes
> • University of Hawaii, Hawaii Center for Volcanology (http://www.soest.hawaii.edu/GG/hcv.html) has general and specific information about the Hawaiian shield volcanoes and links to other volcano sites
> • VolcanoWorld (http://volcano.und.nodak.edu/vw.html)
> • NOAA - VENTS Program (http://www.pmel.noaa.gov/vents/) has information about black smokers and activity along mid-ocean ridges
> • RIDGE Program (http://ridge.oce.orst.edu/) has information about activity along mid-ocean ridges and links to further information.

Resources

The web page for each *EarthComm* chapter provides a list of relevant web sites, maps, videos, books, and magazines. Specific links to sources of these materials are often provided.

Managing Collaborative Group Learning

Working in small collaborative groups is seen as an important part of scientific inquiry, and is reinforced by the *National Science Education Standards* and *Benchmarks for Science Literacy*. Scientists, and others, frequently work in teams to investigate things and solve problems. However, there are times when it is important to work alone. You may have students who are more comfortable working on their own. Traditionally, the competitive nature of school curricula has emphasized individual effort through grading, "honors" classes, and so on. Many parents will have been through this experience themselves as students and will be looking for comparisons between their children's performance and other students as a result. Managing collaborative groups may therefore present some initial problems, especially if you have not organized your class in this way before and the idea is new for your students. Below are some key points to keep in mind as you develop a group approach.

- Arrange your classroom furniture into small group areas.
- Explain to students ahead of time how and why they are going to work in groups.
- Stress the responsibility each group member has to the others in the group.
- Choose student groups carefully to ensure each group has a balance of ability, special talents, gender, ethnicity, and so on.
- Make it clear that groups are not fixed for all time and that their composition will change from time to time.
- Promote the idea of fair work-sharing within groups, where everyone is contributing.
- Help students see the benefits of learning with and from each other.
- Ensure that there are some opportunities for students to work alone.
- Provide students with a copy of any rubrics that address group work and discuss the rubrics with them.

There are assessment rubrics provided in this Teacher's Edition to help you and your students manage and evaluate student collaboration (Student Evaluation of Group Participation, page 668; Student Ratings and Self Evaluation, pages 669 and 670).

Enhancing *EarthComm's Earth System Evolution* with *EarthView Explorer*

There are many similarities between *EarthComm Earth System Evolution (ESE)* and *EarthView Explorer (EView)* that will be useful to teachers using both sets of resources in their Earth Science classes. For many of the topics in *ESE* there are many opportunities for more detailed investigation using *EView*. The usage notes for these correlations should be considered carefully before classroom use because the context of the material in the two resources can be significantly different and a proper transition is necessary.

The **Chapter Challenge** at the beginning of each chapter of *Earth System Evolution* is similar to the processes and major data products table in the *EarthView* Teacher's Manual (Table 1). In *EarthComm*, the **Scenario** introduces what in *EarthView* are the Science, Technology and Society issues. Students can use the Scenario to relate their existing understanding of the Earth Science topics to their experiences and begin to comprehend how the topic relates to their lives. In *EarthView*, maps of population help students visualize how phenomena may or may not affect people. In *EarthComm* the **Assessment Criteria** are presented to students before they begin their activities so that they can plan their study at a high level. In *EarthView*, this function is performed by the **Basecamp**, in which students can keep a journal of the purpose and results of each of their activities. These notes are intended to help students to construct a more extensive familiarity with the material.

Most importantly, the *ESE* activities share a similar inquiry-based approach to the *EView* activities. The *ESE* materials have several different approaches to inquiry. They use hands-on experimentation, charts and data analyses. *EView* focuses on the latter two, focusing on in-depth quantitative inquiry.

What follows is a cross-index between *ESE* and *EView* that attempts to use the strengths of each. For example, using specific data investigations in *EView* to dig deeper on various *ESE* topics. Of the three *ESE* chapters, the correlation to *EView* is best for *Chapter 2 – Climate Change*. However, there are many useful correlations in the other chapters as well. The good content matchup arises from the fact that both sets of materials are derived from current Earth Science research topics and thus by necessity share many common elements.

Enhancing *EarthComm's* *Earth System Evolution* with *EarthView Explorer*

Earth System Evolution (*ESE*) Content	Page no.	Matching EarthView (*EView*) Content	Usage Notes
— **CHAPTER 1: ASTRONOMY...AND YOUR COMMUNITY**			
General: *EView's* content covers the upper atmosphere and therefore overlaps with many *ESE* investigations. Most of the celestial mechanics discussed in *ESE* affect the solar input term of *EView's* climate model.			
ACTIVITY 1 The History and Scale of the Solar System			
Getting Started	E2	Climate unit – Geosphere Atmosphere – Climate zones	The impact of large meteorites and comets would impact the albedo in a manner similar to that shown for volcanic eruptions.
ACTIVITY 3 The Earth–Moon System			
Preparing for the Chapter Challenge	E36	Atmosphere unit – Climate Zones Ozone Climate unit – all activity	Students should understand that celestial mechanics affect solar intensity on the surface of the Earth. This can be expanded in the Atmosphere unit on Ozone because total UV radiation will also change with solar intensity. Also, the intensity of solar radiation is a major term in the atmospheric temperature model. Thus, these *EarthView* extensions can be used here.
ACTIVITY 4 Impact Events and the Earth System			
Investigate, Understanding and Applying	E39	Geosphere unit – Earthquakes Climate unit – Geosphere	The concept of a Richter scale of earthquake magnitudes is introduced in *EView's* activity on Earthquakes and thus this activity could be used as additional backround relevant to the comparison of impact energies and Richter magnitudes. The *EView* Climate section ties into **Question 3** in **Understanding and Applying** in that the impact of large meteorites and comets would impact the albedo in a manner similar to that shown for volcanic eruptions.
ACTIVITY 5 The Sun and Its Effects on Your Community			
Digging Deeper	E50	Climate unit – particularly Atmosphere and Hydrosphere Atmosphere – Ozone	The flow of solar energy through the atmosphere is the core process in the *Eview* Climate unit. The Atmosphere and Hydrosphere activities contain diagrams of the energy flow similar to that on page E50. The ozone hole is the topic of the Atmosphere activity on Ozone.
ACTIVITY 6 The Electromagnetic Spectrum and Your Community			
Digging Deeper	E64	Atmosphere unit – Climate Zones Climate unit – Atmosphere	The concept of the electromagnetic spectrum is complex and is relevant to *EView* activities listed. Both *ESE* and *EView* use the spectrum to explain the greenhouse effect.

Enhancing *EarthComm's Earth System Evolution* with *EarthView Explorer*

Earth System Evolution (*ESE*) Content	Page no.	Matching EarthView (*EView*) Content	Usage Notes
CHAPTER 2: CLIMATE CHANGE...AND YOUR COMMUNITY			
General: Because climate is a theme that runs throughout *Earth View*, this Chapter correlates with activities in all of the units.			
ACTIVITY 1 Present-Day Climate in Your Community			
Investigate Parts A and B; Digging Deeper	E85	Atmosphere unit – Climate zones Hydrosphere unit – Rivers & Rain Biosphere unit – Land biomes Climate unit – main model	The maps of biomes and their association with climate patterns are presented in *EView's* Land Biomes activity. The seasonal air temperatures are mapped in the Climate zones activity. The average rainfall is mapped in the Rivers & Rain activity. The Climate model shown in all of the Climate unit's Exploration screens uses glaciers to denote temperature changes and this ties into the **Digging Deeper** discussion of ice ages and glaciers.
ACTIVITY 2 Paleoclimates			
Investigate Part B; Understanding and Applying	E98	Geosphere – Sea-Floor Spreading Climate unit – main model	The Sea-Floor Spreading activity contains the map of sea floor sediments. The advance and retreat of ice sheets in response to temperature change is one of the main features of the *EView* Climate model screen.
ACTIVITY 3 How do Earth's Orbital Variations Affect Climate?			
Investigate Part B	E107	Atmosphere – Climate Zones	The Climate zones activity illustrates the relationship between the climate zones and the angle of sunlight.
ACTIVITY 4 How Do Plate Tectonics and Ocean Currents Affect Global Climate?			
Investigate; Digging Deeper	E118	Hydrosphere – Ocean Temperature Geosphere – Sea-Floor Spreading	The *EView* Ocean Temperature activity focuses on the same current pattern as shown in the figure on page E119 to illustrate the redistribution of heat by currents. The **Information** section of the Ocean Temperature activity also contains a diagram of deep water formation similar to that in *Figure 2* of the **Digging Deeper** section. The discussion of plate movements after Pangea is also addressed in the Sea-Floor Spreading activity.
ACTIVITY 5 How Do Carbon Dioxide Concentrations in the Atmosphere Affect Global Climate?			
Investigate parts A-C; Digging Deeper	E126	Atmosphere – Greenhouse Gases Biosphere – Land Plants Climate – Atmosphere, Biosphere	The *EView* activity on Greenhouse gases reviews the changes in atmospheric carbon dioxide over the last century for an expanded view of the table and graph on page E126. The Climate/Atmosphere activity focuses on the greenhouse effect on temperature. All listed *EView* activities are relevant to the **Digging Deeper** section. In particular, the Greenhouse gases and Land Plants activities introduce the carbon cycle and anthropogenic emissions of carbon dioxide.
ACTIVITY 6 How Might Global Warming Affect Your Community?			
Digging Deeper	E138	Climate – all activities	The interactive effects of albedo and greenhouse gases on global temperature comprise the core model in the Climate unit. The effect of climate change on sea level is also depicted in all Climate model screens.

Enhancing *EarthComm's Earth System Evolution* with *EarthView Explorer*

Earth System Evolution (*ESE*) Content	Page no.	Matching EarthView (*EView*) Content	Usage Notes
CHAPTER 3: CHANGING LIFE...AND YOUR COMMUNITY			
General: The Biosphere unit in *EView* correlates well to the material in this Chapter.			
ACTIVITY 1 The Fossil Record and Your Community			
Investigate Part B; Digging Deeper	E150	Biosphere – Ocean Plants	The Ocean Plants activity also addresses the concepts of food webs and nutrient cycling.
ACTIVITY 2 North American Biomes			
Investigate; Digging Deeper	E157	Biosphere – Land Biomes	The *EView* Land Biomes activity also contains a map of global terrestrial biomes. The overlays of climate zones and precipitation help students visualize the detailed connections of concepts introduced in the **Digging Deeper** section.
ACTIVITY 3 Your Community and the Last Glacial Maximum			
Digging Deeper	E170	Biosphere – Land Biomes	The overlays of climate zones and precipitation on the Land Biomes help students determine how biomes can change.
ACTIVITY 4 The Mesozoic-Cenozoic Boundary Event			
Digging Deeper	E179	Biosphere – Land Plants; Land Biomes Climate – Geosphere	*EView* distribution maps in the Biosphere section, coupled with the illustration of dust effect on climate in the Climate – Geosphere section, can be used to supplement the **Digging Deeper** section.
ACTIVITY 5 How Different Is Your Community Today from that of the Very Deep Past?			
Digging Deeper	E184	Biosphere – Land Plants; Land Biomes Climate – Geosphere	*EView* distribution maps in the Biosphere section, coupled with the illustration of dust effect on climate in the Climate – Geosphere section, can be used to supplement the **Digging Deeper** section.

GETIT™ Geoscience Education Through Interactive Technology
for Grades 6-12

Earthquakes, volcanoes, hurricanes, and plate tectonics are all subjects that deal with energy transfer at or below the Earth's surface. The GETIT CD-ROM uses these events to teach the fundamentals of the Earth's dynamism. GETIT contains 63 interactive—inquiry-based—activities that closely simulate real-life science practice. Students work with real data and are encouraged to make their own discoveries—often learning from their mistakes. They use an electronic notebook to answer questions and record ideas, and teachers can monitor their progress using the integrated class-management module. The Teacher's Guide includes Assessments, Evaluation Criteria, Scientific Content, Graphs, Diagrams and Blackline Masters. GETIT conforms to the National Science Education Standards and the American Association for the Advancement of Science benchmarks for Earth Science.

Enhancing *EarthComm's Earth System Evolution* with GETIT

Earth System Evolution				
Chapter	Activity	Page	Goals	GETIT Activity
1. Astronomy	1. The History and Scale of the Solar System.	E13	**Inquiring Further:** 1. Solar-system walk	• Science Showtime episode: What's the big idea?
	4. Impact Events and the Earth System.	E37	Goal: Compare natural and man-made disasters to the impact of an asteroid.	• How much energy is released by an earthquake? • How much damage is done?
	5. The Sun and its Effects on your Community.	E50	**Digging Deeper:** The Sun and its effects.	• See the light.
	6. The Electromagnetic Spectrum and Your Community.	E58	Goal: Explain electromagnetic radiation and the electromagnetic spectrum in terms of wavelength, speed, and energy. Goal: Understand that some forms of electromagnetic radiation are essential and beneficial to us on Earth, and others are harmful.	• See the light.
		E59	**Investigate Part B:** Scaling the electromagnetic spectrum.	• Scientific notation. • See the light.
	7. Our Community's Place Among the Stars	E69	Study stellar structure and the stellar evolution (the life history of stars).	• Hertzsprung-Russell diagram (located on one of the laptops in the volcanology lab).
		E71	**Investigate Part B:** Luminosity and temperature of stars.	• Hertzsprung-Russell diagram (located on one of the laptops in the volcanology lab).
		E73	**Digging Deeper:** Earth's stellar neighbors.	• Hertzsprung-Russell diagram (located on one of the laptops in the volcanology lab).
2. Climate Change	1. Present-Day Climate in Your Community	E85	**Investigate Part A:** Physical features and climate in your community.	• World geographic map (What's This Map? button)
	4. How do Plate Tectonics and Ocean Currents Affect Global Climate?	E117	Model present and ancient land masses and oceans to determine current flow.	• Plate reconstruction. • Coriolis effect.
		E121	**Digging Deeper:** How Plate Tectonics affects Global Climate	• Earthquakes, volcanoes, and plate margins.

Module Overview: *Earth System Evolution*

A great many changes have occurred in the Earth system over the vastness of geologic time. Each of these changes has contributed to the evolution of the different spheres of the earth system and the creation of the modern Earth. *Earth System Evolution* uses a student-based inquiry approach to help students gain an understanding of how many changes in the Earth system have contributed to the creation of the present physical environment and have influenced the evolution of life. The module consists of three chapters: *Astronomy and Your Community, Climate Change and Your Community,* and *Changing Life and Your Community.* Each looks at change through time and focuses on concepts necessary to understand some of the major events that affect such change. *Astronomy and Your Community* deals with the formation of Earth and its role as a subsystem in a dynamic universe. *Climate Change and Your Community* examines the relationships between the physical environment and climate on Earth through time. In *Changing Life and Your Community* students investigate how and why the organisms found in their community have changed over time.

Themes

Through their inquiry in this module, students develop understandings of the evolution of Earth's systems. The major themes addressed include the following concepts that relate to the *National Science Education Standards* for Grades 9-12:

- Earth is a part of the solar system, which in turn is part of the galaxy, and on a larger scale, the universe.
- Earth's external energy depends on varying factors, including distance from the Sun, solar variability, and gravitational forces.
- The Earth's physical environment includes landmasses, atmosphere, and oceans, which have all contributed to its formation and continued changes.
- Fossil evidence has shown how organisms have evolved from ancient environments and continue to change as our environment evolves.

Astronomy... and Your Community

This chapter gives students a feeling of their place as a member of not only of their local community, but also of the Earth community, the solar system, and the universe. In *Astronomy and Your Community*, students are presented with a hypothetical situation in which an asteroid is predicted to pass very close to Earth. Students are asked to prepare a booklet for their community that discusses some of the possible hazards from outer space and the benefits of living in our solar system. Students begin the chapter by putting the scale of the solar system into perspective and learning how the solar system formed. Students then examine the relationship between the Earth and Moon and the Earth and the Sun. Students investigate how an asteroid collision would affect their community. Finally, students examine Earth's place within the Milky Way Galaxy. By the end of the chapter students should understand that the history and size of their planetary community and the processes and events that occur within the solar system influence the Earth. They should also develop an appreciation of what the chances are of those events happening.

Climate Change... and Your Community

The **Chapter Challenge** for *Climate Change and Your Community* asks students to write a series of newspaper articles that explore global climate change. As students move through the chapter they gain the knowledge needed to write the articles concerning how global climate has changed over time, what causes global climate change, and the meaning of "global warming" and its possible affect on their community. Finally, students are asked to use the knowledge they have acquired to write an editorial piece that states whether or not their community should be concerned about global warming, and what steps the community should take in response to the possibility of global warming.

Changing Life... and Your Community

In this chapter, students are challenged to produce a display that illustrates the biological changes that their community has experienced over several scales of geologic time. Students begin the chapter by exploring the process of fossilization and determining which organisms in their community are most likely to be preserved in the fossil record. Students look at how climate influences the types of organisms found in a specific place. Students use this information to hypothesize about how organisms in their community might have differed 20,000 years ago, when glaciers covered portions of North America. Extinction events and how they fit into the evolutionary scheme are explored. Finally, students investigate how their community has changed throughout geologic time, and what effect this might have had on the types of organisms found in their community.

1

Astronomy
...and Your Community

EARTH SYSTEM EVOLUTION
CHAPTER I

ASTRONOMY...
AND YOUR COMMUNITY

Chapter Overview

Chapter 1 takes students beyond their local community and gives them a sense of their place in the Earth community, the solar system, and the universe. **Astronomy and Your Community** presents students with a hypothetical situation in which an asteroid is predicted to pass very close to Earth. The **Chapter Challenge** requires them to prepare a booklet for their community that discusses some of the possible hazards from outer space.

Students begin the chapter by putting the scale of the solar system into perspective and learning how our solar system formed. They examine the relationships between the Earth and the Moon and between the Earth and the Sun. Students investigate how an asteroid collision would affect their community. Finally, they examine Earth's place in the Milky Way Galaxy. By the end of **Chapter 1**, students understand:
- the history of our planetary community
- the size of our planetary community
- the processes and events in the solar system that influence the Earth
- the chances of those events happening

Chapter Goals for Students

- Understand how extraterrestrial processes affect the Earth systems.

- Participate in scientific inquiry and construct logical conclusions based on evidence.

- Appreciate the value of Earth science information in improving the quality of life, globally and in the community.

Chapter Timeline

Chapter 1 takes about three weeks to complete, assuming one 45-minute period per day, five days per week. Adjust this guide to suit your school's schedule and standards. Build flexibility into your schedule by manipulating homework and class activities to meet your students' needs.

A sample outline for presenting the chapter is shown on the following pages. This plan assumes that you assign homework at least three nights a week, and that you assign **Understanding and Applying What You Have Learned** and **Preparing for the Chapter Challenge** as group work to be completed during class. This outline also assumes that **Inquiring Further** sections are reserved as additional, out-of-class activities. This is only a sample, not a suggested or recommended method of working through the chapter. Adjust your daily and weekly plans to meet the needs of your students and your school.

Note that part of **Activity 2** asks students to observe the Moon for at least four weeks. You may wish to have them begin collecting their observations before starting **Chapter 1**, so that the lunar observation chart has been mostly filled in. Students can add additional information as they complete the investigation.

Day	Activity	Homework
1	Getting Started; Activity; Chapter Challenge; Assessment Criteria	
2	Activity 1 – Investigate	Investigate continued; Digging Deeper; Check Your Understanding
3	Activity 1 – Review; Understanding and Applying; Preparing for the Chapter Challenge	
4	Activity 2 – Investigate; Parts A to D	Digging Deeper; Check Your Understanding
5	Activity 2 – Review; Understanding and Applying; Preparing for the Chapter Challenge	
6	Activity 3 – Investigate	Digging Deeper; Check Your Understanding
7	Activity 3 – Review; Understanding and Applying; Preparing for the Chapter Challenge	
8	Activity 4 – Investigate	Digging Deeper; Check Your Understanding
9	Activity 4 – Review; Understanding and Applying; Preparing for the Chapter Challenge	
10	Activity 5 – Investigate	Digging Deeper; Check Your Understanding

Day	Activity	Homework
11	**Activity 5 – Review; Understanding and Applying; Preparing for the Chapter Challenge**	
12	**Activity 6 – Investigate, Parts A, B, and C**	**Digging Deeper; Check Your Understanding**
13	**Activity 6 – Review; Understanding and Applying; Preparing for the Chapter Challenge**	
14	**Activity 7 – Investigate, Parts A and B**	**Digging Deeper; Check Your Understanding**
15	**Activity 7 – Review, Understanding and Applying; Preparing for the Chapter Challenge**	
16	**Complete Chapter Report**	**Finalize Chapter Report**
17	**Present Chapter Report**	

National Science Education Standards

Preparing a booklet to inform citizens about possible hazards from outer space sets the stage for the **Chapter Challenge**. Students learn the place of the Earth in the solar system, the galaxy, and the universe. Through a series of activities, students begin to develop the content understandings outlined below.

CONTENT STANDARDS

Unifying Concepts and Processes
- Systems, order, and organization
- Evidence, models, and explanation
- Constancy, change, and measurement
- Evolution and equilibrium

Science as Inquiry
- Identify questions and concepts that guide scientific investigations
- Design and conduct scientific investigations
- Use technology and mathematics to improve investigations
- Formulate and revise scientific explanations and models using logic and evidence
- Communicate and defend a scientific argument
- Understand scientific inquiry

Earth and Space Science
- Energy in the Earth system
- Geochemical cycles
- Origin and evolution of the Earth system
- Origin and evolution of the universe

Science and Technology
- Communicate the problem, process, and solution
- Understand science and technology

Science in Personal and Social Perspectives
- Personal and community health
- Natural and human-induced hazards

History and Nature of Science
- Science as a human endeavor
- Nature of scientific knowledge
- Historical perspectives

Key Science Concepts and Skills

Activities Summaries	Earth Science Principles
Activity 1: The History and Scale of the Solar System To develop an understanding of the size of our solar system, students make models of the solar system using different scales.	• Astronomical distance and time (astronomical units, light-years, parsecs) • Nebular theory, birth of the planets • Relationship of our solar system in the Milky Way Galaxy
Activity 2: The Earth–Moon System Students complete an exercise to understand what causes the different lunar phases. They observe the lunar phases for one month. Students examine data on tides and lunar phases to understand the relationship between the tides and phases of the Moon. They complete calculations to determine how the Moon has influenced the length of a year on Earth.	• Lunar phases • Formation of the Moon • Tides
Activity 3: Orbits and Effects Students draw a series of ellipses to understand the relationship between distance between foci and eccentricity of an ellipse. They study the eccentricity of the Earth's orbit around the Sun. They consider how the shape of the Earth's orbit is related to changes in climate through time.	• Eccentricity • Obliquity • Precession • Orbital inclination
Activity 4: Impact Events and The Earth System Students calculate the energy released when a hypothetical asteroid collides with Earth. They calculate the energy released during known impact events in the Earth's past. Students compare these calculations with the energy released through natural and human-made phenomena like earthquakes and bombs.	• Asteroids • Meteors • Impact events
Activity 5: The Sun and Its Effects on Your Community Students plot sunspot activity from 1899 to 1998 and determine the pattern of this activity. They examine data on strength of solar flares. They correlate this data with sunspot activity to understand that the two are related. Students then consider the effect of solar activity on Earth.	• Structure of the Sun • Sunspots • Solar flares • Earth's energy budget
Activity 6: The Electromagnetic Spectrum and Your Community Students use a spectroscope to determine how the visible spectrum looks in natural sunlight, fluorescent light, and incandescent light. They prepare a scale model of the electromagnetic spectrum to get a sense of the relative sizes of the different electromagnetic bands. Students then research space science missions to understand how astronomers use electromagnetic radiation to study objects and events in the solar system.	• Electromagnetic radiation • Using electromagnetic radiation in astronomy • Frequency, wavelength
Activity 7: Our Community's Place Among the Stars Students complete an exercise to understand how brightness varies with distance. To help them understand how astronomers classify stars, they use luminosity and surface temperature to plot stars on a Hertzsprung-Russell diagram.	• Hertzsprung-Russell diagram • Classification of stars • Luminosity • Life cycles of stars

Equipment List for Chapter One:

Materials needed for each group per activity.

Activity 1
- Calculator
- Metric ruler
- Tape measure

Activity 2 Part A
- Pencil or small wood dowel
- Styrofoam® ball, at least 5 cm in diameter
- Lamp with a 150-W bulb and no lampshade (or a flashlight)

Activity 2 Part B
- Copy of lunar observation chart (see **Blackline Master 2.1, Lunar Observation Chart**)

Activity 2 Part C
- Graph paper
- Internet access (or printouts of tidal data for several cities near your community)*

Activity 2 Part D
- Graph paper
- Calculator

Activity 3
- Piece of paper
- Straightedge
- Tape
- Thick sheet of cardboard (8.5 x 11 in.), Styrofoam, or newspaper
- Two pushpins
- Piece of string – 30 cm long
- Metric ruler
- Calculator

Activity 4
- Calculator
- Metric ruler

Activity 5
- Graph paper

Activity 6 Part A
- Spectroscope (or materials to make one: cardboard tube, aluminum foil, scissors, piece of diffraction grating – 2 cm x 2 cm, two small pieces of overhead transparency, tape)
- Fluorescent light
- Incandescent light

Activity 6 Part B
- Tape
- Metric ruler
- Colored pencils: red, orange, yellow, green, blue, violet
- Calculator
- Paper

Activity 6 Part C
- Internet access

Activity 7 Part A:
- Three lamps with 40-W, 60-W, and 100-W bulbs with frosted glass envelopes
- Light meter (optional)
- Graph paper
- Meter stick

Activity 7 Part B
- Copy of blank HR diagram (**Blackline Master 7.1, HR Diagram**)
- Copy of HR diagram with star categories (**Blackline Master 7.2, HR Diagram**)
- Copy Table 1-Selected Properties of Fourteen Stars (**Blackline Master 7.3, Selected Properties of Fourteen Stars**)
- **Blackline Master Astronomy 7.4, Selected Properties of Fourteen Stars**

*See the *EarthComm* web site for information about how to obtain these resources at http://www.agiweb.org/earthcomm

Astronomy ...and Your Community

Getting Started

Throughout time, all systems in the universe are affected by processes and outside influences that change them in some way. This includes Earth and the solar system in which it exists. You have years of experience with life on the third planet from the Sun, and you know a lot about your tiny corner of the universe. Think about the Earth in relation to its neighbors in the solar system.

• What objects make up the solar system?

• How far is the Earth from other objects in the solar system?

• Which objects in the solar system can influence the Earth?

• Can you think of any objects or processes outside the solar system that might affect the Earth?

Write a paragraph about Earth and its place in this solar system. After that, write a second paragraph about processes or events in the solar system that could change Earth. Describe what they do and how Earth is, or might be, affected. Try to include answers to the questions above.

Scenario

Scientists recently announced that an asteroid 2-km wide, asteroid 1997XF11, would pass within 50,000 km of Earth (about one-eighth the distance between the Earth and the Moon) in October 2028. A day later, NASA scientists revised the estimate to 800,000 km. News reports described how an iron meteorite blasted a hole more than 1 km wide and 200 m deep, and probably killed every living thing within 50 km of impact. That collision formed Arizona's Meteor Crater some 50,000 years ago. Such a collision would wipe out a major city today. These reports

Getting Started

Uncovering students' conceptions about Astronomy and the Earth System

Use **Getting Started** to elicit students' ideas about the main topic. The goal of **Getting Started** is not to seek closure (i.e., the "right" answer) but to provide you, the teacher, with information about the students' starting point and about the diversity of ideas and beliefs in the classroom. By the end of the chapter, students will have developed a more detailed and accurate understanding of how the processes and events in the solar system affect the Earth system.

Ask students to work independently or in pairs and to exchange their ideas with others. Avoid labeling answers right or wrong. Accept all responses, and encourage clarity of expression and detail.

The first question will help you to explore students' conceptions about objects in the solar system. They might name the nine planets, note which planets have moons, and note which planets are rocky and which planets are gaseous. Some will take a more general approach and note that the solar system is made up of different kinds of objects, including the Sun, planets, moons, meteorites, asteroids, and comets.

The second question should elicit ideas about conceptions of scale of the solar system, and where Earth is located relative to other objects. Again, students may take different approaches to this question. Some may have specific ideas about distances. ("Earth is 93 million miles from the Sun. Our Moon is about 250,000 miles away from Earth".) Other students may take a more general approach by using relative scales ("The Moon is closer to Earth than any other object in the solar system" or "Earth is much closer to the Sun than it is to Pluto.") A few students will be aware that distance between Earth and other objects changes over time or in cycles.

The third question on the list explores students' ideas about how Earth is impacted by matter in the solar system. Common responses include that the Moon (or the Moon and the Sun) affect tides on Earth, and that the Sun provides a source of energy for life on Earth. Some students will note that meteorites can strike the surface of the Earth.

The last question on the list explores students' ideas about whether and how Earth might be impacted by objects or processes outside the solar system. It will be interesting to see what kinds of ideas students come up with here. Do they view our solar system as a self-contained system, or as a system that is connected to and impacted by other systems within our galaxy and universe?

have raised concern in your community about the possibility of a comet or asteroid hitting the Earth. Your class will be studying outer space and the effects that the Sun and other objects in the solar system can have on the Earth. Can you share your knowledge with fellow citizens and publish a booklet that will discuss some of the possible hazards from outer space?

Chapter Challenge

In your publication, you will need to do the following:

- Describe Earth and its place in the universe. Include information about the formation and evolution of the solar system, and about the Earth's distance from and orbit around the Sun. Be sure to mention Earth's place in the galaxy, and the galaxy's place in the universe.

- Describe the kinds of solar activities that influence the Earth. Explain the hazardous and beneficial effects that solar activity (sunspots and radiation, for example) have on the planet. Discuss briefly the Sun's composition and structure, and that of other stars.

- Discuss the Earth's orbital and gravitational relationships with the Sun and the Moon.

- Explain what comets and asteroids are, how they behave, how likely it is that one will collide with Earth in your lifetime, and what would happen if one did.

- Explain why extraterrestrial influences on your community are a natural part of Earth system evolution.

The booklet should have a model of the solar system that will help citizens understand the relative sizes of and distances between solar-system bodies.

Assessment Criteria

Think about what you have been asked to do. Scan ahead through the chapter activities to see how they might help you to meet the challenge. Work with your classmates and your teachers to define the criteria for assessing your work. Record all of this information. Make sure that you understand the criteria as well as you can before you begin. Your teacher may provide you with a sample rubric to help you get started.

Chapter Challenge and Assessment Criteria

Read (or have a student read) the **Chapter Challenge** aloud to the class. Allow students to discuss what they have been asked to do. Have students meet in teams to begin brainstorming what they would like to include in their **Chapter Challenge** reports. Ask them to summarize briefly in their own words what they have been asked to do. Review the attributes of a high-quality report.

Alternatively, lead a class discussion about the challenge and the expectations. Review the titles of the activities in the Table of Contents. To remind students that the content of the activities corresponds to the content expected for the Chapter Report, ask them to explain how the title of each activity relates to the expectations for the **Chapter Challenge**. Familiarize students with the way activities are structured, pointing out the sections that are common to all activities. Note particularly the section titled **Preparing for the Chapter Challenge**.

Guiding questions for discussion include:

- What do the activities have to do with the expectations of the challenge?
- What have you been asked to do?
- What should a good final report contain?

A sample rubric for assessing the **Chapter Challenge** is shown on the following page. You can copy and distribute the rubric as is, or you can use it as a baseline for developing scoring guidelines and expectations that suit your needs. For example:

- You may wish to ensure that core concepts and abilities derived from your local or state science frameworks also appear on the rubric.
- You may wish to modify the format of the rubric to make it more consistent with your evaluation system.

However you decide to evaluate the Chapter Report, keep in mind that all expectations should be communicated to students and that the expectations should be outlined at the start of their work. Please review **Assessment Criteria** (pages xxiv to xxv of this Teacher's Edition) for a more detailed explanation of the assessment system developed for the *EarthComm* program.

Assessment Rubric for Chapter Challenge on Astronomy

Meets the standard of excellence.
 	
5
	Significant information is presented about _all_ of the following:
	• How the Earth fits into the universe, including:
	  • formation and evolution of the solar system
	  • Earth's distance from, and orbit around, the Sun
	  • Earth's place in the galaxy
	  • the galaxy's place in the universe
	• How solar activity influences the Earth, what the hazardous and beneficial effects of solar radiation include, and how the Sun and the other stars are structured.
	• What the Earth's orbital and gravitational relationships with the Sun and the Moon are.
	• What comets and asteroids are, how they behave, how likely it is that one will collide with Earth, and what would happen if a collision occurs.
	• Why extraterrestrial influences on the community are a natural part of Earth system evolution.
	All of the information is accurate and appropriate.
	The writing is clear and interesting.

Approaches the standard of excellence.
 	
4
	Significant information is presented about _most_ of the following:
	• How the Earth fits into the universe, including:
	  • formation and evolution of the solar system
	  • Earth's distance from, and orbit around, the Sun
	  • Earth's place in the galaxy
	  • the galaxy's place in the universe
	• How solar activity influences the Earth, what the hazardous and beneficial effects of solar radiation include, and how the Sun and the other stars are structured.
	• What the Earth's orbital and gravitational relationships with the Sun and the Moon are.
	• What comets and asteroids are, how they behave, how likely it is that one will collide with Earth, and what would happen if a collision occurs.
	• Why extraterrestrial influences on the community are a natural part of Earth system evolution.
	All of the information is accurate and appropriate.
	The writing is clear and interesting.

Meets an acceptable standard.
 	
3
	Significant information is presented about _most_ of the following:
	• How the Earth fits into the universe, including:
	  • formation and evolution of the solar system
	  • Earth's distance from, and orbit around, the Sun
	  • Earth's place in the galaxy
	  • the galaxy's place in the universe
	• How solar activity influences the Earth, what the hazardous and beneficial effects of solar radiation include, and how the Sun and the other stars are structured.
	• What the Earth's orbital and gravitational relationships with the Sun and the Moon are.
	• What comets and asteroids are, how they behave, how likely it is that one will collide with Earth, and what would happen if a collision occurs.
	• Why extraterrestrial influences on the community are a natural part of Earth system evolution.
	Most of the information is accurate and appropriate.
	The writing is clear and interesting.

Assessment Rubric for Chapter Challenge on Astronomy

Below acceptable standard and requires remedial help. 2	<u>*Limited*</u> information is presented about the following: • How the Earth fits into the universe, including: • formation and evolution of the solar system • Earth's distance from, and orbit around, the Sun • Earth's place in the galaxy • the galaxy's place in the universe • How solar activity influences the Earth, what the hazardous and beneficial effects of solar radiation include, and how the Sun and the other stars are structured. • What the Earth's orbital and gravitational relationships with the Sun and the Moon comprise. • What comets and asteroids are, how they behave, how likely it is that one will collide with Earth, and what would happen if a collision occurred. • Why extraterrestrial influences on the community are a natural part of Earth system evolution. <u>*Most*</u> of the information is accurate and appropriate. Generally, the writing does not hold the reader's attention.
Basic level that requires remedial help or demonstrates a lack of effort. 1	<u>*Limited*</u> information is presented about the following: • How the Earth fits into the universe, including: • formation and evolution of the solar system • Earth's distance from, and orbit around, the Sun • Earth's place in the galaxy • the galaxy's place in the universe • How solar activity influences the Earth, what the hazardous and beneficial effects of solar radiation include, and how the Sun and the other stars are structured. • What the Earth's orbital and gravitational relationships with the Sun and the Moon comprise. • What comets and asteroids are, how they behave, how likely it is that one will collide with Earth, and what would happen if a collision occurred. • Why extraterrestrial influences on the community are a natural part of Earth system evolution. <u>*Little*</u> of the information is accurate and appropriate. The writing is difficult to follow.

ACTIVITY 1 — THE HISTORY AND SCALE OF THE SOLAR SYSTEM

Background Information

Students began **Chapter 1** by reading the scenario of an Earth-crossing asteroid that will come very close to our planet. The **Chapter Challenge** expects them to gather information about extraterrestrial threats to Earth and to make a report to their community explaining the risks and dangers. Before they can write their report, students need to understand more about the solar system and the objects that constitute it, and learn to assess what constitutes a threat to the Earth and their community. **Activity 1** familiarizes them with the scale and history of the solar system.

Comprehending Distances in the Universe
It sometimes surprises people to learn just how far apart solar system objects really are and what their sizes are relative to each other. Popular science fiction television shows like *Star Trek* often show their crews riding on spacecraft between planets and star systems in very short periods of time. In reality, with current methods of propulsion, it would take a Mars-bound crew more than a year to reach the Red Planet safely. A trip to Jupiter would last several more years, and a trip to Pluto might be a lifelong quest.

Likewise, the sizes of solar system objects are often not well understood by those who don't work with them very often. For the purposes of this chapter, a good understanding of relative sizes is important. Asteroids are very small compared to the size of our planet,

and there are many of them in the solar system. To assess the threat from an asteroid hitting Earth, it's important to know that the Earth is more than 12,000 km in diameter whereas a typical asteroid might be only a kilometer or two across. However, an even more crucial bit of information is necessary to understand the problem fully: how close the asteroid's orbit will bring it to the Earth. (We deal more fully with orbits in **Activity 3**.)

To get a handle on the issue of distances in the solar system, we start by exploring the scale of our solar system. This is like walking through your neighborhood and learning the distance to the nearest houses, the closest fire hydrant, the local gas station, the school, and the shopping mall. These all represent different relative sizes and distances of structures that constitute a student's familiar surroundings.

Measurements in Astronomy: Scaling
In the **Investigate** section, students are presented with tabular data about distances and diameters of the Sun and planets. They are asked to scale these distances. Scaling is one way of understanding how far apart things are. Using scales also prepares the student for some commonly used measurements in astronomy.

On Earth, we use the metric system to measure things like weight, mass, and distance. Children's bodies grow a few centimeters per year. Within our homes, objects are a few meters apart at most. When we step outside, we walk several meters across our yards, and school might be a few kilometers away. If we cross the country or go overseas, we travel thousands of kilometers. The most distant destination that humans have ever experienced is the Moon (in 1969 and during the 1970s), some 400,000 km away!

Measurement in Astronomy: Astronomical Units and Light-Years

In the **Digging Deeper** section of **Activity 1**, students are introduced to the systems of measurement that astronomers use. The terms "astronomical units," "light-years," "parsecs," "kiloparsecs," and "megaparsecs" help us quantify the enormous distances that are used in astronomy. The astronomical unit (AU) is based on the average distance between the Earth and Sun; it has been quantified at 149,597,870 km (or 92,955,730 mi.). The AU works well for discussing distances inside the solar system and extending out through the cloud of cometary nuclei, called the Oort Cloud, that orbits some 50,000 AU from the Sun.

When we start to talk about distances to the nearest stars, however, the AU system gets cumbersome. For example, the star nearest the Sun is called Proxima Centauri. It has two companions called Alpha and Beta Centauri. This stellar system lies more than 265,000 AU away. The more distant the stars get, the bigger the AU number gets. Because we are studying the light from distant stars, we can use one property of light—its speed—to help us define the distance to those stars. The distance that light travels in a year at a speed of nearly 300,000 km/s comes out to be 9.4605×10^{12} km—a "light-year." Alpha Centauri's light traveled 4.2 years to reach the Earth; it is thus 4.2 light-years away. A more distant star, like the bright winter star, Sirius, shines across 8.4 light-years of space. A ninth-grader looking for the light that left a star when he or she was born would need to look for 14 or 15 light-years away.

Measurement in Astronomy: Triangulation

The way that astronomers measure the distance to a star is similar to the methods that surveyors use. It involves basic triangulation. First, surveyors establish a baseline. By drawing an imaginary line from both ends of the baseline to the object being measured, a triangle is formed. Surveyors can then measure the angle to the object from either end of the baseline. That gives them two angles of the triangle and the length of one side (the baseline). From that information, the distance to the object can be determined. The farther the object is away, the longer the baseline needs to be. For very distant objects like stars, astronomers need the biggest baseline that they can get, so they use the orbit of the Earth about the Sun (in this case, the baseline equals 2 AU).

Measurement in Astronomy: Parallax and the Parsec

In measuring the distance to the stars, astronomers take advantage of the phenomenon of parallax. Parallax is the apparent change in the position of an object due to a change in the observer's position. An everyday example of this phenomenon is how our thumb, held at arm's length, appears to shift position against a distant background when first we look with only one eye and then with the other. The farther the distance to an object, the less the parallax shift. By taking a photograph of a star and retaking the photograph six months later when the Earth is halfway through its orbit, the distance to the star can be measured. In the two pictures, the star will appear to have shifted relative to the most distant stars in the picture. The angle of that shift is used to determine the distance. (Actually, stellar parallax is defined as half the total shift observed, because that is the value needed to calculate the distance.)

Professional astronomers also use a unit called the parsec. A parsec is the distance a star has if its parallax is 1 second of arc. Astronomers have determined that a parsec contains 3.0857×10^{13} km, or 206,205 AU, or 3.26 light-years. A kiloparsec is 1000 parsecs; a megaparsec is 1,000,000 parsecs.

More Information – on the Web
Visit the *EarthComm* web site
www.agiweb.org/earthcomm/ to access a
variety of links to web sites that will help
you to deepen your understanding of content
and prepare you to teach this activity. Many
of the sites also contain images that you can
download and make into overheads for
incorporation into class discussions.

Goals and Assessment

Clarify that the goals indicate what students should understand and be able to do as a result of the activity. Make sure students understand that Chapter Assessments are based upon these goals.

Goal	Location in Activity	Assessment Opportunity
Produce a scale model of the solar system.	**Investigate; Understanding and Applying What You Have Learned** Question 2	Scale conversions are done correctly; responses to questions are reasonable. Model of solar system is drafted correctly.
Identify some strengths and limitations of scale models.	**Investigate; Digging Deeper**	Responses are reasonable and demonstrate understanding.
Calculate distances to objects in the solar system in astronomical units (AU), light-years, and parsecs.	**Digging Deeper; Check Your Understanding** Question 2 **Understanding and Applying What You Have Learned** Questions 1 and 5	Calculations are done correctly; responses are reasonable and demonstrate understanding.
Explain, in your own words, the nebular theory of the formation of the solar system.	**Check Your Understanding** Question 3	Responses are reasonable and demonstrate understanding.
Explain the formation of the universe.	**Digging Deeper; Check Your Understanding** Question 4	Responses are reasonable and demonstrate understanding.

Earth System Evolution Astronomy

Activity 1

The History and Scale of the Solar System

Goals
In this activity you will:

- Produce a scale model of the solar system.

- Identify some strengths and limitations of scale models.

- Calculate distances to objects in the solar system in astronomical units (AU), light years, and parsecs.

- Explain, in your own words, the nebular theory of the formation of the solar system.

- Explain the formation of the universe.

Think about It

Earth is part of a large number of objects that orbit around a star called the Sun.

- What objects make up the solar system? Where are they located in relation to Earth?

What do you think? Record your ideas in the form of a diagram of the solar system in your *EarthComm* notebook. Without looking ahead in this book, draw the Sun and the planets, and the distances from the Sun to the planets, as nearly to scale as you can. Be prepared to discuss your diagram with your small group and the class.

Activity Overview

Using two different scales, students calculate the distances from the Sun and the diameters of the planets in our solar systems. They consider the limitations associated with the use of each scale. Using the scale of their choice, students then construct a model of the solar system. **Digging Deeper** introduces the various units used by astronomers to measure distances in the universe. The reading also reviews the nebular theory for the birth of our solar system, discusses the birth of the planets, and examines the relationship of our solar system to the Milky Way Galaxy.

Preparation and Materials Needed

No advance preparation is required for this activity.

Materials
- Calculator
- Metric ruler
- Tape measure

Think about It

Student Conceptions

Students are likely to be familiar with the planets of the solar system. They may or may not know the correct order of the planets relative to the Sun. It is unlikely that students have a strong understanding of the relative sizes of the planets and the Sun.

Students are asked to draw a picture of the solar system showing the planets and the Sun, and to try to make the distances between the planets to scale. Remind them not to look ahead in the book because you want to know what preconceived ideas they have. Select drawings that show a range of ideas and hold a brief discussion. Do not label students' drawings or responses correct or incorrect. Rather, use this opportunity to point out that there are differences in thinking.

You can add a demonstration to the **Think about It** discussion. Have one person hold a basketball to represent the Sun. Use a marble to represent Earth. Ask students to predict which of the following distances represents the correct distance from the Sun to Earth:

- Position "A" (about 1 m from the basketball)
- Position "B" (3 m from the basketball)
- Position "C" (6 m from the basketball)
- Position "D" (a point outside the window about 20 m away)

Students should record their prediction in their notebooks. When they complete **Investigate**, they should discover that "D" is closest to scale. Return to students' predictions as you close the activity to see whether they would change their original answer and, if so, why.

Answer for the Teacher Only

In addition to the nine planets listed in *Table 1* on page E5 of the student textbook, there are numerous asteroids (minor planetary bodies), which orbit the Sun mostly in the region between Mars and Jupiter. There are also numerous comets, most of which have highly elliptical orbits. Many of the planets have satellite moons as well.

Assessment Tool

Think about It Evaluation Sheet
Use this evaluation sheet to help students understand and internalize the basic expectations for the warm-up activity.

NOTES

_____

_____

_____

_____

_____

_____

_____

_____

_____

_____

_____

_____

_____

_____

_____

_____

_____

_____

_____

Investigate

1. Use the data in *Table 1* to make a scale model of the solar system. Try using the scale 1 m = 150,000,000 km.

   a) Divide all the distances in the first column by 150,000,000 (one hundred and fifty million). Write your scaled-down distances in your notebook, in meters.

   b) Divide all the diameters in the second column by 150,000,000. Write your scaled-down diameters in your notebook, in meters.

   c) Looking at your numbers, what major drawback is there to using the scale 1 m = 150,000,000 km?

Table I Diameters of the Sun and Planets, and Distances from the Sun		
Object	Distance from Sun (km)	Diameter (km)
Sun	0	1,391,400
Mercury	57,900,000	4878
Venus	108,209,000	12,104
Earth	149,598,770	12,756
Mars	227,900,000	6794
Jupiter	778,200,000	142,984
Saturn	1,429,200,000	120,536
Uranus	2,875,000,000	51,118
Neptune	4,504,400,000	49,528
Pluto	5,915,800,000	2302

2. Now try another scale: 1 m = 3,000,000 km (three million kilometers).

   a) Divide all the distances in the first column by 3,000,000. Write your scaled-down distances in your notebook in meters.

   b) Divide all the diameters in the second column by 3,000,000. Write your scaled-down diameters in your notebook in meters.

   c) Looking at your numbers, what major drawback is there to using the scale 1 m = 3,000,000 km?

3. Using what you have learned about scaling distances and diameters in the solar system, make models of the Sun and the planets. Each of the planets can be drawn on a different sheet of paper using a ruler to lay out the correct sizes for the different planets and the Sun.

4. To represent the distances from the Sun to the planets you will need to use a tape measure. You may want to measure the size of your stride and use this as a simple measuring tool.

E 5

Investigate

Assessment Tools

EarthComm Notebook Entry-Checklist

This checklist provides a brief summary of important processes, concepts, and skills that you might wish to assess during and after an investigation. Use it as a quick guide for student self-assessment and/or an opportunity to score student work quickly. Add further criteria specific to your classroom needs or to this particular investigation.

Investigate Notebook Entry-Evaluation Sheet

Point out the criteria listed on this evaluation sheet that are relevant to this particular investigation. Encourage students to internalize the criteria by making them part of your "assessment conversations" as you circulate around the classroom. For example, while students are working, ask them criteria-driven questions such as:

- Is your work thorough and complete?
- Are all of you participating in the activity?
- Does each of you have a role to play in solving this problem? And so on.

1. a) - b)

Object	Distance from Sun (m)	Diameter (m)
Sun	0	0.009276
Mercury	0.386	3.25E-05
Venus	0.721393	8.07E-05
Earth	0.997325	8.5E-05
Mars	1.519333	4.53E-05
Jupiter	5.188	0.000953
Saturn	9.528	0.000804
Uranus	19.16667	0.000341
Neptune	30.02933	0.00033
Pluto	39.43867	1.53E-05

c) A drawback to using the scale 1 m = 150,000,000 km is that the planets are too difficult to draw or too small to be seen on the model.

2. a) - b)

Object	Distance from Sun (m)	Diameter (m)
Sun	0	0.4638
Mercury	19.3	0.001626
Venus	36.06967	0.004035
Earth	49.86626	0.004252
Mars	75.96667	0.002265
Jupiter	259.4	0.047661
Saturn	476.4	0.040179
Uranus	958.3333	0.017039
Neptune	1501.467	0.016509
Pluto	1971.933	0.000767

c) A drawback to using the scale 1 m = 3,000,000 km is that the planets are too far apart on the model.

3. Students should make models of the Sun and the planets. They will need to decide what scale they are going to use. Each planet can be drawn on a separate sheet of paper. If time permits, students may enjoy coloring and decorating their planets. The 1 m = 3,000,000 km scale pushes the limit to what can be reasonably drawn (at this scale Pluto has a diameter of < 0.8 mm!), and at this scale the Sun will not quite fit on four pieces of letter-size paper taped together. One alternative is to draw the planets and the Sun at different scales. The planets can be drawn at a scale of 1 m = 750,000 km. At this scale the largest planet (Jupiter) will still fit on a single piece of paper (191 mm diameter) and the smallest (Pluto) will have a diameter of approximately 3 mm. Note, however, that at this scale the Sun will have a diameter of ~ 1.85 m, and it cannot be drawn easily. The fewer scales that are used for this exercise, the better. If more than one scale is used in making the model, the scale should be noted on each element of the model.

4. The distance that is needed to position the planets to scale will vary depending on the scale chosen. From the answers to **Question 1**, one can see that even if Mercury is scaled to be only 39 cm away from the Sun, then Pluto would be over 39 m from the Sun. Given the distances necessary to lay out this model, you may wish to do this as a class. This could be done either in the classroom or in the school gymnasium. Another alternative is to pick a more suitable scale (1 m = 600,000,000 km, for example) and students can either tape sheets of paper together on which to construct their models or use a roll of adding-machine tape.

a) Answers will vary.

b) No, it is not possible to make a model using the same scale for the distances between the bodies and the diameters of the bodies. This is because the scale needed to make the distances between the bodies measurable renders the diameters too small to be measured, and the scale needed to make the bodies large enough to be measured makes the distances too large to illustrate easily.

NOTES

Earth System Evolution Astronomy

To do this, stand behind a line and take five steps in as normal a way as possible and note where your last step ended. Now measure the distance from where you started to the end. Divide by five to determine how far you walk with each step. Knowing the length of your stride is an easy way to determine distances.

a) Explain the scale(s) you decided to use and your reasons for your choices.

b) Is it possible to make a model of the solar system on your school campus in which both the distances between bodies and the diameters of the bodies are to the same scale? Why or why not?

Reflecting on the Activity and the Challenge

In this activity you used ratios to make a scale model of the solar system. You found out that scale models help you appreciate the vastness of distances in the solar system. You also found out

that there are some drawbacks to the use of scale models. Think about how you might use the model you made as part of your **Chapter Challenge**.

Geo Words

astronomical unit: a unit of measurement equal to the average distance between the Sun and Earth, i.e., about 149,600,000 (1.496 x 10^8) km.

light year: a unit of measurement equal to the distance light travels in one year, i.e., 9.46 x 10^{12} km.

Digging Deeper

OUR PLACE IN THE UNIVERSE

Distances in the Universe

Astronomers often study objects far from Earth. It is cumbersome to use units like kilometers (or even a million kilometers) to describe the distances to the stars and planets. For example, the star nearest to the Sun is called Proxima Centauri. It is 39,826,600,000,000 km away. (How would you say this distance?)

Astronomers get around the problem by using larger units to measure distances. When discussing distances inside the solar system, they often use the **astronomical unit** (abbreviated as AU). One AU is the average distance of the Earth from the Sun. It is equal to 149,598,770 km (about 93 million miles).

Stars are so far away that using astronomical units quickly becomes difficult, too! For example, Proxima Centauri is 266,221 AU away. This number is easier to use than kilometers, but it is still too cumbersome for most purposes. For distances to stars and galaxies, astronomers use a unit called a **light-year**. A light-year sounds as though it is a unit of time, because a year is a unit of time, but it is really the distance that light travels in a year. Because light travels

E 6

Reflecting on the Activity and the Challenge

Take advantage of this opportunity to discuss with your students the vast amount of space between planets and the Sun. Students should realize that most of their model is "empty" space. Have them discuss the relative sizes of the different planets and the Sun. They will probably point out that the Sun is very large in comparison to the planets. Students may also realize that the "inner" planets are much smaller than the "outer" planets (with the exception of Pluto).

Digging Deeper

As students read the **Digging Deeper** section, the relevance of the concepts investigated in **Activity 1** will become clearer to them. Assign the reading for homework, along with the questions in **Check Your Understanding** if desired.

Assessment Opportunity

Reword or restructure the questions in **Check Your Understanding** for a brief quiz. Use the quiz (or a class discussion of the questions in the textbook) to assess your students' understanding of the main ideas in the reading and the activity.

extremely fast, a light-year is a very large distance. For example, the Sun is only 8 light *minutes* away from Earth, and the nearest stars are several light-years away. Light travels at a speed of 300,000 km/s. This makes a light year 9.46×10^{12} (9,460,000,000,000) km. Light from Proxima Centauri takes 4.21 years to reach Earth, so this star is 4.21 light years from Earth.

Astronomers also use a unit called the **parsec** (symbol pc) to describe large distances. One parsec equals 3.26 light years. Thus, Proxima Centauri is 1.29 pc away. The kiloparsec (1000 pc) and megaparsec (1,000,000 pc) are used for objects that are extremely far away. The nearest spiral galaxy to the Milky Way galaxy is the Andromeda galaxy. It is about 2.5 million light years, or about 767 kpc (kiloparsecs), away.

The Nebular Theory

As you created a scale model of the solar system, you probably noticed how large the Sun is in comparison to most of the planets. In fact, the Sun contains over 99% of all of the mass of the solar system. Where did all this mass come from? According to current thinking, the birthplace of our solar system was a **nebula**. A nebula is a cloud of gas and dust probably cast off from other stars that used to live in this region of our galaxy. More than 4.5 billion years ago this nebula started down the long road to the formation of a star and planets. The idea that the solar system evolved from such a swirling cloud of dust is called the nebular theory.

You can see one such nebula in the winter **constellation** Orion (see *Figure 1*), just below the three stars that make up the Belt of Orion. Through a pair of binoculars or a small, backyard-type telescope, the Orion Nebula looks like a faint green, hazy patch of light. If you were able to view this starbirth region through a much higher-power telescope, you would be able to see amazing details in the gas and dust clouds. The Orion Nebula is very much like the one that formed our star, the Sun. There are many star nurseries like this one scattered around our galaxy. On a dark night, with binoculars or a small telescope, you can see many gas clouds that are forming stars.

Figure 1 Orion is a prominent constellation in the night sky.

Geo Words

parsec: a unit used in astronomy to describe large distances. One parsec equals 3.26 light-years.

nebula: general term used for any "fuzzy" patch on the sky, either light or dark; a cloud of interstellar gas and dust.

constellation: a grouping of stars in the night sky into a recognizable pattern. Most of the constellations get their name from the Latin translation of one of the ancient Greek star patterns that lies within it. In more recent times, more modern astronomers introduced a number of additional groups, and there are now 88 standard configurations recognized.

Teaching Tip

The constellation Orion, shown in *Figure 1*, is one of the most recognizable patterns of stars in the northern sky. Orion, the hunter, is a figure in Greek mythology. The story of Orion has many different forms.

In one version of the tale, Orion was in love with Merope, one of the Seven Sisters who form the Pleiades. Orion was killed when he stepped on Scorpius, the scorpion. The gods felt sorry for him, and placed him, along with his faithful dogs, Canis Major and Canis Minor, in the sky as constellations. The gods also placed various celestial animals, including Lepus, the rabbit, and Taurus, the bull, in the sky for Orion to hunt. Scorpius, however, was placed on the opposite side of the sky so Orion would never be hurt by the scorpion again.

Another version of the story has Orion accidentally killed by the goddess Artemis and placed in the sky to remain eternally one of the mightiest hunters of the night sky.

Earth System Evolution Astronomy

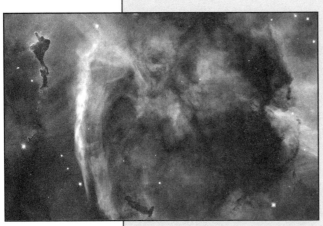

Figure 2 The Keyhole Nebula. Imaged by the Hubble Space Telescope.

In the nebula that gave birth to our solar system, gravity caused the gases and dust to be drawn together into a denser cloud. At the same time, the rate of rotation (swirling) of the entire nebula gradually increased. The effect is the same as when a rotating ice skater draws his or her arms in, causing the rate of rotation to speed up. As the nebular cloud began to collapse and spin faster, it flattened out to resemble a disk, with most of the mass collapsing into the center. Matter in the rest of the disk clumped together into small masses called **planetesimals**, which then gradually collided together to form larger bodies called **protoplanetary bodies**.

At the center of the developing solar system, material kept collapsing under gravitational force. As the moving gases became more concentrated, the temperature and pressure of the center of the cloud started to rise. The same kind of thing happens when you pump up a bicycle tire with a tire pump: the pump gets warmer as the air is compressed. When you let the air out of a tire, the opposite occurs and the air gets colder as it expands rapidly. When the temperature in the center of the gas cloud reached about 15 million degrees Celsius, hydrogen atoms in the gas combined or fused to create helium atoms. This process, called **nuclear fusion**, is the source of the energy from the Sun. A star—the Sun—was born!

Fusion reactions inside the Sun create very high pressure, and like a bomb, threaten to blow the Sun apart. The Sun doesn't fly apart under all this outward pressure, however. The Sun is in a state of equilibrium. The gravity of the Sun is pulling on each part of it and keeps the Sun together as it radiates energy out in all directions, providing solar energy to the Earth community.

The Birth of the Planets

The rest of the solar system formed in the swirling disk of material surrounding the newborn Sun. Nine planets, 67 satellites (with new ones still being discovered!), and a large number of comets and asteroids formed. The larger objects were formed mostly in the flat disk surrounding where the Sun was forming.

Geo Words

planetesimal: one of the small bodies (usually micrometers to kilometers in diameter) that formed from the solar nebula and eventually grew into proto-planets.

protoplanetary body: a clump of material, formed in the early stages of solar system formation, which was the forerunner of the planets we see today.

nuclear fusion: a nuclear process that releases energy when lightweight nuclei combine to form heavier nuclei.

Teaching Tip

Astronomer Sir John Herschel discovered the Keyhole Nebula in the 19th century. The Keyhole is actually a part of a larger region called the Carina Nebula, located 8000 light-years from Earth. In the image shown in *Figure 2*, captured by the Hubble Telescope, the "keyhole" is the circular feature that dominates the photo. It is approximately seven light-years wide and contains filaments of glowing gas and dark clouds of cold molecules and dust. In this region, some of the hottest and most massive stars are known to be growing, each about 10 times as hot and 100 times as large as the Sun.

Four of these planets, shown in *Figure 3*—Mercury, Venus, Earth, and Mars—are called the **terrestrial** ("Earth-like") **planets**. They formed in the inner part of our solar system, where temperatures in the original nebula were high. They are relatively small, rocky bodies. Some have molten centers, with a layer of rock called a mantle outside their centers, and a surface called a crust. The Earth's crust is its outer layer. Even the deepest oil wells do not penetrate the crust.

The larger planets shown in *Figure 3*—Jupiter, Saturn, Uranus, and Neptune—consist mostly of dense fluids like liquid hydrogen. These **gas giants** formed in the colder, outer parts of the early solar nebula. They have solid rocky cores about the size of Earth, covered with layers of hydrogen in both gas and liquid form. They lie far from the Sun and their surfaces are extremely cold.

Pluto is the most distant planet from the Sun. Some astronomers do not even classify Pluto as a planet. Instead, they put it in the category of smaller icy bodies that are found in the outer solar system. Pluto is very different from the terrestrial or the gaseous planets. If anything, it resembles the icy moons of the gas giants. Some scientists think that it may not have been part of the original solar system but instead was captured later by the Sun's gravity, or that it is a moon of an outer planet thrown into a unique tilted orbit around the Sun. Currently, there is a controversy among planetary scientists about whether to include Pluto among the "official" planets! If Pluto were discovered today (instead of in 1930), it probably would not be classified as a planet.

Figure 3 Composite image of the planets in the solar system, plus the Moon.

Geo Words

terrestrial planets: any of the planets Mercury, Venus, Earth, or Mars, or a planet similar in size, composition, and density to the Earth. A planet that consists mainly of rocky material.

gas giant planets: the outer solar system planets: Jupiter, Saturn, Uranus, and Neptune, composed mostly of hydrogen, helium, and methane, and having a density of less than 2 gm/cm^2.

Teaching Tip

Note that *Figure 3* does not show all the planets to the same scale. Rather, two scales are used: the inner planets are shown to scale relative to one another, and the outer planets are shown to scale relative to one another. This presents a good opportunity to reinforce the concepts of the investigation by discussing with students why two different scales are used to show the sizes of the planets. Also, discuss with students why it's not possible to scale the distances from the planets to the Sun accurately. You can tie this discussion to the discussion of *Figure 4* (see next **Teaching Tip**).

Teaching Tip

Use **Blackline Master Astronomy 1.1, Orbits of the Planets** to make an overhead of the diagram in *Figure 4* on page E10. Use the overhead to discuss with students why two diagrams are necessary to show the orbits of the planets to scale. You can tie this discussion to the discussion of *Figure 3* (see previous **Teaching Tip**).

Earth System Evolution Astronomy

There are trillions of **comets** and **asteroids** scattered throughout the solar system. Earth and other solar-system bodies are scarred by impact craters formed when comets and asteroids collided with them. On Earth, erosion has removed obvious signs of many of these craters. Astronomers see these comets and asteroids as the leftovers from the formation of the solar system. Asteroids are dark, rocky bodies that orbit the Sun at different distances. Many are found between the orbits of Mars and Jupiter, making up what is called the asteroid belt. Many others have orbits outside of the asteroid belt. Comets are mixtures of ice and dust grains. They exist mainly in the outer solar system, but when their looping orbits bring them close to the Sun, their ices begin to melt. That is when you can see tails streaming out from them in the direction away from the Sun. Some comets come unexpectedly into the inner solar system. Others have orbits that bring them close to the Sun at regular intervals. For example, the orbit of Halley's comet brings it into the inner solar system every 76 years.

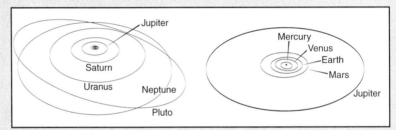

Figure 4 Two diagrams are required to show the orbits of the planets to scale.

Where is the Solar System in Our Galaxy?

Have you ever seen the Milky Way? It is a swath of light, formed by the glow of billions of stars, which stretches across the dark night sky. From Earth, this band of celestial light is best seen from dark-sky viewing sites. Binoculars and backyard-type telescopes magnify the view and reveal individual stars and nebulae. Unfortunately, for those who like to view the night sky, light pollution in densely populated areas makes it impossible to see the Milky Way even on nights when the atmosphere is clear and cloudless.

Galaxies are classified according to their shape: elliptical, spiral, or irregular. Our home galaxy is a flat spiral, a pinwheel-shaped collection of stars held together by their mutual gravitational attraction. Our galaxy shown in *Figure 5* is called the Milky Way Galaxy, or just the galaxy. Our solar system is located in one of the spiral arms about two-thirds of the way out from the center of the galaxy. What is called the Milky Way is the view along the flat part of our galaxy. When you look at the Milky Way, you are looking out through the galaxy parallel to the

Blackline Master Astronomy 1.1
Orbits of the Planets

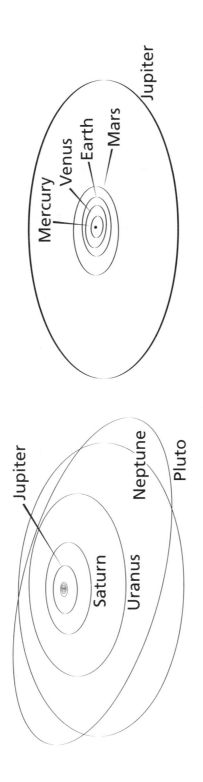

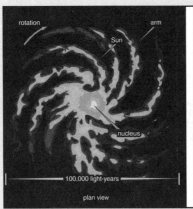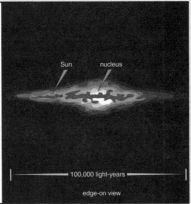

Figure 5 The Milky Way Galaxy. Our solar system is located in a spiral band about two-thirds of the way from the nucleus of the galaxy.

Geo Words

cosmologist: a scientist who studies the origin and dynamics of the universe.

plane of its disk. The individual stars you see dotting the night sky are just the ones nearest to Earth in the galaxy. When you view the Milky Way, you are "looking through" those nearest stars to see the more distant parts of the galaxy. In a sense, you are looking at our galaxy from the inside. In other directions, you look through the nearest stars to see out into intergalactic space!

Our Milky Way Galaxy formed about 10 billion years ago and is one of billions of galaxies in the universe. The universe itself formed somewhere between 12 to 14 billion years ago in an event called the Big Bang. This sounds like the universe began in an explosion, but it did not. In the beginning, at what a scientist would call "time zero," the universe consisted almost entirely of energy, concentrated into a volume smaller than a grain of sand. The temperatures were unimaginably high. Then the universe expanded, extremely rapidly, and as it expanded the temperature dropped and matter was formed from some of the original energy. **Cosmologists** (scientists who study the origin and dynamics of the universe) think that most of the matter in the universe was formed within minutes of time zero! The expansion and cooling that started with the Big Bang continues to this day.

The galaxies and stars are the visible evidence of the Big Bang, but there is other, unseen evidence that it happened. It's called the cosmic background radiation, which is radiation that is left over from the initial moments of the Big Bang. Astronomers using special instruments sensitive to low-energy radio waves have detected it coming in from all directions from the universe. The existence of the cosmic background radiation is generally considered to be solid evidence of how the Big Bang happened.

Check Your Understanding

1. What are the distances represented by a light-year, an astronomical unit, and a parsec?

2. Which of the units in Question 1 would you use to describe each of the following? Justify your choice.

   a) Distances to various stars (but not our Sun)?

   b) Distances to various planets within our solar system?

   c) Widths of galaxies?

3. In your own words, explain the nebular theory for the beginning of our solar system.

4. Briefly describe the origin of the universe.

E 11

EarthComm

Teaching Tip

The Milky Way Galaxy, illustrated in *Figure 5*, is home to our solar system, as well as at least 200 billion other stars and their planets along with thousands of clusters and nebulae. As a galaxy, the Milky Way is actually a giant, inasmuch as its mass is estimated to be between 750 billion and 1 trillion solar masses, and its diameter is approximately 100,000 light-years. The Milky Way Galaxy belongs to the Local Group, a group of three large galaxies and over 30 small galaxies. It is the second largest galaxy in the group, after the Andromeda Galaxy. Visit the *EarthComm* web site to find additional images of the Milky Way Galaxy.

Check Your Understanding

1. A light-year is equal to 9.46×10^{12} km; an AU is equal to 1.49×10^{8} km; a parsec is equal to 3.26 light-years (or 3.08×10^{13} km).

2. **a)** Distances to other stars would best be measured in light-years or parsecs.

    b) Distances to other planets would best be measured in AU.

    c) Widths of galaxies would best be measured in light-years or parsecs.

3. Gravity caused gases and dust to be drawn together into a denser cloud. At the same time, the rate of rotation of the nebula increased. As the nebular cloud began to collapse and spin faster, it flattened out to form a disk with most of the matter concentrated in the center. Matter in the rest of the disk clumped together into planetesimals, which gradually collided together to form larger protoplanetary bodies. As more matter condensed in the center of the nebula, temperature and pressure began to increase until the temperature was high enough to fuse hydrogen atoms to create helium atoms, creating the Sun.

4. The universe formed 12 to 14 billion years ago in an event called the Big Bang, which occurred as an extremely small, concentrated bundle of energy began to expand and cool.

Assessment Tool

Check Your Understanding Notebook Entry–Evaluation Sheet
Use this sheet to evaluate the extent to which students understand the key concepts explored in **Activity 1** and explained in **Digging Deeper**, and to evaluate the students' clarity of expression.

Earth System Evolution Astronomy

Understanding and Applying What You Have Learned

1. Using the second scale (1 m = 3,000,000 km) you used for distance in your model of the solar system:

   a) How far away would Proxima Centauri be from Earth?

   b) How far away would the Andromeda galaxy be on your scale, given that Andromeda is 767 kiloparsecs or 2.5 million light years away?

2. The Moon is 384,000 km from Earth and has a diameter of 3476 km. Calculate the diameter of the Moon and its distance from the Earth using the scale of the model you developed in the **Investigate** section.

3. Refer again to *Table 1*. If the Space Shuttle could travel at 100,000 km/hr, how long would it take to go from Earth to each of the following objects? Assume that each object is as close to the Earth as it can be in its orbit.

   a) The Moon?
   b) Mars?
   c) Pluto?

4. What is the largest distance possible between any two planets in the solar system?

5. Use your understanding of a light-year and the distances from the Sun shown in *Table 1*. Calculate how many minutes it takes for sunlight to reach each of the nine planets in the solar system. Then use the unit "light-minutes" (how far light travels in one minute) to describe the *distances* to each object.

6. Write down your school address in the following ways:

   a) As you would normally address an envelope.

   b) To receive a letter from another country.

   c) To receive a letter from a friend who lives at the center of our galaxy.

   d) To receive a letter from a friend who lives in a distant galaxy.

Preparing for the Chapter Challenge

Begin to develop your brochure for the **Chapter Challenge**. In your own words, explain your community's position relative to the Earth, Sun, the other planets in our solar system, and the entire universe. Include a few paragraphs explaining what your scale model represents and how you chose the scale or scales you used.

Understanding and Applying What You Have Learned

1. a) As given in the **Digging Deeper** reading section, Proxima Centauri is 39,826,600,000,000 km from Earth. Using the scale of 1 m = 3,000,000 km, this is 13,275,533 m. (39,826,600,000,000 ÷ 3,000,000 = 13,275,553). Using scientific notation, this calculation would be $3.98266 \times 10^{13} \div 3 \times 10^{6} = 1.372553 \times 10^{7}$

   b) The Andromeda Galaxy is about 2.5 million light-years away or 767 kiloparsecs. One light-year is 9.46×10^{12} km. Calculating the distance to the Andromeda Galaxy in km yields:

   2.5×10^{6} light-years x 9.46×10^{12} km/light-year = 23.65×10^{18} km

   dividing by 3×10^{6} km/m converts to the scale of 1m = 3,000,000 km (or 3×10^{6} km)

   23.65×10^{18} km ÷ 3×10^{6} km/m = 7.88333×10^{12} m.

   (At this scale, the Andromeda Galaxy would still be ~ 7.88 billion km away!)

2. Answers will vary depending upon the scales that students developed during the investigation. If the scales used were 1 m = 3,000,000 km for the diameter and 1 m = 150,000,000 km for the distance, then the scaled diameter of the model Moon would be 0.001159 m (~ 1 mm), and its scaled distance from the Earth would be 0.00256 m (~ 2.6 mm).

3. It is important to keep in mind that distances between different bodies vary somewhat with time because of the elliptical nature of orbits.

   a) From Question 2, the Moon is 384,000 km away from Earth at their closest approach to one another. At 100,000 km/h the shuttle would take about 3.84 h to go from the Moon to Earth (~ 3 h and 50 min).

   b) Students will need to determine the distance between Mars and Earth at their closest approach to one another. This can be done using *Table 1* on page E5 of the Student Edition. From this table, Mars is 78,301,230 km from the Earth (227,900,000 km – 149,598,770 km). The shuttle would take about 783 h (~ 32.6 days) to go from Earth to Mars.

   c) Students will need to determine the distance between Pluto and Earth at their closest approach to one another. Using the data in *Table 1*, this distance is 5,766,201,230 km (5,915,800,000 km – 149,598,770 km). The shuttle would take about 57,662 h (~ 6.6 years) to go from Earth to Pluto.

4. If Pluto and Neptune were at opposite points in their orbits around the Sun, the greatest distance between any two objects in the solar system would be 10,420,200,000 km, or ~ 69.65 AU (using the distances given in *Table 1*).

5. From the **Digging Deeper** reading section, light travels at 300,000 km/s. Multiplying by 60 s/min yields 18,000,000 for the speed of light in km/m. Once this calculation has been made, students can then divide each of the distances by

the distance light travels in one minute to calculate the time (in minutes) that it takes for light to reach each planet. For example, sunlight travels the 149,598,770 km distance from the Sun to the Earth in ~ 8.31 minutes (149,598,770 ÷ 18,000,000).

Planet	Time Required for Light to Travel to Each Planet (minutes)
Mercury	3.22
Venus	6.01
Earth	8.31
Mars	12.66
Jupiter	43.23
Saturn	79.40
Uranus	159.72
Neptune	250.24
Pluto	328.66

6. Answers will vary depending on your address. Samples follow.

   a) John Doe/123 Address Street/Beverly Hills, CA 90210

   b) John Doe/123 Address Street/Beverly Hills, CA 90210/United States of America

   c) John Doe/123 Address Street/Beverly Hills, CA 90210/United States of America/Earth

   d) John Doe/123 Address Street/Beverly Hills, CA 90210/United States of America/Earth/Milky Way

Preparing for the Chapter Challenge

This section gives students an opportunity to apply what they have learned to the **Chapter Challenge**. They can work on their papers as a homework assignment or during class time within groups.

After completing **Activity 1**, students should be able to explain how a scale model helps them understand objects that are too large or too small to picture easily. They should be able to identify their place in our solar system correctly, and they should know that Earth is only a small part of a larger system that encompasses not only our solar system and galaxy but other galaxies as well. Students also should realize that the largest portion of this vast universe is actually open space that is unoccupied by planets, stars, or other astronomical bodies. Having done the investigation and answered the questions, students should realize that the great distance between Earth and its nearest neighbor leaves us very vulnerable to any extraterrestrial influences. Even though the other planetary bodies in our solar system are close by comparison to the nearest star or galaxy, they are still vastly far away, and most likely offer us little in the way of shielding from extraterrestrial influences. You should help them tie this notion back to the **Chapter Challenge**.

Criteria for assessment (see **Assessment Rubric for Chapter Challenge on Astronomy** on pages 12 and 13) include discussions of how the Earth fits into the universe, including:

- formation and evolution of the solar system
- Earth's distance from, and orbit around, the Sun
- Earth's place in the galaxy
- the galaxy's place in the universe

Review these criteria with your students so that they can be certain to include the appropriate information in their short papers.

Inquiring Further

1. **Solar-system walk**

   Create a "solar-system walk" on your school grounds or your neighborhood. Draw the Sun and the planets to scale on the sidewalk in chalk. Pace off the distances between the Sun and the nine planets at a scale that is appropriate for the site.

2. **Scaling the nearest stars**

   Look up the distances to the five stars nearest to the Sun. Where would they be in your scale model? To show their location, would you need a map of your state? Country? Continent? The world?

3. **Nuclear fusion**

   Find out more about the process of nuclear fusion. Explain how and why energy is released in the process by which hydrogen atoms are converted into helium atoms within the Sun. Be sure to include Albert Einstein's famous equation, $E = mc^2$, in your explanation, and explain what it means.

4. **Star formation**

   Write a newspaper story about star formation. Visit the *EarthComm* web site to find information available on the web sites of the Hubble Space Telescope and the European Southern Observatory to find examples of star-forming nebulae in the galaxy. How are they similar? How are they different? What instruments do astronomers use to study these nebulae?

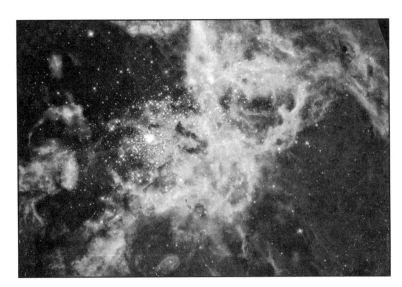

Inquiring Further

1. Solar-system walk

The solar-system walk can be made into a temporary exhibit from which all students in the school could benefit. Walking out the distances to the planets and working with three-dimensional models is also beneficial for many learners. More information about solar system walks can be found on the *EarthComm* web site.

2. Scaling the nearest stars

The five stars closest to the Sun are:

- Proxima Centauri (4.24 light-years)
- Alpha Centauri A (4.34 light-years)
- Alpha Centauri B (4.34 light-years)
- Barnard's Star (5.97 light-years)
- Wolf 359 (7.80 light-years)

Scales will vary with each student. This provides yet another opportunity for students to put the scale of the solar system into a familiar context. It also gives them further practice with mathematics.

3. Nuclear fusion

Fusion is a type of nuclear reaction in which atomic nuclei combine to form a heavier nucleus. During this process large amounts of energy (such as visible light) are produced. Astrophysicists refer to fusion as nucleosynthesis when it occurs in stars because, along with releasing energy, it creates (synthesizes) new elements. Einstein's famous equation $E = mc^2$ explains the equivalence of matter and energy, and it can be used to calculate the energy released when a given mass is converted. $E = mc^2$ is one of the fundamental equations for explaining the primary source of energy in our solar system and beyond. In the equation $E = mc^2$, E represents energy, c^2 stands for the square of the speed of light, and m represents mass. More specifically, m represents the mass difference between the products and reactants in a nuclear reaction. Not all sub-atomic particles (protons, neutrons) have the same mass, and the nucleus of an atom doesn't weigh exactly the same as its constituent parts. Although the mass that is converted to energy during nuclear fusion is very small, the square of the speed of light is great, yielding an enormous amount of energy. In the Sun and other medium-size and smaller stars the fuel for this nucleosynthesis is hydrogen. Nucleosynthesis that normally takes place in stars can produce only elements as heavy as iron. For elements heavier than iron, fusion is an endothermic reaction (energy must be put in to the system to drive the reaction). Elements heavier than iron are relatively rare in the universe and are thought to be synthesized during the explosion of large stars. The *EarthComm* web site provides links to information on nuclear fusion, and reactions important in the nucleosynthesis of elements in stars.

4. Star formation

Answers will vary depending upon which star-forming nebulae students choose to research. Examples of star-forming nebulae include:

- Orion Nebula
- Eagle Nebula
- 30 Doradus Nebula
- Horseshoe Nebula
- Sharpless 106 Nebula

You may want to assign students to specific star-forming regions, to limit the scope of their research and make grading easier for yourself.

Teaching Tip

The photo shown on page E13 of the Student Book was taken using NASA's Hubble Space Telescope. It is a mosaic picture of the 30 Doradus Nebula, which is a vast region of gas and dust where thousands of stars are being born. This fertile star-forming region includes a spectacular cluster of massive stars (the large blue and white cluster just left of center in the picture).

Chapter 1

NOTES

ACTIVITY 2 — THE EARTH-MOON SYSTEM

Background Information

Lunar Effects on Earth

The Moon is so familiar to us that we tend to accept it as part of the scenery. We don't always understand its role in our system. As students will learn in **Activity 2**, Earth's satellite is responsible for a large number of effects on Earth, both past and present. In fact, because of its proximity to Earth and easy availability for study, students can use the Moon as a jumping-off point to investigate the effects of solar-system bodies on their community.

As students investigate the Moon's effect on the Earth, their exploration of the tides gives them indirect proof of the forces of solar and lunar gravity on this planet. This is an important step in fulfilling the **Chapter Challenge**. Although the force of gravity lessens with distance, we see its effect time and time again as we study the interaction of comets with planets, stars with other stars, and even the action of gravity across the gulfs that separate galaxies. It is sometimes difficult for students to grasp the idea that distant objects can have a gravitational effect on Earth processes. The activity on tides gives them evidence for these forces.

Theories about the Origin of the Moon

The Moon is very much bound up with the formation of the Earth. To understand exactly how it formed, we have to take what we know about the Moon and use that to infer its origin. The Moon's average density is 3.3 g/cm³. Compare this to the Earth's average density of 5.5 g/cm³. The difference is that Earth has a large iron core whereas the Moon does not.

Moon rocks have very little water or other volatile gases, whereas many Earth rocks do. However, both Earth and Moon have the same proportions of oxygen isotopes. What does this tell us about the Moon's origin?

Over the years, astronomers have developed many theories about the origin of the Moon. One idea was that the Moon formed in orbit around the Earth, accreting from small planetesimals just as the Earth did. However, the lack of an iron core in the Moon rules out that idea.

Another possibility is that the Earth captured the Moon. This might seem to make sense because the density of the Moon is so different from that of the Earth. However, this scenario requires a long period of time. It also assumes the existence of some effect would help slow the Moon down in order for it to be captured.

Yet another idea is that the Moon might have been "spun off" from the rapidly spinning Earth in its early days. The problem with this idea is that both bodies should show some evidence that this happened, but they don't.

Another theory about the Moon's formation requires an early molten Earth, a Mars-size impactor, and a huge collision. This theory sounds speculative and seems to require a special set of circumstances. But, as it turns out, such a scenario helps explain why the Moon is made mostly of rock that appears to have been heated (and all its water baked out). Because catastrophic collisions were common in the early solar system formation, it is possible that one resulted in the formation of the Moon.

Chapter 1

The basic idea of this particular theory is that about 4.5 billion years ago, when the Earth was about 50 million years old, it collided with another planetary body that had about the same mass as Mars. Enormous amounts of energy were released, and in the process the Mars-size body was destroyed. The upper layers of the Earth's mantle were vaporized and thrown off into Earth's orbit. After some time, this orbiting debris accreted into the body that became our Moon. Today, planetary scientists have created computer models demonstrating that a collision at the right speed and angle could ultimately have produced the debris that could have formed our Moon.

Tidal Forces

The tidal forces exerted on the Earth's oceans by the Sun and the Moon have played a large part in the evolution of the length of the Earth's day and the relative distance between the Earth and the Moon. Soon after the Moon formed, it was probably no more than 24,000 km from the Earth. Earth was spinning rapidly with a day that was probably no more than six hours long. Ocean tides could have been hundreds of meters high, and the pull of this nearby Moon would have flexed the Earth's crust, causing heating and melting.

As the motions of the Earth–Moon system evolved, tidal forces gradually caused the Moon to move farther away from the Earth. As it moved away, it took longer to orbit the Earth. The tidal bulges in the Earth's surface led (that is, were always positioned ahead of the Earth–Moon line) as the Moon orbited the Earth. Although the dynamics are not easy to understand, this produced a torque that slowed down Earth's spin rate as the distance between Earth and Moon increased. This slowdown continues today as the Moon continues to recede from the Earth at the rate of about 3 cm per year. There is further proof of this slowdown in the geologic record. Studies of daily and annual layers in Devonian corals show that 400 million years ago, Earth had a 400-day-long year.

The Moon and the Origins of Life on Earth

There are some scientists who think that life on Earth would not have arisen without the presence of the Moon. The reasoning goes something like this: if the Moon had not formed, particularly in a giant impact, the early Earth might have had much different starting amounts of water, carbon, and nitrogen. Too much of these elements and molecules, combined with heating from the Sun, and the Earth might have ended up in what scientists call a "runaway greenhouse" atmosphere not as conducive to the formation of life. In addition, without a Moon at all, early Earth's atmosphere could have been very unstable. Again, this wouldn't be terribly friendly to life. The formation and ongoing survival of life on this planet needed a stable atmosphere, and possibly the evolution of longer days and shorter years. Until we know more about the earliest epochs of our planet's evolution and all the factors that led to life, however, these arguments about the Moon being necessary for life to emerge on Earth remain interesting avenues of scientific investigation.

Investigating the Moon's Effects

The two most obvious effects of the Moon that we notice today are the phases of the Moon and the ocean tides. For part of the month the Moon appears in the nighttime or early-morning skies. The rest of the month it can be spotted in the daytime sky. Over a period of 28 days, the Moon's shape changes from a thin crescent to a full Moon and back to a crescent. This isn't due to any changes on the Moon's surface or its actual shape, but is an interplay of light and perspective that depends on the positions of the Earth, Moon, and Sun.

Tides are strong and important forces in astronomy and physics. They are defined as the movement of fluids, or the stresses induced in solid objects, by a cyclical change in the net gravitational forces acting upon them. On Earth, ocean tides are the result of the gravitational pull of the Sun and Moon on the Earth's oceans. We see variations in the tides which are caused by the Earth's rotation, the Moon's orbital motion around the Earth, and the Earth's orbital motion around the Sun.

The forces of gravity that produce tidal effects also play a part throughout the solar system. For example, gravitational interactions between Jupiter and its two moons, Io and Europa, produce tides in Io's surface. These heat Io's interior and cause intense volcanic activity in response to the pressures. Around Saturn, the particles that make up its rings orbit in a region where the tidal forces of Saturn and its largest moons are very strong. It's possible that the rings are the remnants of a wandering moon that strayed too close to the planet and was ripped apart. Another theory is that the tidal forces have actually prevented any such moon from forming.

Making Lunar Observations
There are several ways to study Earth's Moon. Students can watch how it appears throughout the month. By making a simple chart of the daily tides, they can observe the effects it has on our planet. They can also study it as a world of its own. **Activity 2** allows students to draw all these things together. **Part A** of **Investigate** offers a particularly useful way to give students a three-dimensional feel for the phases of the Moon. This activity may need to be repeated several times, because not every student can immediately visualize the Earth–Moon–Sun geometry involved.

Observing the Moon is a long-term process that gives the student firsthand experiences in studying another solar-system body and recording observations. No particular equipment is necessary, although students should be encouraged to view the Moon's surface through a pair of binoculars. Craters, valleys, plains, and mountains can all be easily identified.

Lunar Impact Features
Most lunar features are the result of bombardment throughout all the Moon's existence.

The largest impact features are the maria (pronounced "MAH-ree-uh"). They are large, dark plains that formed when impactors punched into the lunar surface. Melted rock—essentially basaltic lavas similar to those that are common on Earth—poured out, forming smooth plains.

Smaller impacts created typical craters. Over the billions of years since it formed, the Moon has been bombarded with millions of smaller impacts. Impactors sometimes obliterated older craters, leaving behind broken walls and canyons that we sometimes describe as mountains. The largest craters have rays of material sprayed out around them.

The smallest impacts come from micrometeoroids. Over geologic time, these little bits of space debris have ground the top few centimeters of the lunar surface into a dusty covering called regolith. If you study videos of the astronauts as they walked across the lunar surface during the Apollo era, you'll see them kicking up this regolith with their boots and spraying it out from under the tires of their lunar dune buggy.

More Information – on the Web
Visit the *EarthComm* web site www.agiweb.org/earthcomm to access a variety of links to web sites that will help you to deepen your understanding of content and prepare you to teach this activity. Many of the sites also contain images that you can download.

Goals and Assessment

Clarify that the goals indicate what students should understand and be able to do as a result of the activity. Make sure students understand that Chapter Assessments are based upon these goals.

Goal	Location in Activity	Assessment Opportunity
Investigate lunar phases using a model and observations in your community.	**Investigate** Parts A and B	Answers to questions are reasonable and reflect an understanding. Lunar observation chart is complete.
Investigate the general idea of tidal forces.	**Investigate** Parts C and D **Digging Deeper; Understanding and Applying What You Have Learned** Questions 2 – 3	Answers to questions are reasonable and reflect an understanding. Graphs of high and low tides and the lunar phases are drawn correctly; both axes are labeled.
Understand the role of the Earth, the Moon, and the Sun in creating tides on Earth.	**Investigate** Part C **Digging Deeper; Check Your Understanding** Questions 2 – 3 **Understanding and Applying What You Have Learned** Questions 1 – 3	Answers to questions are reasonable and reflect an understanding. Graphs are drafted correctly; both axes are labeled. Rate of change of length of year is done correctly; work is shown.
Understand the Earth–Moon system and the Moon's likely origin.	**Investigate** Parts C and D **Digging Deeper; Check Your Understanding** Question 1 **Understanding and Applying What You Have Learned** Question 3	Answers to questions are reasonable and reflect an understanding.
Compare the appearance of the Moon to other solar-system bodies.	**Digging Deeper**	

Earth System Evolution Astronomy

Activity 2   The Earth–Moon System

Goals
In this activity you will:

- Investigate lunar phases using a model and observations in your community.

- Investigate the general idea of tidal forces.

- Understand the role of the Earth, the Moon, and the Sun in creating tides on Earth.

- Understand the Earth–Moon system and the Moon's likely origin.

- Compare the appearance of the Moon to other solar-system bodies.

Think about It

Think about the last time that you gazed at a full Moon.

- What happened to make the Moon look the way it does?
- What is the origin of the Moon?
- How does the Moon affect the Earth?

What do you think? Record your ideas about these questions in your *EarthComm* notebook. Be prepared to discuss your responses with your small group and the class.

E 14

Activity Overview

Students model the phases of the Moon using a Styrofoam® ball and a lamp. They observe the Moon for one month and record their observations on a lunar observation chart. Students examine data for the heights of high and low tides in various coastal cities. To understand how the Moon generates the tides, they correlate the data with information about the lunar phases. Students then use the Internet to collect tidal data for cities near their community. Finally, they complete a series of calculations to illustrate how the rotation of the Earth has changed over time in relation to tidal forces. **Digging Deeper** examines the Earth–Moon system in depth: how it is believed that the Moon formed, what causes the tides, and how tides influence Earth.

Preparation and Materials Needed

Part A

You may wish to prepare the Styrofoam ball on the pencil for the students. Push a very sharp pencil into the foam.

Part B

For this part of the investigation, students are asked to observe the Moon for one month.

NOTE: You may want to have students begin their observations one month before you begin this chapter, so that they have the data to complete the activity.

Students can then fill in any additional information as they complete the investigation. You may wish to make copies of the lunar observation chart, using **Blackline Master Astronomy 2.1, Lunar Observation Chart** on page 54.

Part C

To complete this part of the activity, students will need Internet access. If students collect the data on their own, you may want to standardize responses somewhat by restricting their search to certain cities. If you do not have ready access to the Internet for your students, obtain the data before class and make copies for distribution. Information on obtaining tidal data from different cities can be found on the *EarthComm* web site. **Blackline Master Astronomy 2.2, Graph Paper Master** on page 67 can be used to photocopy graph paper for **Parts C** and **D** of this activity.

Part D

No additional preparation is required to complete this part of the investigation.

Materials

Part A
- Pencil
- Styrofoam ball, at least 5 cm in diameter
- Lamp with a 150-W bulb and no lampshade (or a flashlight)

Part B
- Copy of lunar observation chart (see **Blackline Master Astronomy 2.1, Lunar Observation Chart**)

Part C
- Graph paper (see **Blackline Master Astronomy 2.2**)
- Internet access (or printouts of tidal data for several cities near your community)*

Part D
- Graph paper (see **Blackline Master Astronomy 2.2**)
- Calculator

Think about It

Student Conceptions

Students will most likely have more experience looking at the Moon than at any other extraterrestrial body. Through popular media (or their own pondering) they may realize that the appearance of the Moon's surface has been greatly affected by numerous impact craters. They probably will not understand the details of how such craters are formed or why they are so obvious on the Moon's surface, but not on the Earth's surface. Students are also unlikely to know that the different-colored regions (darker and lighter areas) are the result of differences in the geology of the lunar surface. Most students will not have thought a great deal about the origin of the Moon or its effect on Earth. Some may think that the Moon formed at the same time as the Earth or was captured by the Earth's gravity. Most will probably have heard of tides, and some may even know that they are related to gravity. It is likely, however, that few will understand the details of how the tides correspond to the lunar cycle.

Answer for the Teacher Only

The Moon is scarred by numerous impact craters, which record impacts of smaller bodies with the body of the Moon, most of them early in its history. The dark areas of the Moon, called "maria," are plains covered with basalt lava flows that were generated in connection with major impacts. (See **Background Information: Lunar Impact Features**.)

The general agreement among planetary scientists today is that the Moon originated as a result of a single collision between the Earth and an unusually large planetary body early in the history of solar system. This collision threw off an enormous mass of the outer part of the Earth. The individual masses then gradually coalesced to form the Moon.

The gravitational pull of the Moon on the Earth, together with the weaker gravitational pull of the Sun, produces the tides in the Earth's ocean. There are tides

*The *EarthComm* web site provides suggestions for obtaining these resources.

in the solid Earth as well. Because the material of the solid Earth is much stiffer than the waters of the oceans, however, the amplitude of the solid-Earth tides is much smaller, of the order of centimeters.

Assessment Tool

Think about It Evaluation Sheet
Use this evaluation sheet to help students understand and internalize the basic expectations for the warm-up activity.

Blackline Master Astronomy 2.1
Lunar Observation Chart

	Sunday	Monday	Tuesday	Wednesday	Thursday	Friday	Saturday
Time / Moon Phase	○	○	○	○	○	○	○
Time / Moon Phase	○	○	○	○	○	○	○
Time / Moon Phase	○	○	○	○	○	○	○
Time / Moon Phase	○	○	○	○	○	○	○
Time / Moon Phase	○	○	○	○	○	○	○

NOTES

Investigate

Part A: Lunar Phases

1. Attach a pencil to a white Styrofoam® ball (at least 5 cm in diameter) by pushing the pencil into the foam. Set up a light source on one side of the room. Use a lamp with a bright bulb (150-W) without a lampshade or have a partner hold a flashlight pointed in your direction. Close the shades and turn off the overhead lights.

2. Stand approximately 2 m in front of the light source. Hold the pencil and ball at arm's length away with your arm extended towards the light source. The ball represents the Moon. The light source is the Sun. You are standing in the place of Earth.

   a) How much of the illuminated Moon surface is visible from Earth? Draw a sketch of you, the light source, and the foam ball to explain this.

3. Keeping the ball straight in front of you, turn 45° to your left but stay standing in one place.

   a) How much of the illuminated Moon surface is visible from Earth?

   b) Has the amount of light illuminating the Moon changed?

   c) Which side of the Moon is illuminated? Which side of the Moon is still dark? Draw another diagram in your notebook of the foam ball, you, and the light source in order to explain what you see.

4. Continue rotating counterclockwise away from the light source while holding the ball directly in front of you. Observe how the illuminated portion of the Moon changes shape as you turn 45° each time.

a) After you pass the full Moon phase, which side of the Moon is illuminated? Which side of the Moon is dark?

b) How would the Moon phases appear from Earth if the Moon rotated in the opposite direction?

⚠ Be careful not to poke the sharp end of the pencil into your skin while pushing the pencil into the foam. Use caution around the light source. It is hot. Do not touch the Styrofoam to the light.

Part B: Observing the Moon

1. Observe the Moon for a period of at least four weeks. During this time you will notice that the apparent shape of the Moon changes.

   a) Construct a calendar chart to record what you see and when you see it. Sketch the Moon, along with any obvious surface features that you can see with the naked eye or binoculars.

   b) Do you always see the Moon in the night sky?

   c) How many days does it take to go through a cycle of changes?

   d) What kinds of surface features do you see on the Moon?

   e) Label each phase of the Moon correctly and explain briefly the positions of the Sun, the Earth, and the Moon during each phase.

⚠ Tell an adult before you go outside to observe the Moon.

Investigate

Part A: Lunar Phases

1. Students should be careful not to poke themselves with the sharp end of the pencil. Try to darken the room as much as possible. Students can use either a lamp or a flashlight as the light source for this experiment.

2. a) None of the Moon's illuminated surface is visible from the Earth. This is the "new Moon" phase.

3. a) One quarter of the Moon's illuminated surface is visible from Earth. This is the "first quarter" Moon.

   b) The amount of light illuminating the Moon has not changed, but the amount visible from Earth has.

   c) The side of the Moon facing the light source is visible.

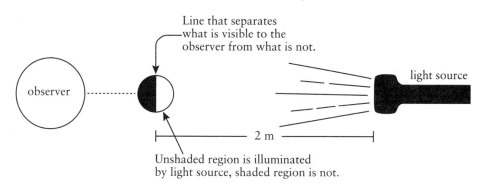

2 a)   VIEW FROM ABOVE

Line that separates what is visible to the observer from what is not.

observer

light source

2 m

Unshaded region is illuminated by light source, shaded region is not.

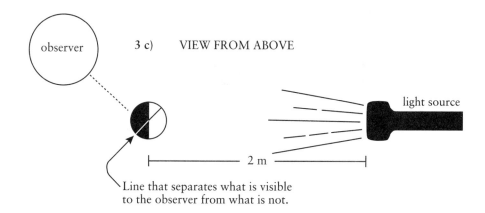

observer

3 c)   VIEW FROM ABOVE

light source

2 m

Line that separates what is visible to the observer from what is not.

4. a) The "full Moon" phase is the phase in which the side of the Moon facing the Earth is entirely illuminated by the Sun. The side away from the Earth is dark. Students may need to duck if their body is in the path of the "sunlight."

    b) The full Moon and new Moon phases would be no different if the Moon revolved around the Earth in the opposite direction. During the intermediate phases, however, the illuminated surface of the Moon visible from Earth would be the opposite.

Part B: Observing the Moon

1. a) Use **Blackline Master Astronomy 2.1, Lunar Observation Chart** if you wish to provide students with a copy of the calendar chart. Charts will vary depending on when activity is completed. A sample chart is provided below.

Sunday	Monday	Tuesday	Wednesday	Thursday	Friday	Saturday
Time Moon Phase	Time Moon Phase	**1** Time 8:00 Moon Phase	**2** Time 7:45 Moon Phase	**3** Time 8:15 Moon Phase	**4** Time 8:30 Moon Phase	**5** Time 8:00 Moon Phase
Time Moon Phase	Time Moon Phase	Time Moon Phase	Time Moon Phase	Time Moon Phase	Time Moon Phase	Time Moon Phase
Time Moon Phase	Time Moon Phase	Time Moon Phase	Time Moon Phase	Time Moon Phase	Time Moon Phase	Time Moon Phase
Time Moon Phase	Time Moon Phase	Time Moon Phase	Time Moon Phase	Time Moon Phase	Time Moon Phase	Time Moon Phase
Time Moon Phase	Time Moon Phase	Time Moon Phase	Time Moon Phase	Time Moon Phase	Time Moon Phase	Time Moon Phase

    b) Ask students to consider whether there were any clear nights on which they could not see the Moon. They should note that it is not always possible to see the Moon.

    c) A complete cycle from one new Moon to the next takes ~ 29.5 days (one synodic month) to complete. Note that this is longer than it takes the Moon to make one revolution around the Earth (~ 27.3 days, which is one sidereal month) because the Earth's position relative to the Sun is also changing during that time.

d) Answers will vary depending upon whether students are viewing the Moon with the naked eye or through binoculars. Students should be able to see craters, and possibly lighter and darker regions that correspond to the lunar highlands and the maria.

e) Students should label each phase of the Moon on their chart. They may wish to compare their completed chart with the Moon phases shown on a calendar.

Earth System Evolution Astronomy

Part C: Tides and Lunar Phases

1. Investigate the relationship between tides and phases of the Moon.

   a) On a sheet of graph paper, plot the high tides for each city and each day in January shown in *Table 1*. To prepare the graph, look at the data to find the range of values. This will help you plan the scales for the vertical axis (tide height) and horizontal axis (date).

   b) On the same graph, plot the Moon phase using a bold line. Moon phases were assigned values that range from zero (new Moon) to four (full Moon).

2. Repeat this process for low tides.

3. Answer the following questions in your *EarthComm* notebook:

   a) What relationships exist between high tides and phases of the Moon?

   b) What relationships exist between low tides and phases of the Moon?

   c) Summarize your ideas about how the Moon affects the tides. Record your ideas in your *EarthComm* notebook.

Table 1 Heights of High and Low Tides in Five Coastal Locations during January 2001 (All heights are in feet.)													
			Breakwater, Delaware		Savannah, Georgia		Portland, Maine		Cape Hatteras, North Carolina		New London, Connecticut	
Date	Moon Phase		Moon Phase	High	Low	High	Low	High	Low	High	Low	High	Low
1/3/01	First Quarter		2	3.6	0.2	7.3	0.5	8.5	1	2.6	0.2	2.4	0.3
1/6/01	Waxing Gibbous		3	4.5	0	8.2	0.5	9.7	0.1	3.4	–0.4	3	–0.2
1/10/01	Full Moon		4	5.6	–0.9	9.4	–1.5	11.6	–1.9	4.2	–0.8	3.5	–0.7
1/13/01	Waning Gibbous		3	5.1	–0.7	8.8	–0.9	11	–1.4	3.7	–0.6	3	–0.5
1/16/01	Last Quarter		2	4.1	–0.1	7.9	–0.2	9.7	0.1	3	–0.2	2.7	0
1/20/01	Waning Crescent		1	4.3	0.1	7.3	0.2	9.4	0.2	3.2	0	2.8	0
1/24/01	New Moon		0	4.6	0	8.1	–0.1	9.7	–0.1	3.3	–0.1	2.8	–0.1
1/30/01	Waxing Crescent		1	3.7	0.1	7.4	0.1	8.7	0.6	2.6	0	2.4	0.2

Part C: Tides and Lunar Phases

1. a) – b)

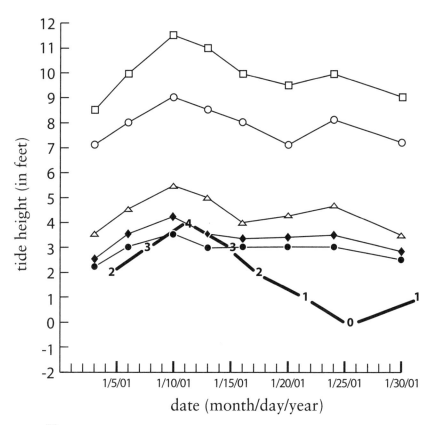

Key

△ Breakwater, Delaware	—1— Lunar Phases
○ Savannah, Georgia	**0** = new moon
□ Portland, Maine	**1** = waxing or waning crescent
◆ Cape Hatteras, North Carolina	**2** = first or last quarter
● New London, Connecticut	**3** = waxing or waning gibbous
	4 = full moon

2. a) – b)

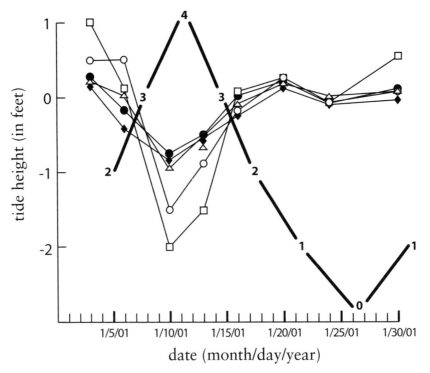

Key

△	Breakwater, Delaware	—**1**— Lunar Phases
○	Savannah, Georgia	**0** = new moon
□	Portland, Maine	**1** = waxing or waning crescent
◆	Cape Hatteras, North Carolina	**2** = first or last quarter
●	New London, Connecticut	**3** = waxing or waning gibbous
		4 = full moon

(**Note:** The heights of the tides in the table in the book have been shown in feet. The authors encourage the use of metric units for scientific data. However, the students are asked to consult tide tables for their area in **Step 4**, and in the tables they will find values are typically given in feet. Therefore, for the sake of comparison, the tables are given in feet in the textbook.)

3. a) The highest high tides occur during the ful Moon and new Moon phases, and the lowest high tides occur during the first-quarter and last-quarter phases.

   b) The lowest low tides occur during the full Moon and new Moon phases, and the highest low tides occur during the first-quarter and last-quarter phases.

   c) Students should refer to their graphs.

Assessment Tools

EarthComm Notebook Entry-Checklist
Use this checklist as a quick guide for student self-assessment and/or an opportunity to quickly score student work. Add further criteria specific to your classroom needs or to this particular investigation.

Investigate Notebook Entry-Evaluation Sheet
Point out the criteria listed on this evaluation sheet that are relevant to this particular investigation. Encourage students to internalize the criteria by making them part of your "assessment conversations" as you circulate around the classroom.

4. *Table 1* shows data from the month of January 2001. At the *EarthComm* web site, you can obtain tidal data during the same period that you are doing your Moon observations. Select several cities nearest your community.

a) Record the highest high tide and the lowest low tide data for each city. Choose at least eight different days to compare. Correlate these records to the appearance of the Moon during your observation period. Make a table like *Table 1* showing high and low tides for each location.

b) What do you notice about the correlation between high and low tides and the appearance of the Moon?

Part D: Tidal Forces and the Earth System

1. Use the data in *Table 2*.

a) Plot this data on graph paper.

Label the vertical axis "Number of Days in a Year" and the horizontal axis "Years before Present." Give your graph a title.

b) Calculate the rate of decrease in the number of days per 100 million years (that is, calculate the slope of the line).

2. Answer the following questions:

a) How many fewer days are there every 10 million years? every million years?

b) Calculate the rate of decrease per year.

c) Do you think that changes in the number of days in a year reflect changes in the time it takes the Earth to orbit the Sun, or changes in the time it takes the Earth to rotate on its axis? In other words, is a year getting shorter, or are days getting longer? How would you test your idea?

Table 2 Change in Rotation of Earth Due to Tidal Forces		
Period	**Date (millions of years ago)**	**Length of Year (days)**
Precambrian	600	424
Cambrian	500	412
Ordovician	425	404
Silurian	405	402
Devonian	345	396
Mississippian	310	393
Pennsylvanian	280	390
Permian	230	385
Triassic	180	381
Jurassic	135	377
Cretaceous	65	371
Present	0	365.25

E 17

Chapter 1

4. a) Answers to these questions will vary depending upon where you live and the cities for which students elect to collect data.

   b) In general, students should notice that the most extreme high and low tides occur during the full Moon and new Moon phases.

Part D: Tidal Forces and the Earth System

1. a)

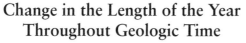

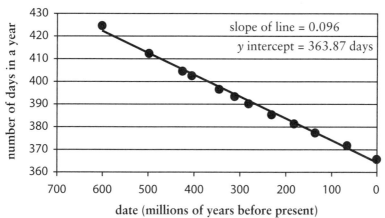

Change in the Length of the Year
Throughout Geologic Time

slope of line = 0.096
y intercept = 363.87 days

number of days in a year

date (millions of years before present)

   b) The slope of the line for this data can be calculated either by doing a linear regression of the data or plotting the data by hand and calculating the rise over run (slope) for the best-fit line. This exercise would be an excellent opportunity to sharpen students' skills at manipulating spreadsheets, and most modern spreadsheets have tools that will calculate the slope of a best-fit line through a set of data. Regardless of how it is done, the graph above shows the data plotted and the parameters for the equation to the best-fit line. The slope of this line is 0.096. Students who have calculated their slopes will have answers that vary slightly from this value (it should be a value around 0.01 or 10%). A slope of 0.096 means that for every one unit change in the *x*-axis variable, there is a change in the *y*-axis variable of ~ 9.6%. It is important to note that the *x*-axis variable is millions of years and the *y*-axis variable is days. Accordingly, for every 100 million years, the length of the year decreases by 9.6 days (100 x .096). Rounding to the nearest day gives an answer of 10 days.

2. a) The length of an Earth year decreases by one day every 10 million years and by 0.1 days every one million years.

   b) Dividing 0.1 days per million years by one million yields a rate of decrease of 0.0000001 days per year or ~ 3.15 seconds per year (assuming 86,400 seconds per day and 365 days per year).

c) Tidal friction acts to slow the rotation of the Earth on its axis. The friction of the Earth's tides does not slow the Earth's revolution around the Sun. As the rotation of the Earth slows, the days grow longer because it takes longer for the Earth to make one complete rotation about its axis. Accordingly, the years are not getting shorter; the days are getting longer. Student's propositions for a way to test their ideas will vary but might include such things as using the laws of physics to analyze the various forces involved, measuring the rotation of the Earth by satellite, or using astronomical observations to chart the Earth's course around the Sun. Note that any solution that is based on making modern measurements without looking at theoretical, historical, or geologic evidence will need to have an extended time series of such observations.

Blackline Master Astronomy 2.2
Graph Paper Master

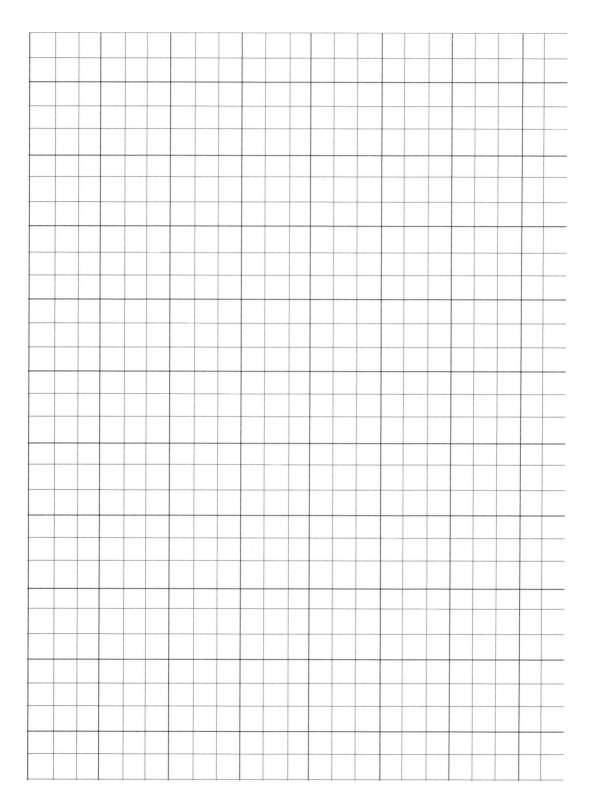

Earth System Evolution Astronomy

Reflecting on the Activity and the Challenge

In this activity you used a simple model and observations of the Moon to explore lunar phases and surface characteristics of the Moon. You also explored the relationship between tides and the phases of the Moon. The tides also have an effect that decreases the number of days in a year over time. That's because tides slow the rotation of the Earth, making each day longer. You now understand that tides slow the rotation of the Earth, and how this has affected the Earth. This will be useful when describing the Earth's gravitational relationships with the Moon for the **Chapter Challenge**.

Geo Words

accretion: the process whereby dust and gas accumulated into larger bodies like stars and planets.

Digging Deeper

THE EVOLUTION OF THE EARTH–MOON SYSTEM

The Formation of the Earth and Moon

Figure I The Moon is the only natural satellite of Earth.

You learned in the previous activity that during the formation of the solar system, small fragments of rocky material called planetesimals stuck together in a process called **accretion**. Larger and larger pieces then came together to form the terrestrial planets. The leftovers became the raw materials for the asteroids and comets. Eventually, much of this material was "swept up" by the newborn inner planets. Collisions between the planets and the leftover planetesimals were common. This was how the Earth was born and lived its early life, but how was the Moon formed?

Reflecting on the Activity and the Challenge

Have students read this brief passage and share their thoughts about the main point of the activity in their own words. Hold a class discussion about how this investigation relates to the **Chapter Challenge**. How does the Moon affect their community? The answer is probably more obvious if you live near a coast, but students should now be able to understand that they are affected by tidal forces even if they do not live in a coastal area. Students can now consider how changes in the number of days in a year (changes in the length of a day) would affect their community.

Digging Deeper

Assign the reading for homework, along with the questions in **Check Your Understanding** if desired.

Assessment Opportunity

Use (or rephrase) the questions in **Check Your Understanding** for a brief quiz to check comprehension of key ideas and skills. Use the quiz (or a class discussion) to assess your students' understanding of the main ideas in the reading and the activity. A few sample questions are provided below:

Question: Use an illustration to explain the difference between a spring tide and a neap tide.

Answer: Student illustrations should resemble *Figure 4* on page E22 in the text.

Question: If the Earth and Moon were both bombarded by incoming planetesimals during their early history, why do we see craters on the Moon but not on Earth?

Answer: The Earth is not as covered with craters as the Moon is because the Earth is geologically a very active place. Geologic and atmospheric (weather) processes act continuously to reshape the surface of the Earth, so very few craters remain. The Moon, however, is geologically inactive, and therefore features like craters are not destroyed over time.

Question: Why do high tides and low tides not occur at the same time from day to day?

Answer: High tides and low tides do not occur at the same time each day because the period of the Moon's orbital cycle is not the same as the length of one day. By the time the Earth has completed one full rotation, the Moon's position relative to the Earth has changed.

Teaching Tip

The Moon, shown in *Figure 1*, is the second brightest object in the sky after the Sun. It is the only extraterrestrial body to have been visited by humans. The presence of ice in the South Pole of the Moon was confirmed in 1999 by the Lunar Prospector mission. There is no atmosphere and no magnetic field. The crust of the Moon averages 68 km thick, but it varies from essentially 0 m under the feature Mare Crisium to 107 km north of the crater Korolev. Below the crust is a mantle and probably a small core. For reasons currently unknown to scientists, the Moon's center of mass is offset from its geometric center by about 2 km in the direction toward the Earth. Also, the crust is thinner on the near side.

NOTES

Scientists theorize that an object the size of Mars collided with and probably shattered the early Earth. The remnants of this titanic collision formed a ring of debris around what was left of our planet. Eventually this material accreted into a giant satellite, which became the Moon. Creating an Earth–Moon system from such a collision is not easy. In computer simulations, the Moon sometimes gets thrown off as a separate planet or collides with the Earth and is destroyed. However, scientists have created accurate models that predict the orbit and composition of both the Earth and Moon from a collision with a Mars-sized object. The Moon's orbit (its distance from the Earth, and its speed of movement) became adjusted so that the gravitational pull of the Earth is just offset by the centrifugal force that tends to make the Moon move off in a straight line rather than circle the Earth. After the Earth–Moon system became stabilized, incoming planetesimals continued to bombard the two bodies, causing impact craters. The Earth's surface has evolved since then. Because the Earth is geologically an active place, very few craters remain. The Moon, however, is geologically inactive. *Figure 2* shows the Moon's pockmarked face that has preserved its early history of collisions.

Figure 2 Impact craters on the Moon.

Teaching Tip

The image in *Figure 2* shows the surface of the Moon. There are two main types of terrain on the Moon: the heavily cratered and very old highlands, and the relatively smooth and younger maria. Maria (the single form is *mare*) literally means "seas." These features are so named because in ancient times it was thought that they were large areas of water. The maria are actually huge impact craters that were later flooded by lava. They constitute about 16% of the Moon's surface. Most of the Moon's surface is covered with regolith, a mixture of fine dust and rocky debris produced by meteorite impacts. Scientists are not certain why the maria are concentrated on the near side. Can your students think of any reasonable explanations?

Teaching Tip

Use **Blackline Master Astronomy 2.3, Schematic Diagram of Tides** to make an overhead of the diagram shown in *Figure 3* on page E20. Incorporate this overhead into a discussion about the astronomical effects that contribute to diurnal inequality in the tides (i.e., why successive diurnal tides are often not of the same magnitude).

Earth System Evolution Astronomy

When the Earth was first formed, its day probably lasted only about six hours. Over time, Earth days have been getting longer and longer. In other words, the Earth takes longer to make one full rotation on its axis. On the other hand, scientists have no reason to think that the time it takes for the Earth to make one complete revolution around the Sun has changed through geologic time. The result is that there are fewer and fewer days in a year, as you saw in the **Investigate** section. Why is the Earth's rotation slowing down? It has to do with the gravitational forces between the Earth, the Moon, and the Sun, which create ocean tides.

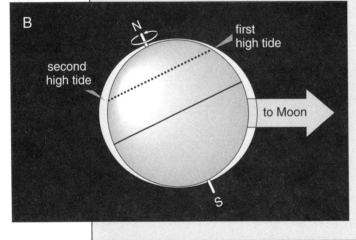

Tides

The gravitational pull between the Earth and the Moon is strong. This force actually stretches the solid Earth about 20 cm along the Earth–Moon line. This stretching is called the Earth tide. The water in the oceans is stretched in the same way. The stretching effect in the oceans is greater than in the solid Earth, because water flows more easily than the rock in the Earth's interior. These bulges in the oceans, called the ocean tide, are what create the high and low tides (see *Figure 3*). It probably will seem strange

Figure 3 Schematic diagram of tides. Diagram A illustrates how the ocean surface would behave without the Moon and the Sun (no tides). Diagram B illustrates, that in the presence of the Moon and the Sun, shorelines away from the poles experience two high tides and two low tides per day.

E 20

Blackline Master Astronomy 2.3
Schematic Diagram of Tides

A

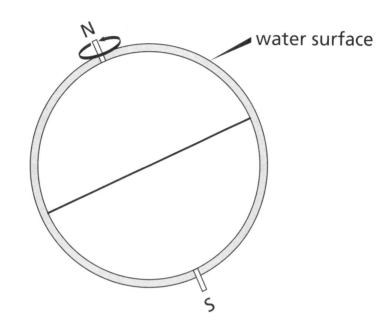

water surface

B

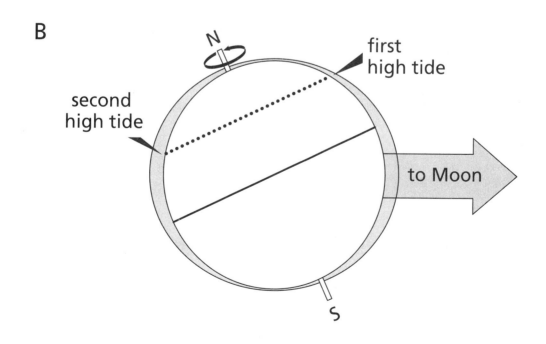

first high tide

second high tide

to Moon

to you that there are two bulges, one pointing toward the Moon and the other away from the Moon. If the tides are caused by the pull of the Moon, why is there not just one bulge pointing toward the Moon? The explanation is not simple. If you are curious, you can pursue it further in the **Inquiring Further** section of this activity.

As the Earth rotates through a 24-h day, shorelines experience two high tides—one when the tidal bulge that points toward the Moon passes by, and once when the tidal bulge that points away from the Moon passes by. The tidal cycle is not exactly 24 h. By the time the Earth has completed one rotation (in 24 h), the Moon is in a slightly different place because it has traveled along about 1/30 of the way in its orbit around the Earth in that 24-h period. That's why the Moon rises and sets about 50 min. later each day, and why high and low tides are about 50 min. later each day. Because there are two high tides each day, each high tide is about 25 min. later than the previous one.

The gravitational pull of the Sun also affects tides. Even though it has much greater mass than the Moon, its tidal effect is not as great, because it is so much farther away from the Earth. The Moon is only 386,400 km away from the Earth, whereas the Sun is nearly 150,000,000 km away. The Moon exerts 2.4 times more tide-producing force on the Earth than the Sun does. The changing relative positions of the Sun, the Moon, and the Earth cause variations in high and low tides.

The lunar phase that occurs when the Sun and the Moon are both on the same side of the Earth is called the new Moon. At a new Moon, the Moon is in the same direction as the Sun and the Sun and Moon rise together in the sky. The tidal pull of the Sun and the Moon are adding together, and high tides are even higher than usual, and low tides are even lower than usual. These tides are called **spring tides** (see *Figure 4*). Don't be confused by this use of the word "spring." The spring tides have nothing to do with the spring season of the year! Spring tides happen when the Sun and Moon are in general alignment and raising larger tides. This also happens at another lunar phase: the full Moon. At full Moon, the Moon and Sun are on opposite sides of the Earth. When the Sun is setting, the full Moon is rising. When the Sun is rising, the full Moon is setting. Spring tides also occur during a full Moon, when the Sun and the Moon are on opposite sides of the Earth. Therefore, spring tides occur twice a month at both the new-Moon and full-Moon phases.

Geo Words

spring tide: the tides of increased range occurring semimonthly near the times of full Moon and new Moon.

Chapter 1

Teaching Tip

Use **Blackline Master Astronomy 2.4, Spring and Neap Tides** to make an overhead of the illustration in *Figure 4* on page E22. Use the overhead to discuss with students the interaction between the Sun, the Moon, and the Earth, and to consider how this interaction affects the tides.

Earth System Evolution Astronomy

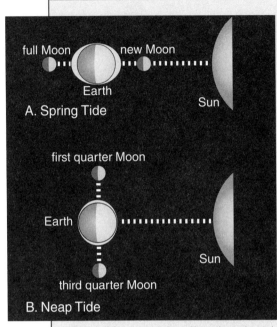

full Moon   new Moon

Earth

A. Spring Tide   Sun

first quarter Moon

Earth   Sun

third quarter Moon

B. Neap Tide

Figure 4 Schematic diagrams illustrating spring and neap tides.

When the line between the Earth and Sun makes a right angle with the line from the Earth to the Moon, as shown in *Figure 4*, their tidal effects tend to counteract one another. At those times, high tides are lower than usual and low tides are higher than usual. These tides are called **neap tides**. They occur during first quarter and third-quarter Moons. As with spring tides, neap tides occur twice a month.

The tide is like a kind of ocean wave. The high and low tides travel around the Earth once every tidal cycle, that is, twice per day. This wave lags behind the Earth's rotation, because it is forced by the Moon to travel faster than it would if it were free to move on its own. That's why the time of high tide generally does not coincide with the time that the Moon is directly overhead. The friction of this lag gradually slows the Earth's rotation.

Another way to look at tides is that the tidal bulges are always located on the sides of the Earth that point toward and away from the Moon, while the Earth with its landmasses is rotating below the bulges. Each time land on the spinning Earth encounters a tidal bulge there is a high tide at that location. The mass of water in the tidal bulge acts a little like a giant brake shoe encircling the Earth. Each time the bulge of water hits a landmass, energy is lost by friction. The water heats up slightly. (This is in addition to the energy lost by waves hitting the shore, which also heats the water by a small fraction of a degree.) Over long periods of time, the tidal bulge has the effect of slowing down the rotation of the Earth, and actually causing the Moon to move away from the Earth. The current rate of motion of the Moon away from the Earth is a few centimeters a year. This has been established by bouncing laser beams off of reflectors on the Moon to measure its distance. Although the Moon's orbit is not circular and is complex in its shape, measurements over many years have established that the

Geo Words

neap tide: the tides of decreased range occurring semimonthly near the times of the first and last quarter of the Moon.

Blackline Master Astronomy 2.4
Spring and Neap Tides

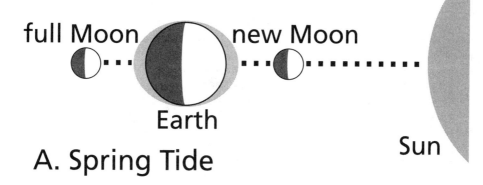

A. Spring Tide

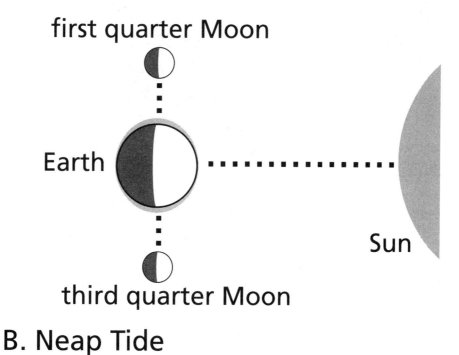

B. Neap Tide

Moon is indeed moving away from the Earth. Special super-accurate clocks have also established that the day is gradually becoming slightly longer as well, because of this same phenomenon, called "tidal friction." The day (one rotation of the Earth on its axis) has gradually become longer over geologic time. As the Earth system evolves, cycles change as well.

In this activity you limited the factors that cause and control the tides to the astronomical forces. These factors play only one part. The continents and their different shapes and ocean basins also play a large role in shaping the nature of the tides. Although many places on Earth have two high tides and two low tides every day (a semidiurnal tide), some places experience only one high tide and one low tide every day (a diurnal tide). There are still other places that have some combination of diurnal and semidiurnal tides (mixed tides). In these places (like along the west coast of the United States) there are two high tides and two low tides per day, but the heights of the successive highs and lows are considerably different from one another.

Figure 5 How do tides affect coastal communities?

Check Your Understanding

1. How did the Moon likely form?

2. Describe the relative positions of the Earth, the Moon, and the Sun for a spring tide and for a neap tide.

3. What effect have tides had on the length of a day? Explain.

Understanding and Applying What You Have Learned

1. Refer back to the graph of the changing length of the day that you produced in the investigation. Think about the causes of tidal friction and the eventual outcome of tidal friction. Predict how long you think the day will eventually be. Explain the reasoning for your prediction.

Teaching Tip

Figure 5 provides an opportunity to discuss the question of why the tidal range is greater in some places than others. (The tidal range is calculated as the difference in sea level between high tide and low tide.) This figure shows a Nova Scotia coastline where an extensive tidal flat has been exposed during low tide because of the relatively large tidal range, which averages 18 feet at this location. Even though the gravitational forces that drive tides vary according to the orbital cycles and geometrical configuration of the Earth–Moon–Sun system, these forces affect the different parts of the Earth equally at any given time. Why, then, do tidal ranges vary between different localities? Besides being affected by gravitational forces, the heights of tides are also strongly affected by the positions of the continents and the shape of the ocean basin and coastline. Some places, like the Bay of Fundy, for example, have very large tidal ranges (as high as 51 feet!). In other places (like the central Pacific Ocean, for example) the difference between high and low tides can be less than 1 foot.

Check Your Understanding

1. The Moon most likely formed when an object the size of Mars collided with and probably shattered parts of the early Earth. The remnants of the collision formed a ring of debris around what was left of the Earth. Eventually, the material accreted to form the Moon.

2. Spring tides occur when the Sun and Moon are aligned such that a single line can be drawn connecting all 3 bodies. Neap tides occur when the line between the Earth and the Sun and the line between the Earth and the Moon are at right angles to one another.

3. Tides have increased the length of a day because tidal friction has slowed the rotation of the Earth while the duration of the year (the time for one complete revolution of the Earth around the Sun) has stayed the same.

Assessment Tool

Check Your Understanding Notebook Entry-Evaluation Sheet
Use this sheet to evaluate the extent to which students understand the key concepts explored in **Activity 2** and explained in **Digging Deeper**, and to evaluate the students' clarity of expression.

Understanding and Applying What You Have Learned

1. a) Student responses to this question will likely vary—it is not an easy concept to reason out. In this case, the correct answer is difficult to predict, and is not as important as the reasoning behind the student's answer. Some students may say that the rotation of the Earth will slow until the day equals a lunar month (because the Moon's gravity plays a major part in creating the tidal bulge). Others may say that the day will eventually become a year long as the rotation of the Earth slows to match the period of the Earth's revolution around the Sun (because the Sun also takes part in creating the tidal bulge).

Teaching Tip

The underlying cause of tidal friction is that the Earth doesn't rotate at the same rate as either the orbit of the Moon around the Sun or the orbit of the Earth around the Sun. Because of this, the portion of the Earth that is receiving the greatest gravitational pull from these two bodies changes, causing the tidal bulge to move relative to the surface of the Earth. The frictional force slows the rotation of the Earth, lessening the movement of the bulge relative to the Earth as this system slowly approaches a state of equilibrium. This lengthens the day. The eventual outcome of this phenomenon is quite complicated because other things are affected by the Earth's rotation slowing down. For example, the slowing of the Earth's rotation will also affect the Moon's orbit. The system will, however, continue to move towards an equilibrium state. In this case, the pondering of the eventual outcome of this phenomenon may be more useful for the student than the knowing of it.

NOTES

Earth System Evolution Astronomy

2. Think about the roles that the Sun and Moon play in causing the ocean tides.

a) If the Earth had no Moon, how would ocean tides be different? Explain your answer.

b) How would the ocean tides be different if the Moon were twice as close to the Earth as it is now?

c) What differences would there be in the ocean tides if the Moon orbited the Earth half as fast as it does now?

3. Look at Figure 3B on page E20. Pretend that you are standing on a shoreline at the position of the dotted line. You stand there for 24 h and 50 min, observing the tides as they go up and down.

a) What differences would do you notice, if any, between the two high tides that day?

b) Redraw the diagram from *Figure 3B*, only this time, make the arrow to the moon parallel to the equator. Make sure you adjust the tidal bulge to reflect this new position of the moon relative to the Earth. What differences would you now see between the two high tides that day (assuming that you are still at the same place)?

c) Every month, the moon goes through a cycle in which its orbit migrates from being directly overhead south of the equator to being directly overhead north of the equator and back again. To complicate things, the maximum latitude at which the moon is directly overhead varies between about

28.5° north and south, to about 18.5° north and south (this variation is on a 16.8 year cycle). How do you think the monthly cycle relates to the relative heights of successive high tides (or successive low tides)?

4. Return to the tide tables for the ocean shoreline that is nearest to your community (your teacher may provide a copy of these to you).

a) When is the next high tide going to occur? Find a calendar to determine the phase of the Moon. Figure out how to combine these two pieces of information to determine whether this next high tide is the bulge toward the Moon or away from the Moon.

b) The tide tables also provide the predicted height of the tides. Look down the table to see how much variation there is in the tide heights. Recalling that the Sun also exerts tidal force on the ocean water, try to draw a picture of the positions of the Earth, Moon, and Sun for:

i) The highest high tide you see on the tidal chart.

ii) The lowest high tide you see on the tidal chart.

iii) The lowest low tide you see on the tidal chart.

5. The questions below refer to your investigation of lunar phases.

a) Explain why the moon looks different in the sky during different times of the month.

E 24

2. a) If the Earth had no Moon, then the tides would be less variable with time because only one significant gravitational force—that of the Sun—would be causing them. There would also be a smaller tidal range.

    b) If the Moon were twice as close to the Earth, the gravitational attraction between the two would be much greater (the gravitational force varies as the square of the distance between the two bodies). So, without considering other factors, the tides would be much larger. Of course, for the closer orbit to be stable, either the Moon's mass or its period of revolution would have to be different than it is now.

    c) If the Moon orbited the Earth half as fast as it did now, there would be less time between successive high tides. Currently, there are 12 hours and 25 minutes between successive high tides. This is because after 12 hours the Moon has moved its position, and it takes an extra 25 minutes for the Earth to "catch up" with the Moon. If the Moon were orbiting half as fast, it would take only an extra 12.5 minutes to "catch up" with the Moon.

3. a) One high tide would be higher than the other high tide.

    b) The two high tides would be of the same height.

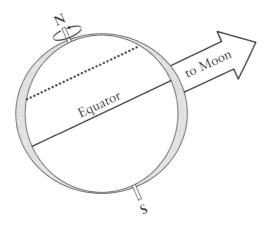

    c) As was seen in **Questions 3(a)** and **3(b)**, the relative position on the Earth where the Moon is directly overhead strongly influences the degree of inequality between successive high tides. When the latitude at which the Moon is directly overhead is at a maximum, the difference between successive high tides is at a maximum. When the Moon is directly overhead at the Equator, this difference is minimized.

4. Answers to these questions will vary depending upon your community and the time in which you complete the activity.

5. a) The different phases of the Moon are seen as variations in the amount of the illuminated part of the Moon that is visible from the Earth. During the new Moon phase, the Moon is between the Sun and the Earth, meaning that no light from the side of the Moon nearest to Earth is reflected from the Moon and the Moon is not visible in the sky. During the full Moon phase, the Moon is on the side of the Earth opposite to the Sun, and the entire illuminated half of the Moon reflects light to Earth. During intermediate stages between the new Moon and the full Moon, variable amounts of the Moon's surface reflect sunlight towards the Earth.

   b) Knowing the phases of the Moon can help understand the timing and variation of the tides, which can be beneficial for people who spend significant time in coastal areas. Looking at tide tables, however, is a more direct and more precise way of knowing the times of high and low tides. Knowing the times when there is a bright Moon in the evening can be of use to hikers and campers.

   c) Because the rate of rotation of the Moon is the same as the time for one revolution of the Moon around the Earth, the same side of the Moon always faces the Earth. As an additional note (students would have no way of knowing this), the combination of the eccentricity and inclination of the Moon's orbit causes the Moon, as seen from the Earth, to "nod" up and down and left and right slightly. These apparent motions, called the "lunar librations," allow us to observe (over a period of time) more than 59% of the Moon's surface from the Earth, rather than only 50%.

NOTES

b) What advantage is there to knowing the phases of the Moon? Who benefits from this knowledge?

c) It takes 27.32166 days for the Moon to complete one orbit around the Earth. The Moon also takes 27.32166 days to complete the rotation about its axis. How does this explain why we see the same face of the Moon all the time?

Preparing for the Chapter Challenge

Write several paragraphs explaining the evolution of the Earth–Moon system, how mutual gravitational attraction can affect a community through the tides, and how the changing length of the day could someday affect the Earth system. Be sure to support your positions with evidence.

Inquiring Further

1. **Tidal bulge**

   Use your school library or the library of a nearby college or university, or the Internet, to investigate the reason why the tidal bulge extends in the direction away from the Moon as well as in the direction toward the Moon. Why does the Earth have two tidal bulges instead of just one, on the side closest to the Moon?

2. **Tidal forces throughout the solar system**

   Tidal forces are at work throughout the solar system. Investigate how Jupiter's tidal forces affect Jupiter's Moons Europa, Io, Ganymede, and Callisto. Are tidal forces involved with Saturn's rings? Write a short report explaining how tidal friction is affecting these solar-system bodies.

3. **Impact craters**

   Search for examples of impact craters throughout the solar system. Do all the objects in the solar system show evidence of impacts: the planets, the moons, and the asteroids? Are there any impact craters on the Earth, besides the Meteor Crater in Arizona?

E 25

Preparing for the Chapter Challenge

In **Activity 2**, students studied the Earth–Moon system and learned that tides actually slow down the rate at which Earth rotates on its axis. Students practiced making and interpreting graphs, and understanding the consequences of the data graphed. They should re-examine the **Chapter Challenge** to consider ways that they can incorporate what they have learned about tides and changes in cycles within the Earth–Sun–Moon system to help local residents understand how the Earth is influenced by its neighbors in space.

For assessment of this section (see **Assessment Rubric for Chapter Challenge on Astronomy** on pages 12 and 13), students should include discussion about:
- what the Earth's orbital and gravitational relationships with the Sun and the Moon are
- why extraterrestrial influences on the community are a natural part of Earth system evolution

Inquiring Further

1. Tidal bulge
The Earth's tides are caused by the difference in the Moon's and the Sun's gravitational pull on the two sides of the Earth—but there is more to the story than just that. You need to think in terms of the combination of two different forces that are felt by the materials of the Earth:
- the gravitational attraction of the Moon
- certain centrifugal forces in the Earth

The origin of the latter is not obvious. Here's how they come about:

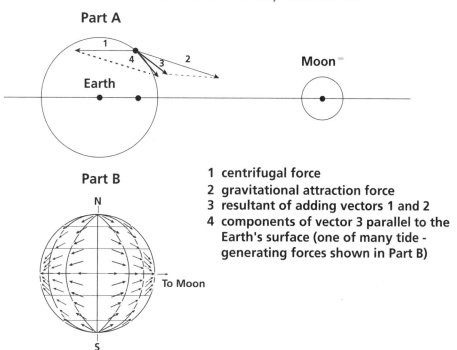

Part A

Earth

Moon

Part B

N

To Moon

S

1 centrifugal force
2 gravitational attraction force
3 resultant of adding vectors 1 and 2
4 components of vector 3 parallel to the Earth's surface (one of many tide-generating forces shown in Part B)

The Earth–Moon system revolves around the center of mass of the system. Because of the much greater mass of the Earth, the center of mass is located within the Earth, about 4600 km from its center.

(**IMPORTANT NOTE:** This centrifugal force is entirely different from—and in addition to—the centrifugal force caused by the Earth's rotation, which has nothing to do with the tides.)

Now look at the sum of the Moon's gravitational force and the Earth's centrifugal force at two points on the Earth's surface: one on the side facing the Moon and the other facing away from the Moon. Some mathematics shows right away that the net force on the Earth at the point facing the Moon is directed toward the Moon, and the net force at the point facing away from the Moon is directed away from the Moon. This is the basic reason for the existence of the double tidal bulge. For your students to follow the actual math involved, they would need to consult an advanced textbook or reference book on oceanography.

2. Tidal forces throughout the solar system

Tidal forces are at work throughout the solar system. Similar to the effect of the Moon on the period of the Earth's rotation, Jupiter is very gradually slowing down because of the tidal drag produced by its satellites. Also, the same tidal forces are changing the orbits of the moons, very slowly forcing them farther from Jupiter. The innermost moon of Jupiter is Io. The surface of Io is covered with volcanoes that have formed as a result of the gravitational force between Jupiter, Io, and Jupiter's other moons. The tidal forces on Io stretch and warm the moon, forming volcanoes that are actually similar to geysers, except that they spew lava rather than water. The other moons of Jupiter feel the effects of its gravitational pull as well. Because they are farther away, however, the effects are not as great.

Below are three theories for the formation of the rings of Saturn, two of which involve tidal forces:

- The first theory suggests that the rings represent a medium-sized moon of Saturn that was torn apart by tidal forces.
- The second theory suggests that the rings are composed of material that was never allowed to condense into moons because of the tidal forces.
- The third theory suggests that the ring material is the remains of a satellite shattered by meteor impacts.

3. Impact craters

The impact of asteroids, comets, and meteorites onto the surfaces of planets and moons is a major geologic process. Craters are the most common features on some planets and moons. On Earth, the workings of the atmosphere, water, and continental movements have hidden most impact craters, and their true number will never be known. Over 150 impacts have been studied on Earth. Visit the *EarthComm* web site to view images of impact craters on Earth and on other planets and moons.

NOTES

Earth System Evolution Astronomy

NOTES

Chapter 1

E 27

NOTES

ACTIVITY 3 — ORBITS AND EFFECTS
Background Information

In the **Chapter Challenge,** your students are asked to consider a threat from an asteroid whose orbit takes it near Earth's orbit. There has been a great deal of speculation in the press in recent years about this scenario. Our popular literature and films have taken advantage of the public's appetite for pseudo-scientific stories by dreaming up such implausibities as UFOs that hide behind comets and solar-system bodies that can readily jump out of their.orbits to head straight for Earth. These make dramatic reading and movie scripts, but they simply aren't supported by science. Anyone with a rudimentary knowledge of how orbits work can easily debunk such stories. The science community has offered straightforward explanations of the mechanics of the situation.

Objects in the solar system move around the Sun in easily predictable paths called orbits. The type of orbit a body has in its trip around the Sun affects not only its likelihood of collision with another body but also the heating it receives from the Sun and, consequently, its climate. Understanding the properties of orbits, how they can be calculated, and what effects they have on Earth moves students one step closer to making common-sense assessments of what constitutes a threat to the Earth and what doesn't.

Understanding Orbits
Throughout much of human history, people believed that everything moved around the Earth. This led to complicated explanations for the motions of the planets as seen from Earth, particularly when they appeared to move backward! It wasn't until the 16th century that the Polish astronomer Nicolaus Copernicus announced a new model of planetary motion to explain the observed motions of the planets. He placed the Sun at the center of a set of circular planetary orbits. His scheme came quite close to correctly predicting planetary motion, but wasn't quite right because it didn't explain such things as Mars' retrograde motion.

German mathematician Johannes Kepler went one step further and explained that planetary orbits are not circles, but ellipses. He devised three laws of planetary motion:

• *First Law:* Each planet orbits the Sun in an elliptical path, with the Sun occupying one focus of the ellipse. The point nearest the Sun is called "perihelion," and the point in the ellipse farthest away is called "aphelion."

• *Second Law:* As a planet moves in orbit around the Sun, it sweeps out equal areas in equal times. When Earth is at perihelion with respect to the Sun, it moves faster than it does when it is at aphelion.

• *Third Law:* The square of the length of time a planet takes to make one complete orbit around the Sun (its period) is equal to the cube of its average distance from the Sun.

Another way to think of this third law is to examine how it describes the orbits of the planets. Earth is relatively close to the Sun and takes one year to complete its orbit. Jupiter is much farther out, moves more slowly around the Sun, and thus takes longer to make a trip around the Sun. Yet another way to look at it is that Jupiter's year is 12 Earth-years long!

The same laws apply throughout the universe, to stars orbiting each other, stars orbiting the center of a galaxy, or galaxies orbiting each other.

Calculating Orbits for Near-Earth Bodies

Most bodies in the solar system follow an elliptical orbit of some kind. Because astronomers know the properties of ellipses, they can calculate such orbits quite accurately. When a new asteroid or comet is discovered, for example, observers are asked to send accurate positions for the object to the International Astronomical Union's Minor Planet Center at Harvard University. Astronomers use these positions to plot points of an orbit. The more positions that are sent in, the more accurate the orbital calculations will be. Once an orbit is calculated, it is published, and observers around the world use it to determine when the object will be visible to them. Comet observers, for example, will know how long they will have to observe and where to look in the sky for their favorite object.

These measurements are extremely important when an object appears to be headed for a close encounter with Earth. You will see these objects called by a variety of names — Earth-crossing asteroids (ECAs), Earth-approaching asteroids or near-Earth objects (NEOs), and potentially hazardous asteroids (PHAs). Many comets and asteroids have orbits that cross Earth's orbit.

How is it that an object can have an orbit that intersects Earth's? Quite simply, it is because many of these objects follow paths that are much more elliptical than ours. Solar-system bodies do not necessarily follow paths that are nested within each other. If you look at the orbits of Neptune and Pluto, for example, you will see that Pluto's path is highly elliptical and has an inclination (a tilt)

that takes it out of the plane of the solar system. For part of its 248-year orbit, Pluto is actually closer to the Sun than Neptune is!

Most comets follow highly elliptical orbits. Some are steeply inclined, too. Some asteroids orbiting in the vicinity of Earth follow such orbits, too. Others may simply be following more Earth-like elliptical orbits but are too small and dim to have been spotted yet by observers. Astronomers haven't mapped the existence of every nearby body and its orbit. So, from a standpoint of threats to Earth, the possibility of a collision from just about any direction (and not necessarily from something orbiting in the same plane) does exist. Astronomers are devising ways to learn more about the near-Earth environment by using more sensitive equipment to detect smaller objects on possible collision courses with our planet.

The Stability of Orbits

You might have been asked the question, "Can an object jump out of its orbit and take up a new orbit?" The answer requires an understanding of the laws of motion, the orbital laws, and the law of gravity. Basically, an object in motion (or at rest) tends to stay in motion (or at rest) unless acted upon by another force or object. This is the gist of Newton's first law of motion. But there is also an element of common sense to answering the question. If a student is asking the question, have them imagine a car parked on a street. It doesn't just start rolling down the street by itself. It will stay stopped until it is acted upon by something else—perhaps another car will smash into it and cause it to move.

In the previous section, we studied the Moon orbiting the Earth. The Moon, as it moves around the Earth, is always falling toward Earth because our planet's gravity is attracting the Moon toward it. But the Moon has sufficient motion that it almost perfectly balances the pull of gravity, and so it stays in orbit—i.e., it tends to stay in motion. It would take the collision of another body to push the Moon out of that orbit and out into space or careening into the surface of the Earth.

Likewise, a comet traveling in an orbit around the Sun tends to stay in that orbit unless some outside force or object changes its path. Now in fact, cometary orbits can be changed by gravitational tugs from nearby planets. This has happened throughout our solar system's history. However, the beauty of determining and calculating the orbital path of a comet or asteroid is that we can use those calculations to predict where the object will go and what it will do—so long as it is not acted upon by an outside force. And no comet or asteroid or moon or planet will simply move out of an orbit without the action of another force. Understand that force and you understand better the laws of motion that govern the movements of all bodies in the universe.

Understanding Orbital Effects – The Example of Uranus

In the **Digging Deeper** reading section of this activity, students read about the combined effects of Earth's axial tilt, the precession of its axis, and the motions of bodies in high-inclination orbits. As they work through their orbital calculations, students should take time to consider different kinds of axial tilts, higher or lower inclinations, and the effects of different orbits on the bodies occupying them.

The gas giant planet Uranus is worth studying for its very different orbital aspects. It lies just over 19 AU (about 2.8 billion km) from the Sun and takes 84 Earth years to complete one orbit. Its axial tilt (as measured from the perpendicular to the plane of the solar system) is an astounding 82.1°! It

rotates backward on its axis in slightly more than 17 hours. In practical terms, these very great differences have great effects on the atmospheric dynamics of the planet. Because it lies on its side with respect to the Sun, Uranus is basically rolling around in its orbit like a giant bowling ball. This means that during parts of its 84-year orbit, its poles end up pointing toward the Sun. During these times, those parts of the Uranian atmosphere are heated, causing atmospheric disturbances. At other times, the equatorial regions are pointed toward the Sun and are heated.

Planetary scientists think that, early in its history, Uranus may have had a much more normal axial tilt. A huge force would have been required to change its axial tilt, and it's possible that another large body may have collided with Uranus, causing it to tip on its side. It would be an interesting thought experiment for your students to consider what conditions on Earth would be like if our planet were to be tipped over on its side by an outside force and made to roll around in its orbit as Uranus does. In their study of Earth's current conditions, they should find the clues to help them formulate different scenarios for our planet and their community as they imagine differences in Earth's orbital characteristics.

Threats and Information

In recent years, the possibility of an asteroid crashing down on Earth has been the subject of many books, movies, and web sites. Getting credible information about such a threat is difficult. There are differing opinions among scientists and political leaders about the likelihood of collisions and what we could do to avoid them. Once your students have come to an understanding of orbits and their effects on Earth, they are ready to begin assessing some of the information they find in the media and on the Internet.

More Information – on the Web

Visit the *EarthComm* web site www.agiweb.org/earthcomm/ to access a variety of links to web sites that will help you to deepen your understanding of content and prepare you to teach this activity. Many of the sites also contain images that you can download.

Goals and Assessment

Clarify that the goals indicate what students should understand and be able to do as a result of the activity. Make sure students understand that Chapter Assessments are based upon these goals.

Goal	Location in Activity	Assessment Opportunity
Measure the major axis and distance between the foci of an ellipse.	Investigate	Ellipses are drawn correctly, axes are measured correctly.
Understand the relationship between the distance between foci and the eccentricity of an ellipse.	Investigate; Digging Deeper; Check Your Understanding Question 2	Calculations of eccentricity are done correctly. Responses to questions are accurate and reflect an understanding.
Calculate the eccentricity of the Earth's orbit.	Understanding and Applying What You Have Learned Question 1	Calculation of eccentricity is done correctly.
Draw the Earth's changing orbit in relation to the Sun.	Understanding and Applying What You Have Learned Question 4	Drawings are scaled correctly; drawings accurately reflect changes in shape of Earth's orbit over time.
Explain how the Earth's changing orbit and its rotation rate affect its climate.	Digging Deeper; Check Your Understanding Question 3 Understanding and Applying What You Have Learned Question 4(c)	Responses to questions are accurate and reflect an understanding.
Draw the orbits of a comet and an asteroid in relation to the Earth and the Sun.	Understanding and Applying What You Have Learned Question 5	Drawing of solar system is correct. Inclinations of asteroid and comet orbits are correctly represented.

NOTES

Earth System Evolution Astronomy

Activity 3   Orbits and Effects

Goals

In this activity you will:

- Measure the major axis and distance between the foci of an ellipse.

- Understand the relationship between the distance between the foci and eccentricity of an ellipse.

- Calculate the eccentricity of the Earth's orbit.

- Draw the Earth's changing orbit in relation to the Sun.

- Explain how the Earth's changing orbit and its rotation rate could affect its climate.

- Draw the orbits of a comet and an asteroid in relation to the Earth and the Sun.

Think about It

The Earth rotates on its axis as it revolves around the Sun, some 150,000,000 km away. The axis of rotation is now tilted about 23.5°.

- What is the shape of the Earth's orbit around the Sun?
- How might a change in the shape of the Earth's orbit or its axis of rotation affect weather and climate?

What do you think? In your *EarthComm* notebook, draw a picture of the Earth's orbit around the Sun, as seen from above the solar system. Record your ideas about how this shape affects weather and climate. Be prepared to discuss your responses with your small group and the class.

Activity Overview

Students use pushpins and string to draw ellipses with varying distances between the foci. They measure the major axes of the ellipses and calculate eccentricity to understand the relationship between eccentricity and distance between the foci. Students then consider how the exercise relates to the orbit of the planets around the Sun. The **Digging Deeper** reading section looks at the eccentricity, axial tilt, and precession of the Earth's orbit, and considers how these parameters affect the Earth's climate. The reading also discusses orbital inclination and its implications for impact events on Earth.

Preparation and Materials Needed

Preparation

You will need about 30 cm of string for each student (or group). They will need to help each other tie the string into a loop that is 12 cm long, or you may pre-tie the loops to allow more time for the activity.

Use cardboard, Styrofoam, or newspaper to stabilize the pushpins. If you think students will have too much difficulty making the points, make the sheet ahead of time and photocopy it for the class. But you may prefer to have them construct the points as a good exercise in measurement.

Try this activity yourself so that you will be able to help students as they work and they can see your example. Trying this investigation ahead of time will also allow you to make up a data table to use as an answer key. It will also help you become familiar with the challenges of making ellipses and calculating the eccentricities.

Materials
- Piece of paper
- Straightedge
- Tape
- Thick sheet of cardboard (8.5 x 11 in.), Styrofoam, or newspaper
- Two pushpins
- Piece of string – 30 cm long
- Metric ruler
- Calculator

Think about It

Student Conceptions

Students are likely to think that the Earth's orbit around the Sun is nearly circular. Ask them to identify the four seasons on their drawings. This should expose students' conceptions about how far we are away from the Sun in winter versus summer (we are actually closer to the Sun in the Northern Hemisphere winter!).

Answer for the Teacher Only

Give students a few minutes to complete this task. Circulate around the room to identify different ideas about the shape of the Earth's orbit. Select several students to share their ideas.

The Earth's orbit around the Sun is an ellipse with a very small eccentricity. The eccentricity is so small that a careful and accurate drawing of the orbit appears to the eye to be a circle. The difference in the Earth–Sun distance varies by about 3% as the Earth revolves around the Sun in its elliptical orbit. Because the flux of solar energy decreases with the square of the distance away from the Sun (an effect known in physics as the inverse-square law), the difference in insolation (the term used for the rate at which the Earth's surface receives solar energy) varies by about 7% between perihelion (when the Earth is closest to the Sun, on 5 January) and aphelion (when the Earth is farthest from the Sun, on 5 July).

A full answer to the second question would have to be very lengthy. The **Digging Deeper** reading section covers much of the relevant material. In short, variations in the Earth's orbit and axial tilt lead to systematic variations in insolation, which by way of various interacting effects lead, in turn, to changes in the Earth's climate.

Assessment Tool

Think about It Evaluation Sheet
Use this evaluation sheet to help students understand and internalize the basic expectations for the warm-up activity.

Blackline Master Astronomy 3.1
Ellipse

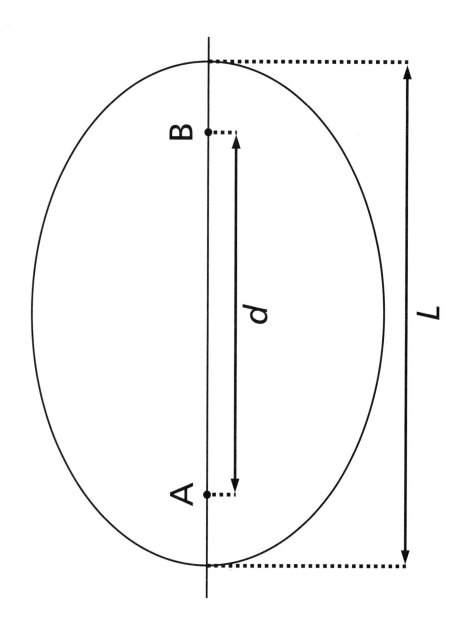

Ellipse with foci AB showing major axis length L and distance between the foci d.

Investigate

1. Draw an ellipse using the following steps:

   • Fold a piece of paper in half.

   • Use a straightedge to draw a horizontal line across the width of the paper along the fold.

   • Put two dots 10 cm apart on the line toward the center of the line. Label the left dot "A" and the right dot "B."

   • Tape the sheet of paper to a piece of thick cardboard, and put two pushpins into points A and B. The positions of the pushpins will be the foci of the ellipse.

   • Tie two ends of a piece of strong string together to make a loop. Make the knot so that when you stretch out the loop with your fingers into a line, it is 12 cm long.

   • Put the string over the two pins and pull the loop tight using a pencil point, as shown in the diagram.

   • Draw an ellipse with the pencil. Do this by putting the pencil point inside the loop and then moving the pencil while keeping the string pulled tight with the pencil point.

   a) Draw a small circle around either point A or point B and label it "Sun."

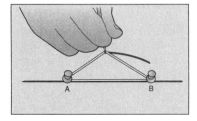

2. Repeat the process using the following measurements and labels:

   • Two points 8 cm apart labeled C and D (1 cm inside of points A and B).

   • Two points 6 cm apart, labeled E and F (2 cm inside of points A and B).

   • Two points 4 cm apart, labeled G and H (3 cm inside points A and B).

   • Two points 2 cm apart, labeled I and J (4 cm inside points A and B).

3. Copy the data table on the next page into your notebook.

   a) Measure the width (in centimeters) of ellipse "AB" at its widest point. This is the major axis, L (see diagram on the next page). Record this in your data table.

   b) Record the length of the major axis for each ellipse in your data table.

   c) Record the distance between the two foci, d (the distance between the two pushpins) for each ellipse in your data table (see diagram).

   d) The eccentricity E of an ellipse is equal to the distance between the two foci divided by the length of the major axis. Calculate the eccentricity of each of your ellipses using the equation $E = d/L$, where d is the distance between the foci and L is the length of the major axis. Record the eccentricity of each ellipse.

   ⚠ Be sure the cardboard is thicker than the points of the pins. If not, use two or more pieces of cardboard.

EarthComm

Investigate

1. Circulate around the classroom, helping students construct the points and hold the string tight. Having a model will help them understand what is expected (see **Preparation**). It is difficult to make the loop exactly 12 cm long. Students can help each other. It may be easiest to loop one end around one of the pushpins and tie a half-knot, then pull the ends until the loop measures 12 cm. Students may also have trouble with the string slipping off the tip of the pencil.

a)

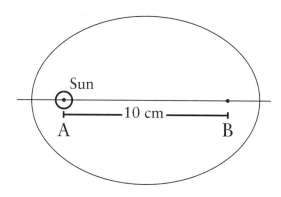

2. Students ellipses will have the measurements given in the following table.

3.

Ellipse	Major Axis (L) (cm)	Distance between Foci (d) (cm)	Eccentricity $E = d/L$
AB	14	10	0.714286
CD	16	8	0.5
EF	18	6	0.333333
GH	20	4	0.2
IJ	22	2	0.090909

(**Note:** These are the "ideal" numbers. When the string is wrapped around the push-pins, a bit of the length of the string will be "lost" and numbers may be slightly smaller. You will need to decide on your own standards for accuracy and precision.)

Teaching Tip

Use **Blackline Master Astronomy 3.1, Ellipse** of the illustration of an ellipse on page E30 to make an overhead transparency. This overhead can be used to discuss any questions regarding the properties of an ellipse.

Earth System Evolution Astronomy

Ellipse	Major Axis (L) (cm)	Distance between the Foci (d) (cm)	Eccentricity E = d/L
AB		10	
CD		8	
EF		6	
GH		4	
IJ		2	

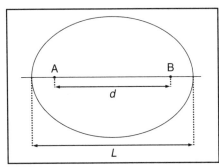

Ellipse with foci AB showing major axis length L and distance between the foci d.

4. Study your data table to find a relationship between the distance between the foci and the eccentricity of an ellipse.

a) Record the relationship between the distance between the foci and the eccentricity in your notebook.

b) Think of your ellipses as the orbits of planets around the Sun. Does the distance to the center of the Sun stay the same in any orbit?

c) Which orbit has the least variation in distance from the Sun throughout its orbit? Which has the most?

5. Earth's orbit has an eccentricity of about 0.017. Compare this value to the ellipse with the lowest eccentricity of those you drew.

a) Why does it make sense to describe Earth's orbit as "nearly circular"?

Reflecting on the Activity and the Challenge

In this activity, you explored a geometric figure called an ellipse. You also learned how to characterize ellipses by their eccentricity. The orbits of all nine planets in our solar system are ellipses, with the Sun at one focus of the ellipse representing the orbit for each planet. As you will see, although Earth's orbit is very nearly circular (only slightly eccentric), the shape of its orbit is generally believed to play an important role in long-term changes in the climate. The shape of the Earth's orbit is not responsible for the seasons. You will need this information when describing the Earth's orbital relationships with the Sun and the Moon in the **Chapter Challenge**.

E 30

4. a) As the distance between the foci increases, the eccentricity of the ellipse increases. When the foci coincide, the eccentricity is zero, and the ellipse becomes a circle. When the distance between the foci is equal to the major diameter, the ellipse becomes a straight line.

    b) Distance to the center of the "Sun" changes with eccentricity. The more eccentric the orbit, the greater the variation in distance.

    c) The orbit "IJ" has the lowest eccentricity; it looks very much like a circle. The orbit "AB" has the greatest eccentricity, and the most variation in its distance from the Sun.

5. a) The eccentricity of the Earth's orbit is one-fifth the value of the least eccentric ellipse drawn: the Earth's orbit is very much like a circle.

Assessment Tools

EarthComm **Notebook Entry-Checklist**
Use this checklist as a quick guide for student self-assessment and/or an opportunity to quickly score student work. Add further criteria specific to your classroom needs or to this particular investigation.

Investigate Notebook Entry-Evaluation Sheet
Point out the criteria listed on this evaluation sheet that are relevant to this particular investigation. Encourage students to internalize the criteria by making them part of your "assessment conversations" as you circulate around the classroom.

Reflecting on the Activity and the Challenge

Students examined the relationship between the distance between the foci and the eccentricity of an ellipse. They should begin to think about how this relationship is tied to the orbits of the planets, and they should consider what role the shape of the Earth's orbit plays in long-term changes in climate. These concepts are explored further in the **Digging Deeper** reading section. Take this time to review the investigation and make sure that students understand the concept of eccentricity.

Digging Deeper

ECCENTRICITY, AXIAL TILT, PRECESSION, AND INCLINATION

Eccentricity

After many years of analyzing observational data on the motions of the planets, the astronomer Johannes Kepler (1571–1630) developed three laws of planetary motion that govern orbits. The first law states that the orbit of each planet around the Sun is an ellipse with the Sun at one focus. The second law, as shown in *Figure 1*, explains that as a planet moves around the Sun in its orbit, it covers equal areas in equal times. The third law states that the time a planet takes to complete one orbit is related to its average distance from the Sun.

Geo Words

eccentricity: the ratio of the distance between the foci and the length of the major axis of an ellipse.

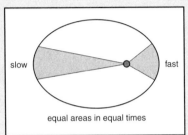

equal areas in equal times

Figure 1 Kepler's Second Law states that a line joining a planet and the Sun sweeps out equal areas in equal intervals of time.

As you saw in the **Investigate** section, the shape of an ellipse can vary from a circle to a very highly elongated shape, and even to a straight line. The more flattened the ellipse is, the greater its eccentricity. Values of **eccentricity** range from zero for a circle, to one, for a straight line. (A mathematician would say that the circle and the line are "special cases" of an ellipse.) The two planets in the solar system with the most elliptical orbits are Mercury, the closest planet, and Pluto, the farthest one. Both have eccentricities greater than 0.2. The orbit of Mars is also fairly elliptical, with an eccentricity of 0.09. In comparison, the Earth's orbit is an ellipse with an eccentricity of 0.017. (This is a much lower value than even the eccentricity of ellipse IJ in the investigation, which looked much like a circle.) If you were to draw an ellipse with an eccentricity of 0.017 on a large sheet of paper, most people would call it a circle. It's eccentric enough, however, to make the Earth's distance from the Sun vary between 153,000,000 km and 147,000,000 km. To make things more complicated, the eccentricity of the Earth's orbit changes over time, because of complicated effects having to do with the weak gravitational pull of other planets in the solar system. Over the course of about 100,000 years, the Earth's orbit ranges from nearly circular (very close to zero eccentricity) to more elliptical (with an eccentricity of about 0.05).

Planetary scientists have found that some solar-system objects have highly elliptical orbits. Comets are a well-known example. As they move closer

Digging Deeper

Assign the reading for homework, along with the questions in **Check Your Understanding** if desired.

Assessment Opportunity

Reword or restructure the questions in **Check Your Understanding** for a brief quiz. Use the quiz (or a class discussion of the questions in the textbook) to assess your students' understanding of the main ideas in the reading and the activity.

Assessment Tool

Check Your Understanding Notebook Entry-Evaluation Sheet
Use this sheet to evaluate the extent to which students understand the key concepts explored in **Activity 3** and explained in **Digging Deeper**, and to evaluate the students' clarity of expression.

Teaching Tip

Figure 1 on page E31 illustrates the principle of Kepler's second law which states that as a planet moves around the Sun in its orbit, it covers equal areas in equal times. In this figure each green-shaded section represents a region of equal area that is formed as the Earth moves in its orbit over a given time. In other words, the time needed for the Earth to move through the distance of the orbit that is represented by each green area is the same. Because the two areas shaded in green are equal, and because the distance from the Sun to the left side of the orbit is greater than the distance from the right side of the orbit (note that this ellipse is greatly exaggerated), the left area is formed by fewer degrees of rotation about the Sun than the right area. Because the time needed to cover both areas is the same, the angular velocity of the Earth is slower on the left than on the right.

Use **Blackline Master Astronomy 3.2, Kepler's Second Law** of *Figure 1* on page E31 to make an overhead transparency. Incorporate this overhead into a discussion about Kepler's laws.

Earth System Evolution Astronomy

to the Sun, the icy mix that makes up a comet's nucleus begins to turn into gas and stream away. The result is a ghostly looking tail and a fuzzy "shroud," that you can see in *Figure 2*. It is called a **coma**, and it forms around the nucleus. When a comet gets far enough away from the Sun, the ices are no longer turned to gas and the icy nucleus continues on its way.

Figure 2 The comet's head or coma is the fuzzy haze that surrounds the comet's true nucleus.

Another good example is the distant, icy world Pluto. Throughout much of its year (which lasts 248 Earth years) this little outpost of the solar system has no measurable atmosphere. It does have a highly eccentric orbit, and its distance from the Sun varies from 29.5 to 49.5 AU. The strength of solar heating varies by a factor of almost four during Pluto's orbit around the Sun. As Pluto gets closest to the Sun, Pluto receives just enough solar heating to vaporize some of its ices. This creates a thin, measurable atmosphere. Then, as the planet moves farther out in its orbit, this atmosphere freezes out and falls to the surface as a frosty covering. Some scientists predict that when Pluto is only 20 years past its point of being closest to the Sun, its atmosphere will collapse as the temperature decreases.

Axial Tilt (Obliquity)

The Earth's axis of rotation is now tilted at an angle of 23.5° to the plane of the Earth's orbit around the Sun, as seen in *Figure 3*. Over a cycle lasting about 41,000 years, the axial tilt varies from 22.1° to 24.5°. The greater the angle of tilt, the greater the difference in solar energy, and therefore temperature, between summer and winter. This small change, combined with other long-term changes in the Earth's orbit, is thought to be responsible for the Earth's ice ages.

Precession

The Earth also has a slight wobble, the same as the slow wobble of a spinning top. This wobble is called the **precession** of the Earth's axis. It is caused by differences in the gravitational pull of the Moon and the Sun on the Earth. It takes about 25,725 years for this wobble to complete a cycle. As the axis wobbles, the timing of the seasons changes. Winter occurs when a hemisphere, northern or southern, is tilted away from the Sun. Nowadays, the Earth is slightly closer to the Sun during winter (January 5) in the Northern Hemisphere. Don't let anybody tell you that winter happens because the Earth is farthest from the Sun! The Earth's orbit is nearly circular! Also, even if a particular hemisphere of the Earth is tilted towards the Sun, it is not significantly closer to the Sun than the other hemisphere, which is tilted away.

Geo Words

coma: a spherical cloud of material surrounding the head of a comet. This material is mostly gas that the Sun has caused to boil off the comet's icy nucleus. A cometary coma can extend up to a million miles from the nucleus.

precession: slow motion of the axis of the Earth around a cone, one cycle in about 26,000 years, due to gravitational tugs by the Sun, Moon, and major planets.

E 32

Blackline Master Astronomy 3.2
Kepler's Second Law

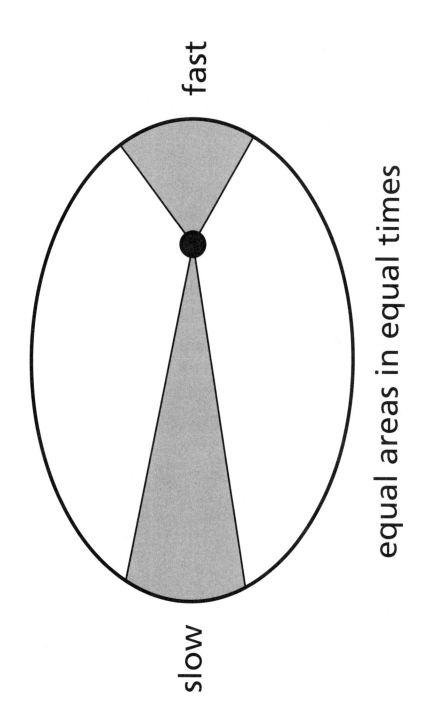

fast

slow

equal areas in equal times

Teaching Tip

The comet's coma or head (as shown in *Figure 2*) and the comet's tail are what we see from Earth. During observation, the shape of the coma can vary from comet to comet and even for the same comet. The shape depends on the distance of the comet from the Sun and the rate of production of dust and gas. For faint comets or bright comets producing little dust, the coma is usually round. A comet that produces significant quantities of dust has a fan-shaped coma. This fan-like shape results from different-sized dust particles being released: the larger particles remain along the comet's orbital path while smaller dust particles are pushed away from the Sun. For comets with fan-shaped or parabolic comae, there is no obvious boundary between the coma and tail.

Teaching Tip

Figure 3 on page E33 of the Student Edition illustrates two important cycles that, in combination with variations of the eccentricity of the Earth's orbit, play a large part in governing the spatial relationship between the Earth and Sun on time scales greater than the year. Use **Blackline Master Astronomy 3.3, Important Earth's Cycles** of the illustration to make an overhead transparency. Incorporate this overhead into a discussion about variations in the Earth's orbital parameters and the spatial relationship between the Earth and the Sun. These concepts are important to considerations about variability in the Earth's climate, and will be revisited in the second chapter of this book.

Brief explanations on each of these cycles are given in the **Digging Deeper** reading section. The first of these cycles, the obliquity cycle, is shown diagrammatically in the upper left-hand illustration of this figure. In this illustration, the yellow shaded region represents the range over which the tilt of the Earth's axis varies. The second of these cycles, the precession of the equinoxes, is shown diagrammatically in the lower right-hand illustration of this figure. Note that the eccentricity of the Earth's orbit is greatly exaggerated in both of the lower illustrations. The precession of the equinoxes is governed by the combined effects of two other phenomena, called the axial precession and the orbital precession. These are shown in the upper-right and lower-left illustrations respectively. In the upper-right illustration the circular arrow above the Earth signifies the rotation (or wobble) of the Earth's axis of rotation. The dashed ellipse in the lower-left diagram shows a future orientation of the Earth's orbit as the orbit itself rotates (actually, precesses) about the Sun in the direction of the black arrow.

NOTES

The precession of the Earth's axis is one part of the precession cycle. Another part is the precession of the Earth's orbit. As the Earth moves around the Sun in its elliptical orbit, the major axis of the Earth's orbital ellipse is rotating about the Sun. In other words, the orbit itself rotates around the Sun! These two precessions (the axial and orbital precessions) combine to affect how far the Earth

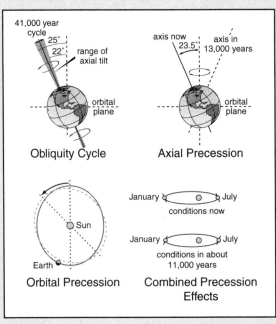

Figure 3 The tilt of the Earth's axis and its orbital path about the Sun go through several cycles of change.

Geo Words

orbital plane: (also called the ecliptic or plane of the ecliptic). A plane formed by the path of the Earth around the Sun.

inclination: the angle between the orbital plane of the solar system and the actual orbit of an object around the Sun.

is from the Sun during the different seasons. This combined effect is called precession of the equinoxes, and this change goes through one complete cycle about every 22,000 years. Ten thousand years from now, about halfway through the precession cycle, winter will be from June to September, when the Earth will be farthest from the Sun during the Northern Hemisphere winter. That will make winters there even colder, on average.

Inclination

When you study a diagram of the solar system, you notice that the orbits of all the planets except Pluto stay within a narrow range called the **orbital plane** of the solar system. If you were making a model of the orbits of the planets in the solar system you could put many of the orbits on a tabletop. However, Pluto's orbit, as shown in *Figure 4*, could not be drawn or placed on a tabletop (a plane, as it is called in geometry). Pluto's path around the Sun is inclined 17.1° from the plane described by the Earth's motion around the Sun. This 17.1° tilt is called its orbital **inclination**.

Blackline Master Astronomy 3.3
Important Earth's Cycles

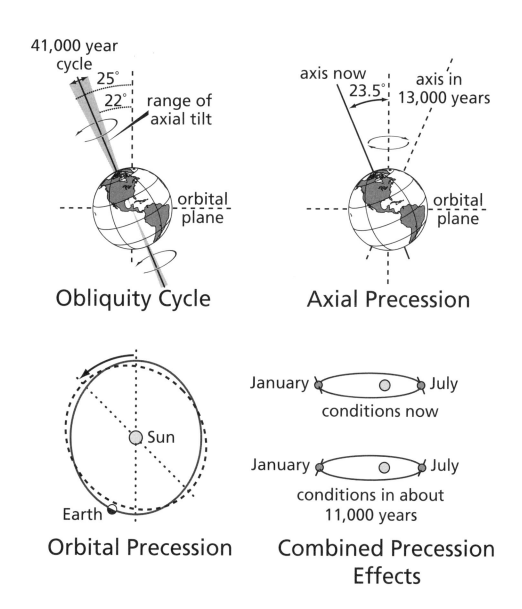

41,000 year cycle

25°

22°

range of axial tilt

orbital plane

Obliquity Cycle

axis now

23.5°

axis in 13,000 years

orbital plane

Axial Precession

Sun

Earth

Orbital Precession

January ○ July
conditions now

January ○ July
conditions in about 11,000 years

Combined Precession Effects

Blackline Master Astronomy 3.4
Inclination of Pluto's Orbit

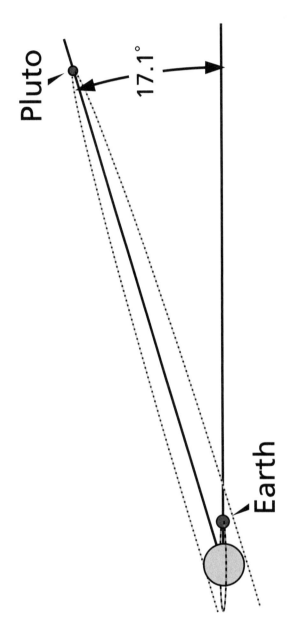

Blackline Master Astronomy 3.5
Orbit of Asteroid 1996 JA$_1$

north ecliptic polar view

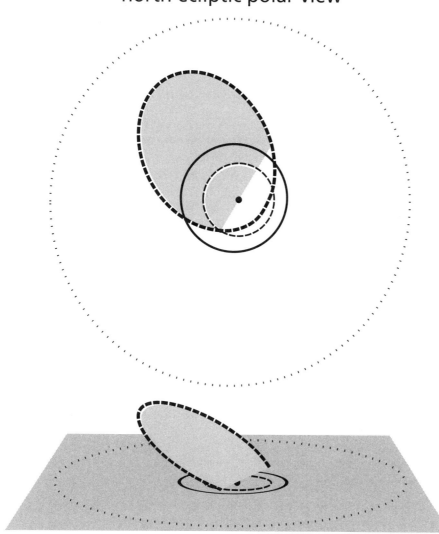

—— orbit of Mars ⋯⋯ orbit of Jupiter
---- orbit of Earth ▪▪▪▪ orbit of asteroid 1996 JA$_1$

Earth System Evolution Astronomy

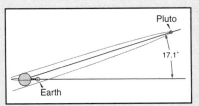

Figure 4 Pluto's orbit is inclined 17.1° to the orbital plane of the rest of the solar system.

What are the orbital planes of asteroids and comets? Both are found mainly in the part of the solar system beyond the Earth. Although some asteroids can be found in the inner solar system, many are found between the orbits of Mars and Jupiter. The common movie portrayal of the "asteroid belt" as a densely populated part of space through which one must dodge asteroids is wrong. The asteroids occupy very little space. Another misconception is that asteroids are the remains of a planet that exploded.

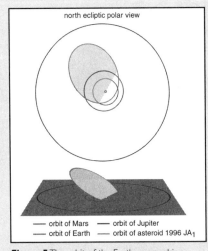

Figure 5 The orbit of the Earth-approaching asteroid 1996 JA, in relation to the Earth.

The orbits of asteroids are more eccentric than the orbits of the planets, and they are often slightly inclined from the orbital plane. As the Earth orbits the Sun, it can cross the orbital paths of objects called Earth-approaching asteroids. There is a great deal of interest in finding Earth-approaching or Earth-crossing asteroids. A collision with an object a few miles across could be devastating, because of its very high velocity relative to the Earth. Astronomers search the skies for asteroids and map their orbits. In this way they hope to learn what's coming toward Earth long before it poses a danger to your community.

Comets are "loners" that periodically visit the inner solar system. They usually originate in the outer solar system. As shown in *Figure 5*, they have very high-inclination orbits—some as much as 30° from the plane of the solar system. In addition, their orbits are often highly eccentric. Astronomers also search the skies for comets. Once a comet is discovered, its orbit is calculated and the comet is observed as it moves closer to the Sun and changes. A collision of a comet's nucleus with the Earth would be serious, but a collision with a comet's tail is much more likely. A collision with the tail would have little, if any, effect on the Earth, because the tail consists mainly of glowing gas with very little mass.

Check Your Understanding

1. In your own words, explain what is meant by the eccentricity of an ellipse.

2. For an ellipse with a major axis of 25 cm, which one is more eccentric; the one with a distance between the foci of 15 cm or with a distance between the foci of 20 cm? Explain.

3. How does the precession of the Earth's axis of rotation affect the seasons? Justify your answer.

4. Why is there a danger that a large asteroid might strike the Earth at some time in the future?

Teaching Tip

Use **Blackline Master Astronomy 3.4, Inclination of Pluto's Orbit** of the illustration in *Figure 4* on page E34 to make an overhead transparency. Incorporate this overhead into a discussion about the orbital inclination of Pluto.

Teaching Tip

The **Blackline Master Astronomy 3.5, Orbit of Asteroid 1996 JA$_1$** *Figure 5* on page E34 can be used to help illustrate the geometry of a highly-inclined orbit by providing a 3-dimensional perspective view (in this case, the orbit of asteroid 1996 JA$_1$). Also note that the ellipticity of this asteroids' orbit is much greater than that of the planets shown. What implications does this have for the likelihood of an Earth impact?

Check Your Understanding

1. An ellipse can be thought of as a "flattened" or "stretched" circle. The degree of flattening or stretching is characterized by eccentricity, which is the relationship between the distance between the two foci of the ellipse and the length of the major axis of the ellipse. The more flattened the ellipse, the more eccentric it is. Although the eccentricity of an ellipse is not defined in **Digging Deeper**, it is defined in the table at the top of page E30 as the ratio of the distance between foci and the length of the major axis of the ellipse.

2. According to the definition of the eccentricity of an ellipse, the greater the ratio of the distance between the foci and the length of the major axis, the greater the eccentricity. Thus, an ellipse with a distance of 20 cm between foci has greater eccentricity than an ellipse with a distance of 15 cm (given that the length of the major axis of both is 25 cm).

3. The precession of the Earth's axis causes changes in the timing of the Earth's seasons relative to the year, as defined by the Earth's orbit around the Sun. This happens because the wobble causes the different hemispheres to be tilted towards or away from the Sun at different times of the year.

4. The Earth will likely cross paths someday with a large asteroid because the orbits of asteroids are more eccentric than the orbits of the planets, and they are often slightly inclined from the orbital plane.

Understanding and Applying What You Have Learned

1. The major axis of the Earth's orbit is 299,200,000 km, and the distance between the foci is 4,999,632 km. Calculate the eccentricity of the Earth's orbit. How does this value compare to the value noted in the **Digging Deeper** reading section?

2. On the GH line on the ellipse that you created for your **Investigate** activity, draw the Earth at its closest position to the Sun and the farthest position away from the Sun.

3. Refer to the table that shows the eccentricities of the planets to answer the following questions:

   a) Which planet would show the greatest percentage variation in its average distance from the Sun during its year? Explain.

   b) Which planet would show the least percentage variation in its average distance from the Sun throughout its year? Explain.

   c) Is there any relationship between the distance from the Sun and the eccentricity of a planet's orbit? Refer to *Table 1* on page E5.

   d) Why might Neptune be farther away from the Sun at times than Pluto is?

   e) Look up the orbital inclinations of the planets and add them to a copy of the table.

Eccentricities of the Planets	
Planet	**Eccentricity**
Mercury	0.206
Venus	0.007
Earth	0.017
Mars	0.093
Jupiter	0.048
Saturn	0.054
Uranus	0.047
Neptune	0.009
Pluto	0.249

4. Draw a scale model to show changes in the Earth's orbit of about the same magnitude as in nature (a cycle of 100,000 years).

   a) Draw the orbit of the Earth with a perfectly circular orbit at 150,000,000 km from the Sun. Use a scale of 1 cm = 20,000,000 km. Make sure that your pencil is sharp, and draw the thinnest line possible.

   b) Make another drawing of the actual shape of the Earth orbit—an ellipse. This ellipse has 153,000,000 km as the farthest distance and 147,000,000 km as the closest point to the Sun.

   c) Does the difference in distance from the Sun look significant enough to cause much difference in temperature? Explain.

5. Draw the solar system as viewed from the plane of the ecliptic (orbital plane).

Understanding and Applying What You Have Learned

1. 4,999,632 ÷ 299,200,000 = 0.01671. When rounded, this is the value given in the **Digging Deeper** reading section (0.017).

2.

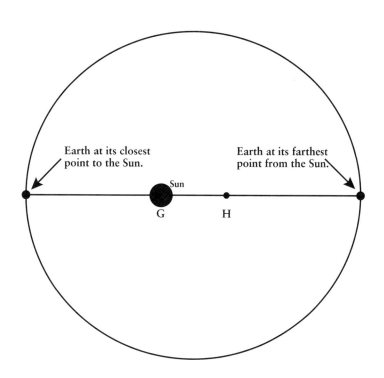

Note: Remind students that the ellipse GH does not approximate the actual eccentricity of Earth's orbit around the Sun. By using an ellipse with an exaggerated eccentricity, this question simply helps students to understand that distance from the Sun varies during one revolution when the orbit is elliptical.

3. a) Pluto has the greatest eccentricity, so it should show the greatest percentage variation in distance from the Sun during the year.

    b) Venus has the lowest eccentricity, thus its orbit is most like a circle and has the lowest percentage change in distance from the Sun during the year.

    c) When students refer to the table on page E5, they will note that there is no general direct relationship between distance from the Sun and eccentricity of orbit.

    d) Neptune is farther away from the Sun than Pluto at times, because Pluto has a highly eccentric orbit and Neptune's orbit crosses inside that of Pluto.

e) Direct students to the *EarthComm* web site or the local or school library to find information on the planets' orbital inclinations.

Planet	Eccentricity	Orbital Inclination (degrees)
Mercury	0.206	7.00
Venus	0.007	3.39
Earth	0.017	0.000
Mars	0.093	1.850
Jupiter	0.048	1.304
Saturn	0.054	2.485
Uranus	0.047	0.772
Neptune	0.009	1.769
Pluto	0.249	17.16

4. **a) – c)** From their diagrams students should note that the difference does not appear significant enough to cause any difference, but the small difference is in fact significant. In part, this is because the difference in insolation varies as the square of the Earth–Sun distance. It is generally agreed that seemingly small differences in insolation can have large effects on climate.

5. The intent of this step is for students to draw a representation of the solar system as it would be viewed looking edge-on into the plane of the ecliptic (Earth's orbital plane). In this depiction the diagram should look similar to *Figure 4* on page E34, only with all of the planets and other requested features included. A sample diagram is given on the following page.

    a) Pluto's orbit will be inclined with respect to the plane of the elliptic such that when it is to the right of the Sun (as depicted in the diagram) Pluto will reside above the plane of the ecliptic.

    b) From the data compiled in **Step 3**(e) it should be clear that none of the planets' orbits lie precisely in the plane of the ecliptic relative to the Earth. Accordingly, the orbits of the other planets would look similar to Pluto's orbit, only having variable and lower eccentricities and inclinations.

c) – d)

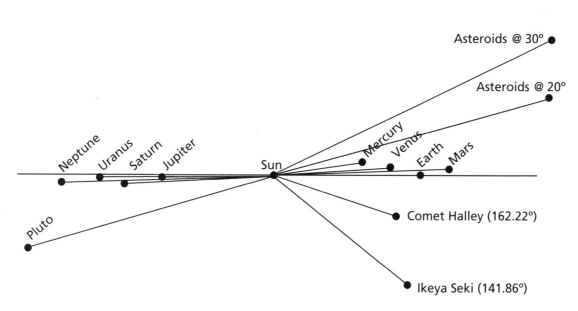

Earth System Evolution Astronomy

a) How will Pluto's orbit look with its 17° inclination?
b) How will the orbits of the other planets look?
c) Draw in the orbits of Earth-crossing asteroids with inclinations of 20° and 30° to the orbital plane.
d) Draw in the orbits of several comets with high inclinations. Some typical high-inclination comets are Comet Halley (162.22°) and Ikeya-Seki (141.86°).

6. Now that you know that the Earth's orbit is elliptical (and the Moon's is too), you can think about a third astronomical factor that controls the nature of tides. Tidal forces are stronger when the Moon is closer to the Earth, and when the Earth is closer to the Sun.

a) Draw a diagram to show the positions of the Moon, Earth, and Sun that would generate the highest tidal ranges (difference in height between high and low tides) of the year.
b) Draw a diagram to show the positions of the Moon, Earth, and Sun that would generate the lowest tidal ranges (difference in height between high and low tides) of the year.

Preparing for the Chapter Challenge

1. In your own words, explain the changes in the Earth's orbital eccentricity, and how it might have affected your community in the past. What effect might it have in the future?

2. Describe the orbits of comets and asteroids and how they are different from those of the planets. What effect could comets and asteroids have on your community if their orbits intersected the Earth's orbit and the Earth and the comet or asteroid were both at that same place in their orbits?

Inquiring Further

1. **The gravitational "slingshot" effect on spacecraft**

   The gravitational tug of the Sun and the planets plays a role in shaping the orbits of solar-system bodies. NASA has used the gravitational pull of Jupiter and Saturn to influence the paths of spacecraft like Pioneer and Voyager. Investigate how this gravitational "slingshot" effect works and its role in moving small bodies from one orbit to another.

2. **Investigate the orbits of comets and asteroids**

   Look up the orbital information for some typical comets and asteroids. Try to include some with high inclinations and orbital eccentricities.

E 36

6. a) – b)

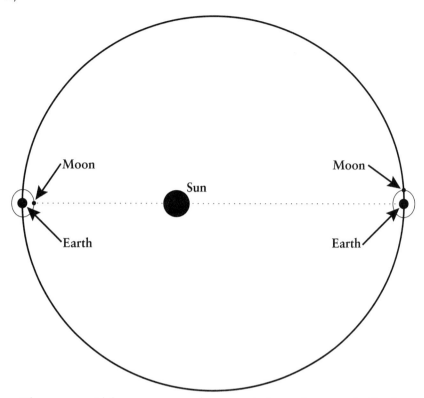

The greatest tidal range occurs when: a path drawn between the Earth, Moon, and Sun forms a line, the Moon is at its closest point to the Earth, and the Earth is at its closest point to the Sun (configuration on left side of the diagram above).

The smallest tidal range is produced when: a path drawn between the Earth, Moon, and Sun forms a right angle, the Earth is at its farthest distance from the Sun, and the Moon is at its farthest distance from the Earth (configuration on right side of the diagram above).

Preparing for the Chapter Challenge

After completing **Activity 3**, students should be able to calculate the eccentricity of an ellipse. They should also understand that the eccentricity of the Earth's orbit is not constant, but varies over time between almost zero and about 0.05. They should also understand that the Earth's axial tilt follows a cycle and that the Earth wobbles on its axis. Additionally, students should understand that interactions between the axial wobble and the precession of the Earth's orbit about the Sun cause changes in the timing of the seasons relative to the Earth's position in its orbit. Students should be able to explain in their own words how a combination of these factors can produce climate changes. They should also understand how the orbits of comets and asteroids affect the likelihood that an object will strike the Earth. Students may find it helpful to construct physical models to explore the relationships explained in the text more deeply.

Criteria for assessment (see **Assessment Rubric for Chapter Challenge on Astronomy** on pages 12 and 13) include discussions of:
- the Earth's distance from, and orbit around, the Sun
- the Earth's orbital relationships with the Sun and the Moon
- what comets and asteroids are, how they behave, how likely it is that one will collide with Earth in the future, and what would happen if one did
- why extraterrestrial influences on the community are a natural part of Earth system evolution

Review these criteria with your students so that they can be certain to include the appropriate information in their short papers.

Inquiring Further

1. The gravitational "slingshot" effect on spacecraft
Some space probes, like Galileo, were deliberately aimed close to planets in order to increase their velocity through a gravitational slingshot effect. When a small body passes close by a large planet, the planet's gravitational attraction exerts a sideways force on the moving small body, causing it to change its direction. The path of the small body has to be adjusted very carefully, though: if it passes too close to the planet, it could be captured as a satellite, or even caused to crash into the planet. Students can learn more about the slingshot effect and its use by visiting the *EarthComm* web site.

2. Investigate the orbits of comets and asteroids
Students can refer to the *EarthComm* web site to help them find orbital information for some typical comets and asteroids. Detailed information is available for a large number of comets and asteroids, and student answers will vary depending upon the selections they make. You may wish to limit student searches and also to standardize results by choosing several specific comets and asteroids for the students to study.

NOTES

Chapter I

ACTIVITY 4 — IMPACT EVENTS AND THE EARTH SYSTEM

Background Information

Impacts are some of the most dramatic events in the solar system. An impact at the end of the Cretaceous Period some 65 million years ago is considered a likely culprit in the mass extinction that led to the vanishing of the dinosaurs. Such collisions make for great drama and exciting science fiction stories, but they can be readily explained. To understand the dynamics of such collisions, **Activity 4** asks students to calculate the explosive and seismic effects of different-sized impactors and to examine the aftereffects of impacts on Earth and other solar-system bodies. When they combine this knowledge with an understanding of the most likely objects to be involved in impacts, the students will have another tool for assessing the likelihood and magnitude of threats to their planet and community.

The History of Solar-System Impacts
When astronomers try to calculate the chances of an impact on Earth by other bodies, they look for examples of previous impacts on the planet. Today there are only a few surviving impact sites. The most famous is Meteor Crater in Arizona, but there are others scattered across the globe. In North America, Haughton Crater on Devon Island inside the Arctic Circle is a prominent ancient crater. It is currently the site of a unique experiment by Mars enthusiasts to set up a research station similar to what is being planned for the Martian surface. Buried under the northern coast of Mexico's

Yucatan peninsula lie the cratered remains of the impactor that devastated life at the end of the Cretaceous. This crater is named Chicxulub (pronounced Cheek-shoo-loob), after the nearby town of Puerto Chicxulub. In all, there are more than 145 known impact sites around the world.

The Mechanics of Solar-System Impacts
The usual suspects for impactors are comets and asteroids. As your students have learned in the previous activities, the orbits of these objects can be unusual and can bring them very close to Earth. This increases the chances for a collision. The power of such an impact is determined by such factors as the size of the impacting body and its structure— i.e., whether its composition is mostly rock, or a combination of ice and rock (in the case of comets). A rocky body is far more likely to survive the ride through our planetary atmosphere to reach the ground and do damage. A comet, on the other hand, may lose much or all of its mass on the way to Earth's surface, depending on how much ice it had before it began its trip.

As an impacting mass moves closer, it sets up shock waves in our atmosphere. When it finally hits the ground, it blasts out huge amounts of debris, leaving behind a round depression called a crater. Most craters have raised rims and flat floors, and some have a central peak at their centers. Some of the excavated material falls back down as ejecta, along with any remnants of the original impactor.

Material that is thrown out of the crater at high speed falls back to Earth, forming rays that point away from the origin of the impact. (The lunar surface has several examples of impact craters surrounded by rays.) Dust from the event is kicked up into the atmosphere and circulated around the globe. If enough dust is raised, it blocks sunlight and global temperatures fall. It is this scenario that describes what may have happened in the end–Cretaceous mass extinction.

Impacts in the early solar system figured significantly in a process called accretion. This is how planets and moons are formed when smaller particles of material come together to form larger ones. Gravitational attraction is also part of the process. As the largest worlds began to take shape, they were targets for leftover bits of solar-system debris that had not yet been swept up. Their surfaces were pelted by this debris during a period that planetary scientists call the "early heavy bombardment." This continued even as the larger bodies in the solar system went through a process called differentiation. Heavier elements gravitated toward the cores of these worlds and lighter ones formed their outer layers. Differentiation thus eliminated the early impact craters from the surfaces of these worlds.

However, the rain of impactors continued to pound the newborn planets. During a period that planetary scientists call the "late heavy bombardment," the worlds continued to sweep up leftover bits of debris still moving in planet-crossing orbits. These impacts left their mark. The evidence of a period in solar-system history that ended more than 3.8 billion years ago can still be seen today in crater fields scattered across our Moon, Mars, Mercury, and the icy moons of the outer solar system.

The Probability of Modern-Day Solar-System Impacts

The rate at which comets and asteroids strike the Earth in our time is fairly low. However, all we need to do is look around at the Moon, Mars, Mercury, some of the larger asteroids, and the moons of the outer solar system to realize that impacts were much more numerous in the past. These collisions have changed every hard surface in the solar system. In 1994, many people watched as Comet Shoemaker-Levy 9 collided with Jupiter, so we can take it for granted that impacts are still happening and will continue to occur.

Assessing the Danger from Modern-Day Impacts

Although the likelihood of impacts today is far less than it was 3.8 billion years ago, the threat still exists. Why is this? Because we have not accurately accounted for all the bodies in near-Earth space that could pose a danger to us. This is not due to lack of observation, but because many bodies are simply too small to observe until they get closer to us. In addition, comets following high-inclination, eccentric orbits simply aren't seen until the Sun's heat begins to melt their ices, forming a cloudy coma around the nucleus and causing them to grow a pair of tails that stretch out in the solar wind.

It's important to know, however, that no near-Earth objects that we presently know about (i.e., those we have detected from Earth) are in danger of colliding with Earth. Your students can find data on these objects on the Internet if they want to see for themselves.

In recent years, astronomers have made great advances in finding Earth-crossing asteroids, but a great many more remain out there to be discovered and charted. Because of the great many unknown objects—undiscovered asteroids or previously unknown comets—that could potentially cross Earth's path, no one knows when the next major impact will happen.

Astronomers have, however, calculated the odds of an impact event occurring. What they have found can help your students understand the danger and the odds of an impact affecting their community.

Our greatest threat comes from objects that are about 2 km in diameter and that could produce an energy equivalent to about one million megatons of conventional explosives. Statistically, one of these has a chance of colliding with Earth once or twice every million years. The resulting devastation would be worldwide—a global catastrophe that would wipe out a large fraction of the Earth's human population. What this means for your students is that each one has about one chance in 6000 to 12,000 of witnessing an impact of this size.

What about impacts by smaller objects? Of course the Earth is hit every day by the very smallest particles, called meteoroids. These often do little more than streak across the sky, or in some cases, leave meteorites scattered across the ground. Occasionally they have been known to hit cars or even people, but mainly they fall harmlessly to Earth.

What about larger objects, like the 20-meter-wide piece of space debris that created Meteor Crater some 50,000 years ago? Here we are talking about impacts that cause a great deal of local damage and kick up some atmospheric dust but do not necessarily put the entire world at risk. The last such impact was in 1908 in Tunguska, in Siberia. The power of the collision blasted across a forest, scattering trees in a radial pattern out from the impact site across more than 2000 km². At the center of the impact, a forest fire burned for weeks, and ash and tundra fragments were blown skyward and distributed around the world high in the atmosphere. People 800 km away could hear the sound of the blast. Today, such a blast might cause people to think that a huge bomb had been detonated!

When we start to discuss objects 1 km in diameter, we are getting into the realm of global catastrophic effects. There are perhaps a hundred known potentially hazardous asteroids ranging in size from about 0.5 km up to nearly 9 km. These are not always easy to spot until they are relatively close and bright enough to be observed. Groups like the Spaceguard Survey and some search and tracking programs look for such near-Earth asteroids. Scientists are asking world governments to spend time and money building early detection systems that could give people a chance to prepare themselves for the effects of the impacts. They argue that our first and best defense against such an occurrence is to know where the 1-km bodies are. Then we can assess the true risks and take action to protect ourselves against possible impacts.

For now, it is extremely difficult to detect and defend against the smaller objects. Perhaps we will be able to do so in the very near future. In any case, the risk that an individual could be affected by an impact like the one that shook Tunguska is smaller than the risk of being affected by a larger impact. If you look at it in comparison to more commonly occurring events like earthquakes, volcanic eruptions, or severe weather, then the risk of being affected by an impact is very small indeed.

Lunar Craters: Impacts or Volcanoes?
As you and your students observe the Moon over a four-week period, here is an interesting historical fact about craters to consider. Until the 1930s, many astronomers classified the Moon's craters as the calderas of ancient, extinct volcanoes. If you examine a volcanic crater and a well-formed impact crater, there are some similarities. But modern studies in high-impact physics can reproduce craters that are nearly identical to Meteor Crater and all of the craters seen on

the other solid bodies. The Apollo missions to the Moon and later missions by spacecraft like Voyager to the outer solar system show impact craters everywhere. In the case of the lunar surface, it is true that volcanism played a part in creating the huge maria basins. But those volcanic deposits were the direct result of an impactor colliding with the surface, melting rock and releasing molten lavas from deep below the lunar surface. Today the Moon is geologically nearly dead, but it continues to be pelted by space debris.

Among the outer planets, it is likely that impact cratering continues to occur at the icy moons, sometimes allowing ice volcanoes to ooze their slushy water-ice mixtures across the frozen surfaces.

More Information – on the Web
Visit the *EarthComm* web site www.agiweb.org/earthcomm/ to access a variety of links to web sites that will help you to deepen your understanding of content and prepare you to teach this activity. Many of the sites also contain images that you can download.

Goals and Assessment

Clarify that the goals indicate what students should understand and be able to do as a result of the activity. Make sure students understand that Chapter Assessments are based upon these goals.

Goal	Location in Activity	Assessment Opportunity
Investigate the mechanics of an impact event and make scale drawings of an impact crater.	**Investigate; Digging Deeper; Understanding and Applying What You Have Learned** Question 1	Drawing of Chicxulub crater is scaled correctly.
Calculate the energy (in joules) released when an asteroid collides with Earth.	**Investigate; Understanding and Applying What You Have Learned** Question 4	Calculations are done correctly; work is shown.
Compare natural and human-made disasters to the impact of an asteroid.	**Investigate**	Calculations of the Richter scale equivalent of impact events is done correctly; work is shown. Comparisons are accurate.
Understand the consequences to your community should an impact event occur.	**Digging Deeper; Understanding and Applying What You Have Learned** Questions 2 – 3	Responses are reasonable.
Investigate the chances for an asteroid or comet collision.	**Digging Deeper; Understanding and Applying What You Have Learned** Question 5	Responses are reasonable.

NOTES

Activity 4   Impact Events and the Earth System

Goals

In this activity you will:

- Investigate the mechanics of an impact event and make scale drawings of an impact crater.

- Calculate the energy (in joules) released when an asteroid collides with Earth.

- Compare natural and man-made disasters to the impact of an asteroid.

- Understand the consequences to your community should an impact event occur.

- Investigate the chances for an asteroid or comet collision.

Think about It

Meteor Crater in Arizona is one of the best-preserved meteor craters on Earth. It is 1.25 km across and about 4 km in circumference. Twenty football games could be played simultaneously on its floor, while more than two million spectators observed from its sloping sides.

- How large (in diameter) do you think the meteor was that formed Meteor Crater?
- How would the impact of the meteor have affected living things near the crater?

What do you think? Record your ideas about these questions in your *EarthComm* notebook. Be prepared to discuss your responses with your small group and the class.

Activity Overview

Students calculate the energy that would be released if a hypothetical asteroid were to collide with the Earth. They calculate the amount of energy released by objects that are known to have made impact with the Earth in the past. Students then compare these calculations to estimates of natural and human-made phenomena on Earth, like earthquakes and nuclear explosions. They make a scale drawing of the Chicxulub crater. Finally, students convert their energy calculations to the Richter scale equivalent and compare impact events to the five greatest earthquakes in the world between 1900 and 1998. The **Digging Deeper** section of **Activity 4** introduces asteroids and comets, assesses the likelihood that one of these objects will strike the Earth, and discusses the consequences if it were to strike the Earth.

Preparation and Materials Needed

No advance preparation is required to complete this activity.

Materials
- Calculator
- Metric ruler

Teaching Tip

The movie *Armageddon* has several scenes you may want to show your students. Examples include the scene in which advisors tell the president about the approach of the asteroid, and the scene in which pieces of the asteroid are striking Earth. The latter scene will be most meaningful to students, because it shows damage to communities from the asteroid. The scene in which the miners are passing through the large chunks of asteroids to land on the main piece may also be of interest.

Think about It

Student Conceptions

Students are likely to say that the meteor was similar in size to the crater that it left behind. They are likely to say that the impact event would kill all of the organisms living near the crater. Ask students to consider how far beyond the actual impact site they would expect organisms to be affected.

Answer for the Teacher Only

Scientists have estimated that the meteor that formed Meteor Crater was about 20 m in diameter. Because of its very high velocity upon impact, the energy available to form the crater was extremely large relative to the actual size of the meteor. The impact probably affected living things only in the immediate vicinity of the impact site, in an area with a radius of perhaps ten to a few hundreds of kilometers. It would have taken a much larger impact to disrupt environments on a continental or global scale.

Assessment Tool

Think about It Evaluation Sheet
Use this evaluation sheet to help students understand and internalize the basic expectations for the warm-up activity.

NOTES

Earth System Evolution Astronomy

Investigate

1. Given the following information, calculate the energy released when an asteroid collides with Earth:

   • The spherical, iron–nickel asteroid has a density of 7800 kg/m³.

   • It is 40 m in diameter.

   • It has a velocity of 20,000 m/s relative to the Earth.

   Note: It is very important to keep track of your units during these calculations. You will be expressing energy with a unit called a "joule." A joule is 1 kg m²/s².

   a) Find the volume of the asteroid in cubic meters. The equation for the volume of a sphere is as follows:

   $$V = \frac{4}{3}\pi r^3$$

   where V is volume of the sphere, and r is the radius of the sphere.

   b) Multiply the volume by the density to find the total mass of the asteroid.

   c) Calculate the energy of the asteroid. Because the asteroid is moving, you will use the equation for kinetic energy, as follows:

   $$KE = \frac{1}{2}mv^2$$

   where KE is kinetic energy, m is the mass, and v is the velocity.

   Express your answer in joules. To do this express mass in kilograms, and velocity in meters per second.

   For some perspective, a teenager uses over 10,000 kJ (kilojoules) of energy each day. (There are 1000 J (joules) in a kilojoule.)

2. The combination of calculations that you just performed can be summarized as:

   $$\text{Energy} = \frac{2}{3}\pi\rho r^3 v^2$$

   where r is the radius,
   ρ is the density, and
   v is the velocity of the object.

   a) Suppose an object makes an impact with the Earth at 10 times the velocity of another identical object. By what factor will the energy of the object increase?

   b) Suppose an object makes an impact with the Earth, and it has 10 times the radius of another object traveling at the same speed. By what factor will the energy of the object increase?

   c) How do these relationships help to explain how small, fast-moving objects can release a tremendous amount of energy as well as larger yet slower-moving objects?

3. The asteroid described in **Step 1** above was the one responsible for Meteor Crater in Arizona.

   a) Copy the following table into your notebook. Enter your calculation for Meteor Crater.

   b) Calculate the energy released by the impacts shown in the table.

Investigate

1. a) The radius of the asteroid is equal to 20 m (half of the diameter). Therefore, the volume is 33,510 m^3 ($V = 4/3 \times \pi \times 20^3$).

    b) The mass of the asteroid is 261,378,000 kg ($m = 33{,}510$ m^3 $\times$ 7800 kg/m^3). If the initial calculation of volume is not rounded to the nearest whole number, then the mass calculation comes out to 261,380,508 kg.

    c) Using the formula for kinetic energy ($\frac{1}{2}mv^2$) and a mass of 261,378,000 kg, the energy of the asteroid is [$\frac{1}{2} \times 261{,}378{,}000$ kg $\times (20{,}000$ m/s$)^2$] = 5.228 $\times$ 10^{16} J (kg$\cdot$m/s^2)

2. a) The energy will increase by a factor of 100 (i.e., 10^2).

    b) The energy will increase by a factor of 1000 (i.e., 10^3).

    c) Because the total kinetic energy of a moving object increases as a function of the square of its speed (velocity) and as the cube of its radius, (mass equals density multiplied by volume and volume increases as the cube of its radius) increases in either of these parameters will greatly increase the total kinetic energy of a moving object.

3. a) - b) Note that calculations of the Richter-scale magnitude shown in the table below refer to **Step Six** of this investigation and are not required for the answer to **Step 3**.

Object	Radius (m)	Density (kg/m^3)	Impact Velocity (m/s)	Energy (joules)	Richter-Scale Magnitude Equivalent
Asclepius	100	3000	30,000	5.655 $\times$ 10^{18}	6.69
Comet Swift-Tuttle	1000	5000	60,000	3.770 $\times$ 10^{22}	9.26
Chicxulub impactor	5000	3000	32,000	8.042 $\times$ 10^{23}	10.15
SL9 Fragment	2150	1000	60,000	7.493 $\times$ 10^{22}	9.46
Meteor Crater	20	7800	20,000	5.228 $\times$ 10^{16}	5.33

Object	Radius (m)	Density (kg/m³)	Impact Velocity (m/s)	Energy (joules)	Richter Scale Magnitude Equivalent
Asclepius	100	3000	30,000		
Comet Swift-Tuttle	1000	5000	60,000		
Chicxulub impactor	5000	3000	32,000		
SL9 Fragment Q	2150	1000	60,000		
Meteor Crater	20	7800	20,000		

Note:

- Asclepius is an asteroid that passed within 690,000 km of Earth in 1989.

- Comet Swift-Tuttle is a future threat to the Earth–Moon system, having passed Earth in 1992 and being scheduled for return in 2126.

- SL9 Fragment Q is a fragment of Comet Shoemaker-Levy that impacted Jupiter in 1994.

- Chicxulub impactor is the name of the asteroid that triggered the extinction of the dinosaurs 65 million years ago.

4. Use the table below to compare the energy from all these events to known phenomena.

a) In your notebook, explain how the energies of these four impact events compare to some other known phenomena.

Phenomena	Kinetic Energy (joules)
Annual output of the Sun	10^{34}
Severe earthquake	10^{18}
100-megaton hydrogen bomb	10^{17}
Atomic bomb	10^{13}

5. Make a scale drawing of the Chicxulub impactor compared with Earth. The diameter of the Earth is 12,756 km.

a) If you made the diameter of the Chicxulub impactor 1 mm, what would the diameter of the Earth be?

b) If you made the diameter of the Chicxulub impactor 0.5 mm, which is probably about as small as you can draw, what would the diameter of the Earth be?

E 39

EarthComm

4. **a)** None of the energies released by the impacts is larger than the annual energy released by the Sun. All of the energies released by the impacts, with the exception of the energy released by the asteroid that formed Meteor Crater, are larger than the energy released by a severe earthquake, a 100 megaton hydrogen bomb, or an atomic bomb.

5. **a)** The Chicxulub impactor had a diameter of 10,000 m. If a circle 1 mm in diameter represented it, then this equates to being scaled down by a factor of 10^7. Scaling the diameter of the Earth (12,756,000 m) by an equivalent factor yields a model Earth with a diameter of 1275.6 mm.

   b) The diameter of the Earth would be 637.8 mm (1257.6 ÷ 2).

Chapter 1

Earth System Evolution Astronomy

6. How do these impact events compare with the energy released in an earthquake? If you have a calculator capable of handling logarithms, answer the following questions:

   a) Calculate the Richter scale equivalent of the energy released by the four impact events. Use the following equation:

   $$M = 0.67 \log_{10}E - 5.87$$

where M is the equivalent magnitude on the Richter scale, and E is the energy of the impact, in joules.

   b) How do your results compare with the table below, which shows the five greatest earthquakes in the world between 1900 and 1998? Which impacts exceed the world's greatest earthquakes?

Location	Year	Magnitude
Chile	1960	9.5
Prince William Sound, Alaska	1964	9.2
Andreanof Islands, Aleutian Islands	1957	9.1
Kamchatka	1952	9.0
Off the Coast of Ecuador	1906	8.8

Reflecting on the Activity and the Challenge

You have calculated the energy released when asteroids of different sizes hit the Earth's surface, and you have compared these to other energy-releasing events. This comparison will be helpful as you explain the hazards associated with an impact in your **Chapter Challenge** brochure.

6. **a)** See the table in **Step 3** on page 141. This calculation is made using the formula given on page E40 of the Student Edition. An example calculation for the Meteor Crater follows.

   The energy release during the Meteor Crater impact has been calculated above to be 5.228×10^{16} J. The $\log_{10}$ of this number is the $\log_{10}$ of 5.228 plus 16, or 16.72. Multiplying this by 0.67 yields 11.20. Subtracting 5.87 gives a Richter-scale value of 5.33.

   b) In each case, the energy released during impacts of the Asclepius and Meteor Crater impactors was less than the smallest of the five earthquakes listed. The energy released during the Comet Swift-Tuttle and SL9 fragment impacts was greater than all of the earthquakes listed except the Chilean earthquake of 1960, and the impact of Chicxulub impactor released more energy than any of the earthquakes listed.

Assessment Tools

EarthComm **Notebook Entry-Checklist**
Use this checklist as a quick guide for student self-assessment and/or an opportunity to quickly score student work. Add further criteria specific to your classroom needs or to this particular investigation.

Investigate Notebook Entry-Evaluation Sheet
Point out the criteria listed on this evaluation sheet that are relevant to this particular investigation. Encourage students to internalize the criteria by making them part of your "assessment conversations" as you circulate around the classroom.

Reflecting on the Activity and the Challenge

Have a student read this section aloud to the class, and discuss the major points raised. Students should recognize that asteroids pose a nonnegligible threat to the Earth community. Take this opportunity to discuss the repercussions of an impact for their community. Students should realize that even if their community is not close to the impact site, the effects of an asteroid hitting Earth would still be felt, because clouds of dust and debris would block sunlight.

Digging Deeper

ASTEROIDS AND COMETS

Asteroids

Asteroids are rocky bodies smaller than planets. They are leftovers from the formation of the solar system. In fact, the early history of the solar system was a period of frequent impacts. The many scars (impact craters) seen on the Moon, Mercury, Mars, and the moons of the outer planets are the evidence for this bombardment. Asteroids orbit the Sun in very elliptical orbits with inclinations up to 30°. Most asteroids are in the region between Jupiter and Mars called the asteroid belt. There are probably at least 100,000 asteroids 1 km in diameter and larger. The largest, called Ceres, is about 1000 km across. Some of the asteroids have very eccentric orbits that cross Earth's orbit. Of these, perhaps a few dozen are larger than one kilometer in diameter. As you learned in the activity, the energy of an asteroid impact event increases with the cube of the radius. Thus, the larger asteroids are the ones astronomers worry about when they consider the danger of collision with Earth.

The closest recent approach of an asteroid to Earth was Asteroid 1994 XM 11. On December 9, 1994, the asteroid approached within 115,000 km of Earth. On March 22, 1989 the asteroid 4581 Asclepius came within 1.8 lunar distances, which is close to 690,000 km. Astronomers think that asteroids at least 1 km in diameter hit Earth every few hundred million years. They base this upon the number of impact craters that have been found and dated on Earth. A list of asteroids that have approached within

Figure 1 Image of the asteroid Ida, which is 58 km long and 23 km wide.

two lunar distances of Earth (the average distance between Earth and the Moon) is provided in *Table 1* on the following page. Only close-approach distances less than 0.01 AU for asteroids are included in this table.

Digging Deeper

Assign the reading for homework, along with the questions in **Check Your Understanding** if desired.

Assessment Opportunity

Use a quiz to assess student understanding of the concepts presented in this activity. Some sample questions are listed below:

Fill in the missing word or words in the sentences below.

1. A _____ is a small rock in space. When this rock survives through the Earth's atmosphere, it is called a _____.
(*meteoroid, meteorite*)

2. Most asteroids are found in the region between Jupiter and Mars, which is called the _____.
(*asteroid belt*)

3. One of the clues scientists use to support that the theory that the extinction of the dinosaurs was caused by the impact of an asteroid is a layer of _____ sediment that is _____ years old and found worldwide.
(*iridium-rich, 65 million*)

4. A _____ is a mass of frozen gas and rocky dust particles that has a tail, which is caused by the action of _____.
(*comet, solar wind*)

Teaching Tip

Figure 1 is an image of asteroid Ida, photographed by the robot spacecraft Galileo. Ida is the first asteroid to have been discovered to have a moon. The tiny moon, named Dactyl, is about 1 mi. across. As with many names in astronomy, *Ida* and *Dactyl* are from Greek mythology. The name Dactyl is derived from the Dactyli, a group of mythological beings who lived on Mount Ida. Here, the infant Zeus was hidden by the nymph Ida and protected by the Dactyli. Visit the *EarthComm* web site to view additional images of the asteroid Ida and its moon.

Earth System Evolution Astronomy

Table 1 Asteroids with Close-Approach Distances to Earth			
Name or Designation	**Date of Close Earth Approach**	**Distance**	
		(AU)	**(LD)**
1994 XM1	1994–Dec–09	0.0007	0.3
1993 KA2	1993–May–20	0.001	0.4
1994 ES1	1994–Mar–15	0.0011	0.4
1991 BA	1991–Jan–18	0.0011	0.4
1996 JA1	1996–May–19	0.003	1.2
1991 VG	1991–Dec–05	0.0031	1.2
1999 VP11	1982–Oct–21	0.0039	1.5
1995 FF	1995–Apr–03	0.0045	1.8
1998 DV9	1975–Jan–31	0.0045	1.8
4581 Asclepius	1989–Mar–22	0.0046	1.8
1994 WR12	1994–Nov–24	0.0048	1.9
1991 TU	1991–Oct–08	0.0048	1.9
1995 UB	1995–Oct–17	0.005	1.9
1937 UB (Hermes)	1937–Oct–30	0.005	1.9
1998 KY26	1998–Jun–08	0.0054	2.1

(AU) – Astronomical distance Unit: 1.0 AU is roughly the average distance between the Earth and the Sun.
(LD) – Lunar Distance unit: 1.0 LD is the average distance from the Earth to the Moon (about 0.00257 AU).

Most (but not all) scientists believe that the extinction of the dinosaurs 65 million years ago was caused by the impact of an asteroid or comet 10 km in diameter. Such a large impact would have sent up enough dust to cloud the entire Earth's atmosphere for many months. This would have blocked out sunlight and killed off many plants, and eventually, the animals that fed on those plants. Not only the dinosaurs died out. About 75% of all plants and animals became extinct. One of the strong pieces of evidence supporting this hypothesis is a 1-cm-thick layer of iridium-rich sediment about 65 million years old that has been found worldwide. Iridium is rare on Earth but is common in asteroids.

NOTES

Our planet has undergone at least a dozen mass-extinction events during its history, during which a large percentage of all plant and animal species became extinct in an extremely short interval of geologic time. It is likely that at least some of these were related to impacts. It is also likely that Earth will suffer another collision sometime in the future. NASA is currently forming plans to discover and monitor asteroids that are at least I km in size and with orbits that cross the Earth's orbit. Asteroid experts take the threat from asteroids very seriously, and they strongly suggest that a program of systematic observation be put into operation to predict and, hopefully avoid an impact.

Comets

Comets are masses of frozen gases (ices) and rocky dust particles. Like asteroids, they are leftovers from the formation of the solar system. There are many comets in orbit around the Sun. Their orbits are usually very eccentric with large inclinations. The orbits of many comets are very large, with distances from the Sun of greater than 20,000 astronomical units (AU). The icy head of a comet (the nucleus) is usually a few kilometers in diameter, but it appears much larger as it gets closer to the Sun.

That's because the Sun's heat vaporizes the ice, forming a cloud called a coma. Radiation pressure and the action of the **solar wind** (the stream of fast-moving charged particles coming from the Sun) blow the gases and dust in the coma in a direction away from the Sun. This produces a tail that points away from the Sun even as the comet moves around the Sun. Halley's Comet, shown in *Figure 2*, is the best known of these icy visitors. It rounds the Sun about every 76 years, and it last passed by Earth in 1986.

Figure 2 Halley's Comet last appeared in the night sky in 1986.

Comets have collided with the Earth since its earliest formation. It is thought that the ices from comet impacts melted to help form Earth's oceans. In 1908 something hit the Earth at Tunguska, in Siberian Russia. It flattened trees for hundreds of miles, and researchers believe that the object might have been a comet. Had such an event occurred in more recent history in a more populated area, the damage and loss of life would have been enormous. A list of comets that have approached within less than 0.11 AU of Earth is provided in *Table 2* on the following page.

Geo Words

solar wind: a flow of hot charged particles leaving the Sun.

Teaching Tip

Comet Halley, shown in *Figure 2*, is named after the scientist Edmund Halley. In 1705, Halley used Newton's laws of motion to predict that the comet that had been seen in 1531, 1607, and 1682, and would return in 1758. The comet did return as predicted and was later named in his honor.

The average period of Comet Halley's orbit is 76 years. However, the gravitational pull of the major planets alters the orbital period of the comet, making exact dates of the comet's reappearance difficult to predict. Its next period of visibility will be in 2061.

Like all comets, Halley has a highly eccentric orbit. Its orbit is retrograde, meaning that it orbits in a clockwise direction when viewed from our North Pole. The orbit is inclined 18° to the plane of the ecliptic. The nucleus of Comet Halley is approximately 16 x 8 x 8 km. Visit the *EarthComm* web site for additional images and information about Comet Halley.

Earth System Evolution Astronomy

Geo Words

meteoroid: a small rock in space.

meteor: the luminous phenomenon seen when a meteoroid enters the atmosphere (commonly known as a shooting star).

meteorite: a part of a meteoroid that survives through the Earth's atmosphere.

Table 2 Close Approaches of Comets				
Name	**Designation**	**Date of Close Earth Approach**	**Distance**	
			(AU)	**(LD)**
Comet of 1491	C/1491 B1	1491–Feb–20	0.0094	3.7*
Lexell	D/1770 L1	1770–Jul–01	0.0151	5.9
Tempel-Tuttle	55P/1366 U1	1366–Oct–26	0.0229	8.9
IRAS-Araki-Alcock	C/1983 H1	1983–May–11	0.0313	12.2
Halley	1P/ 837 F1	837–Apr–10	0.0334	13
Biela	3D/1805 V1	1805–Dec–09	0.0366	14.2
Comet of 1743	C/1743 C1	1743–Feb–08	0.039	15.2
Pons-Winnecke	7P/	1927–Jun–26	0.0394	15.3
Comet of 1014	C/1014 C1	1014–Feb–24	0.0407	15.8*
Comet of 1702	C/1702 H1	1702–Apr–20	0.0437	17
Comet of 1132	C/1132 T1	1132–Oct–07	0.0447	17.4*
Comet of 1351	C/1351 W1	1351–Nov–29	0.0479	18.6*
Comet of 1345	C/1345 O1	1345–Jul–31	0.0485	18.9*
Comet of 1499	C/1499 Q1	1499–Aug–17	0.0588	22.9*
Schwassmann-Wachmann 3	73P/1930 J1	1930–May–31	0.0617	24

* Distance uncertain because comet's orbit is relatively poorly determined.
(AU) – Astronomical distance Unit: 1.0 AU is roughly the average distance between the Earth and the Sun.
(LD) – Lunar Distance unit: 1.0 LD is the average distance from the Earth to the Moon (about 0.00257 AU).

Meteoroids, Meteors, and Meteorites

Meteoroids are tiny particles in space, like leftover dust from a comet's tail or fragments of asteroids. Meteoroids are called **meteors** when they enter Earth's atmosphere, and **meteorites** when they reach the Earth's surface. About 1000 tons of material is added to the Earth each year by meteorites, much of it through dust-sized particles that settle slowly through the atmosphere. There are several types of meteorites. About 80% that hit Earth are stony in nature and are difficult to tell apart from Earth rocks. About 15% of meteorites consist of the metals iron and nickel and are very dense. The rest are a mixture of iron–nickel and stony material. Most of the stony meteorites are called chondrites. Chondrites may represent material that

E 44

Chapter 1

NOTES

was never part of a larger body like a moon, a planet, or an asteroid, but instead are probably original solar-system materials.

Figure 3 Lunar meteorite.

Check Your Understanding

1. Where are asteroids most abundant in the solar system?

2. How might a major asteroid impact have caused a mass extinction of the Earth's plant and animal species at certain times in the geologic past?

3. Why do comets have tails? Why do the tails point away from the Sun?

4. What are the compositions of the major kinds of meteorites?

Understanding and Applying What You Have Learned

1. Look at the table of impact events shown in the **Investigate** section. Compare the densities of the object that formed Meteor Crater and SL9 Fragment Q from the Shoemaker-Levy Comet. Use what you have learned in this activity to explain the large difference in densities between the two objects.

2. If an asteroid or comet were on a collision course for Earth, what factors would determine how dangerous the collision might be for your community?

3. How would an asteroid on a collision course endanger our Earth community?

4. Comets are composed largely of ice and mineral grains. Assume a density of 1.1 g/cm^3:

   a) How would the energy released in a comet impact compare to the asteroid impact you calculated in the **Investigate** section? (Assume that the comet has the same diameter and velocity as the asteroid.)

   b) Based upon your calculation, are comets dangerous if they make impact with the Earth? Explain your response.

5. From the information in the **Digging Deeper** reading section, and what you know about the eccentricities and inclinations of asteroid orbits, how likely do you think it is that an asteroid with a diameter of 1 km or greater will hit the Earth in your lifetime? Explain your reasoning. Can you apply the same reasoning to comets?

6. Add the asteroid belt to the model of the solar system you made in the first activity. You will need to think about how best to represent the vast number of asteroids and their wide range of sizes. Don't forget to add in some samples of Earth-approaching asteroids and the orbit of one or two comets.

E 45

Check Your Understanding

1. Most asteroids are in the region between Jupiter and Mars called the asteroid belt.

2. A major impact could send up enough dust to cloud the entire Earth's atmosphere for many months, blocking out sunlight and killing plants and animals. An event like this is thought by many scientists to have occurred 65 million years ago.

3. Comets appear to have tails because the Sun's heat vaporizes the ice around the nucleus to form the coma. Radiation pressure and the action of the solar wind blow gases and dust in the coma away from the Sun, producing a tail.

4. Estimates for the relative abundance of different kinds of meteorites that fall to the Earth (as opposed to those found) show that the majority of meteorites are stony meteorites (80% is quoted in the **Digging Deeper** reading section, although some estimates are as high as 95%). Iron meteorites comprise about 15%, estimates vary and can be as low as 5%. The remainder is comprised of mixed stony-iron meteorites (a few percent). Most of the stony meteorites are called chondrites, although a small fraction of them are very different; these are called achondrites. Chondrites may represent material that was never part of a larger body like a moon, a planet or an asteroid, but instead are probably original solar-system materials. More information about different kinds of meteorites can be found on the *EarthComm* web site.

> **Assessment Tool**
>
> **Check Your Understanding Notebook Entry-Evaluation Sheet**
> Use this sheet to evaluate the extent to which students understand the key concepts explored in **Activity 4** and explained in **Digging Deeper**, and to evaluate the students' clarity of expression.

Understanding and Applying What You Have Learned

1. The density of the object that formed Meteor Crater is much greater than the density of the SL9 Fragment Q. Presumably, this is because the object that formed Meteor Crater was an iron meteorite rather than a comet fragment like SL9 Fragment Q. Being made of ices and rocky materials, comets have a lower density than stony or iron meteorites.

2. Responses may include the diameter of the object, the velocity of the object relative to Earth, and the density of the object.

3. The students should recognize that asteroids pose a considerable threat to their Earth community. This would be a good opportunity to discuss the repercussions

for their community. Even if students themselves were not close to the impact site, they would still be affected as clouds of dust and debris block sunlight.

4. a) Students will need to convert the density from g/cm^3 to kg/m^3 to make the units consistent throughout. In the equation for calculating energy, the density (used to calculate mass) is not raised to any power, so changes in density of the impactor are directly proportional to changes in the resultant energy of the impact (all other factors being equal, twice the density yields twice the energy of impact).

A density of 1.1 g/cm^3 is equivalent to a density of 1100 kg/m^3 (because there are 1000 gm in a kilogram, and 10^6 cm^3 in 1 m^3). This is 14.1% of the density of the Meteor Crater impactor. Multiplying the Meteor Crater impact energy by 0.141 yields 7.37 x 10^{15} J. Substituting the volume, density, and velocity values into the equations given on page E38 yields the same answer.

This number is less than the energy released by the asteroid. This should be expected, because the density of the comet is less than that of the asteroid, whereas everything else is the same.

   b) Student opinions on whether or not comets are dangerous will vary. Encourage students to refer to the table they completed in the investigation to substantiate their responses. By comparison, the SL9 comet fragment given in the table had the second largest impact of all the impactors investigated.

5. Answers will vary. Some students may say, given that scientists estimate that asteroids at least 1 km in diameter hit Earth every few hundred million years, that it is unlikely that one will hit during their lifetime.

6. Students add the asteroid belt to their model of the solar system.

NOTES

Earth System Evolution Astronomy

Preparing for the Chapter Challenge

Assume that scientists learn several months before impact that a large asteroid will hit near your community. Assume that you live 300 km from the impact site. What plans can your family make to survive this disaster? What are some of the potential larger-scale effects of an asteroid impact? Work with your group to make a survival plan. Present your group's plan to the entire class. Be sure to record suggestions made by other groups. This information will prove useful in completing the **Chapter Challenge**.

Inquiring Further

1. **Impact craters on objects other than the Earth**

   In an earlier activity you studied impact sites on the Moon. Look at Mercury, Mars, and the moons of Saturn, Uranus, and Neptune to see other examples of impact craters in the solar system. How are these craters similar to Meteor Crater? How are they different?

2. **Modeling impact craters**

   Simulate an asteroid or comet hitting the Earth. Fill a shoebox partway with plaster of Paris. When the plaster is almost dry, drop two rocks of different sizes into it from the same height. Carefully retrieve the rocks and drop them again in a different place, this time from higher distance. Let the plaster fully harden, then examine, and measure the craters. Measure the depth and diameter and calculate the diameter-to-depth ratio. Which is largest? Which is deepest? Did the results surprise you?

3. **Earth-approaching asteroids**

   Do some research into current efforts by scientists to map the orbits of Earth-approaching asteroids? Visit the *EarthComm* web site to help you get started with your research. How are orbits determined? What is the current thinking among scientists about how to prevent impacts from large comets or asteroids?

4. **Barringer Crater**

   Research the Barringer Crater (Meteor Crater). The crater has been named for Daniel Moreau Barringer, who owned the property that contains the crater. Explain how scientists used Barringer Crater to understand how craters form. Study the work of Dr. Eugene Shoemaker, who was one of the foremost experts on the mechanics of impact cratering.

 Wear goggles while modeling impact craters. Work with adult supervision to complete the activity.

 E 46

Chapter 1

Preparing for the Chapter Challenge

Students should now have a good idea of the risks associated with impact events. To get an even better sense of the risks, students may enjoy reading the history of the Tunguska event. Scientists think that this may have been a giant fireball from a comet that burst over Siberia. There are also several of examples of comets and meteorites striking Earth. The *EarthComm* web site provides additional information on impact events and their possible consequences. The work that students complete here can be inserted into their **Chapter Challenge** projects.

Relevant criteria for assessing this section (see **Assessment Rubric for Chapter Challenge on Astronomy** on pages 12 and 13) include discussions on:

- what comets and asteroids are, how they behave, how likely it is that one will collide with Earth in the future, and what would happen if one did
- why extraterrestrial influences on the community are a natural part of Earth system evolution

Inquiring Further

1. Impact craters on objects other than the Earth

Unlike Earth, Mercury has no atmosphere to disintegrate meteoroids before they hit the planet. Therefore, impact craters dominate the surface of Mercury. Also, an absence of water or tectonics means that craters on Mercury are not eroded (except by other impacts). The largest feature on the surface of Mercury is the circular Caloris Basin, which resulted from a collision with an asteroid. It is 1000 km across.

Terrain on Mars shows evidence of erosion by running water, and the largest craters or impact basins on Mars are thought to be buried in the northern part of the planet. There are several impact basins in the southern hemisphere of Mars, the largest of them being the Hellas Basin, with a diameter of approximately 1800 km. The Argyre Basin is the second largest, measuring 900 km across.

The outer gas planets do not have solid surfaces, but their moons do, and therefore the moons have impact craters. The largest known impact basin in the solar system is found on Jupiter's moon, Callisto. The impact basin is Valhalla. It has a bright central region that is 600 km in diameter, and its rings extend to 3000 km in diameter. The second largest impact basin on Callisto is Asgard. It measures about 1600 km in diameter. Saturn's innermost moon, Mimas, has been impacted. Its largest crater, Herschel, has walls that are approximately 5 km high. Parts of its floor measure 10 km deep, and its central peak rises 6 km (3.7 mi.) above the crater floor.

2. Modeling impact craters

Small-scale model studies like this one serve an important purpose, because they give a first indication of some of the important effects. The results have to be interpreted with caution, however, because the models are not true scale models. It is usually impossible to adjust the properties of the materials in such a way that the model results are an undistorted representation of the full-scale phenomenon.

3. Earth-approaching asteroids

Scientists recognize the hazards of Earth-approaching asteroids. Congress recently asked NASA to develop a plan for monitoring these asteroids. Current technology allows scientists to discover and track most asteroids larger than 1 km in diameter that might strike the Earth. These objects can be detected with ground-based telescopes of moderate size. Most of what is known about Earth-crossing asteroids (ECAs) has been obtained through studies using small ground-based telescopes. Currently, several new ECAs are discovered each month. Programs like NASA's Spaceguard Survey or IMPACT (International Monitoring Programs for Asteroids and Comet Threats) are working towards the development of an international network for the discovery, confirmation, and follow-up observations of ECAs.

4. Barringer Crater

Early theories regarding the origins of Barringer Crater (also known as Meteor Crater) centered on two conflicting hypotheses. The first proposed that the crater was the result of a steam explosion caused by volcanic activity, and the second proposed that it was the result of a meteorite impact. In 1906 and 1909 Daniel Barringer, who had secured mining patents for the crater and the land around it, presented strong arguments that supported the impact hypothesis. Shortly thereafter, the impact hypothesis was championed by a number of geologists, and has since become the accepted explanation for the origins of the crater. Scientists now believe that the crater was formed by the impact of a nickel-iron meteorite approximately 50,000 years ago.

Since Barringer's early studies, Meteor Crater has been the site of much scientific study into the process of meteoric impact. A number of geologic materials that have since been found to be characteristic of impact sites were discovered at Meteor Crater. Additionally, in 1963 Eugene Shoemaker published a landmark paper that analyzed similarities between the Barringer crater and craters created by nuclear test explosions in Nevada. Shoemaker demonstrated that the nuclear craters and the Barringer crater were structurally similar in nearly all respects. His paper provided the clinching arguments in favor of an impact, finally convincing most of the last doubters.

The *EarthComm* web site provides links that will help students get started on their research into the history of scientific discovery at Barringer Crater, and Eugene Shoemaker.

NOTES

ACTIVITY 5 — THE SUN AND ITS EFFECTS ON YOUR COMMUNITY

Background Information

Every day, we are exposed to radiation from a nuclear power plant. No, that isn't meant to frighten you or your students. It is simply a statement of the reality in which the Earth and its systems exist—that "nuclear power plant" is the Sun!

The nuclear power plant we call the Sun is distant enough to spare us from instant radiation-induced death but close enough to warm the Earth and promote the continuation of life. It turns out that we are far more at risk from solar radiation, flares, and other solar effects than we are from more rarely occurring solar-system impacts. This is because solar output is a 24-hour-a-day operation; a ceaseless bath of radiation that affects our planet more directly than Earth-crossing asteroids.

In **Activity 5**, your students will be investigating the Sun as part of their assessment of threats against their communities. Studying the strength and amount of the Sun's prodigious output of energy should allow them to assess the kinds of events that are most likely to affect their community.

The Sun as a Nuclear Furnace

From Earth's relatively safe average distance of about 150 million kilometers, the Sun appears as a smooth, yellow-white ball of gas. It follows the laws of gravity, temperature, and pressure as it goes about turning hydrogen into helium. As we study the Sun through specialized instruments like those aboard the Solar and Heliospheric Observatory (SOHO), Transition Region and Coronal Explorer (TRACE), and Yohkoh satellites, we find it to be a complex structure. The surface of the Sun is in constant turmoil. Dark spots exhibiting extreme magnetic fields rampage across its boiling face in tandem with changes in its overall magnetic activity. Flares and huge eruptions called coronal mass ejections send streams of ionized particles blasting past the Earth, affecting everything from communication systems to weather satellites and Earth-orbiting astronauts.

The designation of the Sun as a nuclear furnace is an apt one. To understand how this furnace works, we need to drive straight to its core. Here is where this star—which is more than 300,000 times the mass of the Earth—sustains a 16-billion-degree furnace. In the Sun's core, even complete atoms cannot exist because of the extremely high temperatures. Hydrogen atoms are dissociated and nuclear protons and electrons move throughout the core in a soup of electrified particles called plasma.

The core's temperatures are so high because it is under enormous pressure from the mass of the solar layers above it. That pressure is estimated to be some 233 billion times the Earth's atmosphere at sea level. At high temperatures and pressures, the protons and electrons are packed so tightly together that their density is 151 g/cm^3. This is 13 times denser than solid lead!

In this alien environment, nuclear fusion—the source of all the Sun's power—takes place. At the end of the fusion chain, a photon is emitted and begins to work its way out of the core, through the Sun's radiative and convective zones. The photon undergoes many interactions along its path out of the Sun that greatly slow its progress. The photon finally escapes through the solar atmosphere some 170,000 years after it was created (and remember, it's traveling at the speed of light, 300,000 km/s) and emerges as some form of radiation.

The Sun's Influence on Earth

The consequences of the Sun's activity are all around us. Life on Earth exists, in part, because of warming from the Sun. Our upper atmosphere forms ozone to shield us from the most dangerous solar radiation. Our magnetic field is shaped and buffeted by the stream of particles blowing away from the Sun, and occasionally we see auroral displays as a result of variations in that stream.

The Sunspot Cycle

Probably the best-known solar variation (to the public at least) is the sunspot cycle. It varies on an approximately 11-year cycle. The stage in the cycle when sunspots are at their highest numbers is called the solar maximum. At this time, the frequency of solar flares and flame-like structures called prominences increases. This is also a time of increased auroral activity. Sunspot activity gradually slows down to a time of few spots, called the solar minimum. The sunspot cycle has been known to go as long as 15 years. For a period of about 70 years between 1645 and 1715, a period called the Maunder Minimum, the Sun showed no spots at all!

Overlaid atop the sunspot cycle is a variation in the polarity of the Sun's entire magnetic field. Apparently, for one 11-year cycle, the polarity of field lines connecting pairs of sunspots point one way. At solar minimum, the fields reverse, and point the opposite way for the next 11 years. Thus, there is a 22-year cycle overlaying the 11-year cycle.

The Causes of Sunspots

The causes of sunspots are not well understood, although the best hypothesis involves magnetic fields. The Sun's convective layer (which lies under its outer atmospheric layers) seems to generate weak magnetic fields that point in a north–south direction. At the same time, the Sun is rotating on its axis, although not all parts rotate at the same rate. The Equator rotates much faster than the polar regions do. This differential rotation stretches and drags the weak magnetic field lines. It has been hypothesized that these distorted field lines could be generating sunspots and other solar activity.

The Effects of Sunspots on Earth

For us on Earth, the arrival of a sunspot maximum (and its associated flares and mass ejections) means increased auroral activity. It also brings increased chances of power outages, damage to orbiting satellites, and disruption of communication systems. In March, 1989, a powerful solar flare knocked out communication with more than 2000 satellites. The hydroelectric system in the Canadian province of Quebec was shut down when its ground circuits were subjected to powerful currents resulting from the solar storm.

There is some evidence to suggest that sunspot variations may also affect Earth's weather. For example, during the Maunder Minimum, Europe and North America experienced some of their lowest temperatures. Historians called this time the "Little Ice Age." Some weather researchers have suggested that drought and storm cycles seem to occur along with the 11- or 22-year cycles of solar activity. These hypothesized links need much more data before they can be rigorously tested. Part of the problem is determining just what (if any) mechanism in the solar radiation cycle links solar and terrestrial weather cycles. There is a larger concern here: is global warming affected by solar radiation? We are still making links between that issue and the activities of humans, particularly with respect to pollution.

Recent Discoveries about the Sun

The last years of the 20th century and the opening years of the 21st are rewarding solar observers with unprecedented looks into the interior of the Sun. For example, the SOHO spacecraft has shown vast rivers of hot gas circulating inside the Sun. Other studies have revealed that the outer layer of the Sun is moving in a lazy 85-km/h flow toward the poles. Helioseismologists—scientists who study the sound waves that reverberate throughout the Sun in a manner reminiscent of the waves that earthquakes generate—have used their work to fashion complex and detailed models of the Sun's interior.

SOHO observations have also shown that our star is a violent place throughout all its cycles, and not just during times of maximum sunspot activity. Currently, solar astronomers are trying to understand the complex mechanisms that heat the Sun's corona. This region of rarefied gas that stretches out tens of thousands of kilometers from the Sun has a temperature of more than a million kelvins. It is not necessarily heated from below, because it lies just over the photosphere, which is a cool 5700 K.

Studying the Sun and Other Stars

As your students safely make their solar observations, they will be exposed to more than its light reflected onto a piece of paper. They will absorb its heat through their skin, and quite possibly get sunburned from exposure to its ultraviolet light. An interesting thought experiment, which will help your students get ready for **Activity 6**, is to consider that our eyes have evolved to be most sensitive to a very narrow range of sunlight—what we call visible light. This seems logical enough—but think about how the world would look to us if our eyes were sensitive only to the Sun's ultraviolet light.

The Sun is not alone in exhibiting spots and variations in its behavior. In recent years, astronomers using Hubble Space Telescope have observed spots on other stars. These spots likely formed by the same mechanism as sunspots form. Determining how it happens with our star will help astronomers understand the activities experienced by other stars. Because our star is part of a larger collection of stars, it is easy to assume that all stars will behave in the same manner.

Collecting starspot data and evidence of flares and coronal mass ejections from other stars will go a long way toward helping astronomers understand all the complex mechanisms that govern star life. Although most stars are too far away to affect Earth directly, understanding how they live can also help us predict which stars might be a long-term threat.

More Information – on the Web

Visit the *EarthComm* web site www.agiweb.org/earthcomm/ to access a variety of links to web sites that will help you to deepen your understanding of content and prepare you to teach this activity. Many of the sites also contain images that you can download.

Goals and Assessment

Clarify that the goals indicate what students should understand and be able to do as a result of the activity. Make sure students understand that Chapter Assessments are based upon these goals.

Goal	Location in Activity	Assessment Opportunity
Explore the structure of the Sun and describe the flow of solar energy in terms of reflection, absorption, and scattering.	**Digging Deeper; Check Your Understanding** Question 3	Responses are reasonable.
Understand that the Sun emits charged particles called the solar wind. Understand how this wind affects space weather.	**Digging Deeper; Check Your Understanding** Questions 1 – 2 **Understanding and Applying What You Have Learned** Question 1	Responses are reasonable and reflect an understanding.
Explain the effect of solar wind on people and communities.	**Digging Deeper; Check Your Understanding** Question 2 **Understanding and Applying What You Have Learned** Question 3	Responses are reasonable and reflect an understanding.
Understand the effects on Earth of sunspots, solar flares, and other solar activity.	**Investigate; Digging Deeper; Check Your Understanding** Question 1 **Understanding and Applying What You Have Learned** Questions 1 – 3	Sunspot activity and solar flare events are correctly graphed; axes are properly labeled. Responses to questions are reasonable and reflect an understanding.
Learn to estimate the chance that solar activity will affect your community.	**Understanding and Applying What You Have Learned** Questions 1 – 2	Predictions are reasonable.

Activity 5

The Sun and Its Effects on Your Community

Goals

In this activity you will:

- Explore the structure of the Sun and describe the flow of solar energy in terms of reflection, absorption, and scattering.

- Understand that the Sun emits charged particles called the solar wind, and how this wind affects "space weather."

- Explain the effect of solar wind on people and communities.

- Understand sunspots, solar flares, and other kinds of solar activities and their effects on Earth.

- Learn to estimate the chances for solar activity to affect your community.

Think about It

Every day of your life you are subjected to radiation from the Sun. Fortunately, the Earth's atmosphere and magnetic field provides protection against many of the Sun's strong outbursts.

- In what ways does solar radiation benefit you?
- In what ways can solar radiation be harmful or disruptive?

What do you think? Record your ideas about these questions in your *EarthComm* notebook. Be prepared to discuss your response with your small group and the class.

Activity Overview

Students plot sunspot activity by year from 1899 to 1998. They use their graphs to identify any patterns in sunspot activity. Students plot solar flare events by strength from 1978 to 2001. They compare their graphs of sunspot activity to their graphs of solar flare events to determine the relationship between the two. **Digging Deeper** takes a closer look at the Sun and its effects on Earth, including solar radiation, sunspots, and solar flares.

Preparation and Materials Needed

No special preparation is required for this investigation. Given the large number of points to be plotted in this investigation, you may wish to use this as an opportunity for students to practice using computer spreadsheet and graphics programs. Alternatively, to save class time, you may wish to plot the data ahead of class and provide copies to your students.

Materials
• Graph paper

Think about It

Student Conceptions

Students are likely to say that the benefits of solar radiation include warmth and light. They are likely to recognize that ultraviolet rays, which can cause skin cancer, are one of the harmful aspects of solar radiation.

Answer for the Teacher Only

With the exception of some deep-ocean ecosystems where the basis for the food chain starts with chemosynthetic bacteria (i.e., bacteria that derive their energy from chemicals in deep-ocean hydrothermal vents and seeps), all life on Earth derives its food and energy from the Sun. In other words, it is the Sun's energy that makes (almost) all life on Earth possible.

Ultraviolet radiation is an example of a harmful higher-energy radiation that comes from the Sun. There are other aspects of solar and cosmic radiation that are harmful to life. Fortunately, the Earth's magnetosphere and atmosphere shield the Earth's surface from much of this high-energy radiation, reducing the harmful effects of solar radiation.

Assessment Tool

Think about It Evaluation Sheet
Use this evaluation sheet to help students understand and internalize the basic expectations for the warm-up activity.

Earth System Evolution Astronomy

Investigate

1. Use the data in *Table 1* to construct a graph of sunspot activity by year.

   a) Plot time on the horizontal axis and number of sunspots on the vertical axis.

   b) Connect the points you have plotted.

   c) Look at your graph. Describe any pattern you find in the sunspot activity.

Table 1 Sunspot Activity 1899 to 1998							
Year	Number of Sunspots	Year	Number of Sunspots	Year	Number of Sunspots	Year	Number of Sunspots
1899	12.1	1924	16.7	1949	134.7	1974	34.5
1900	9.5	1925	44.3	1950	83.9	1975	15.5
1901	2.7	1926	63.9	1951	69.4	1976	12.6
1902	5.0	1927	69.0	1952	31.5	1977	27.5
1903	24.4	1928	77.8	1953	13.9	1978	92.5
1904	42.0	1929	64.9	1954	4.4	1979	155.4
1905	63.5	1930	35.7	1955	38.0	1980	154.6
1906	53.8	1931	21.2	1956	141.7	1981	140.4
1907	62.0	1932	11.1	1957	190.2	1982	115.9
1908	48.5	1933	5.7	1958	184.8	1983	66.6
1909	43.9	1934	8.7	1959	159.0	1984	45.9
1910	18.6	1935	36.1	1960	112.3	1985	17.9
1911	5.7	1936	79.7	1961	53.9	1986	13.4
1912	3.6	1937	114.4	1962	37.6	1987	29.4
1913	1.4	1938	109.6	1963	27.9	1988	100.2
1914	9.6	1939	88.8	1964	10.2	1989	157.6
1915	47.4	1940	67.8	1965	15.1	1990	142.6
1916	57.1	1941	47.5	1966	47.0	1991	145.7
1917	103.9	1942	30.6	1967	93.8	1992	94.3
1918	80.6	1943	16.3	1968	105.9	1993	54.6
1919	63.6	1944	9.6	1969	105.5	1994	29.9
1920	37.6	1945	33.2	1970	104.5	1995	17.5
1921	26.1	1946	92.6	1971	66.6	1996	8.6
1922	14.2	1947	151.6	1972	68.9	1997	21.5
1923	5.8	1948	136.3	1973	38.0	1998	64.3

The number of sunspots on the visible solar surface is counted by many solar observatories and is averaged into a single standardized quantity called the sunspot number. This explains the fractional values in the table.

E 48

Investigate

1. a) -b)

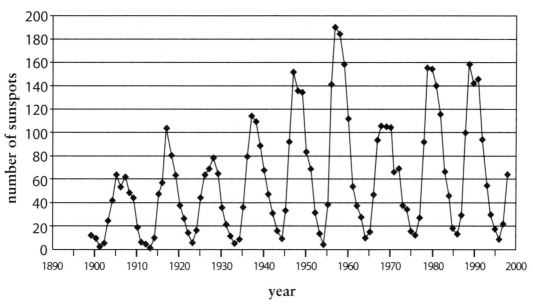

Sunspot Activity from 1899 to 1998

Teaching Tip

As the students are graphing the sunspots, circulate through the classroom to
ensure that the axes are correctly labeled. Assist with any problems graphing data.
You might suggest that students use a colored pencil to connect the points in order
to see the cycle more easily. Keep them focused on the activity by asking such
questions as:

• In what year was the smallest number of sunspots recorded?
• In what year was the largest number of sunspots recorded?
• What year showed the greatest increase?

Once they are finished, they can compare graphs and discuss results.

c) The number of sunspots follows a fairly consistent pattern, cycling between a
    peak about every 10 years and then falling to below 10 sunspots per year.

Chapter 1

2. *Table 2* contains a list of solar flares that were strong enough to disrupt terrestrial communications and power systems.

a) Plot the data from *Table 2* onto a histogram.

b) What pattern do you see in the activity of solar flares?

3. Compare the two graphs you have produced.

a) What pattern do you see that connects the two?

b) How would you explain the pattern?

Table 2 Strongest Solar Flare Events 1978–2001			
Date of Activity Onset	**Strength**	**Date of Activity Onset**	**Strength**
August 16, 1989	X20.0	December 17, 1982	X10.1
March 06, 1989	X15.0	May 20, 1984	X10.1
July 07, 1978	X15.0	January 25, 1991	X10.0
April 24, 1984	X13.0	June 09, 1991	X10.0
October 19, 1989	X13.0	July 09, 1982	X 9.8
December 12, 1982	X12.9	September 29, 1989	X9.8
June 06, 1982	X12.0	March 22, 1991	X9.4
June 01, 1991	X12.0	November 6, 1997	X9.4
June 04, 1991	X12.0	May 24, 1990	X9.3
June 06, 1991	X12.0	November 6, 1980	X9.0
June 11, 1991	X12.0	November 2, 1992	X9.0
June 15, 1991	X12.0		

The X before the number is a designation of the strongest flares.
Source: IPS Solar Flares & Space Service in Australia.

Reflecting on the Activity and the Challenge

In this activity you used data tables to plot the number of sunspots in a given year and to correlate strong solar-flare activity with larger numbers of sunspots. You found out that the number of sunspots varies from year to year in a regular cycle and that strong solar flares occur in greater numbers during high-sunspot years. In your **Chapter Challenge,** you will need to explain sunspots and solar flares, their cycles, and the effects of these cycles on your community.

2. **a)** Students may need guidance to help them to determine the best way to graphically represent the solar flare data. One possible example, shown below, is a stacked histogram. This method has the advantage of showing both the total number of events (total height of the stacked column) and the relative numbers of flares of different intensities.

Strongest Solar Flare Events 1978 to 1997

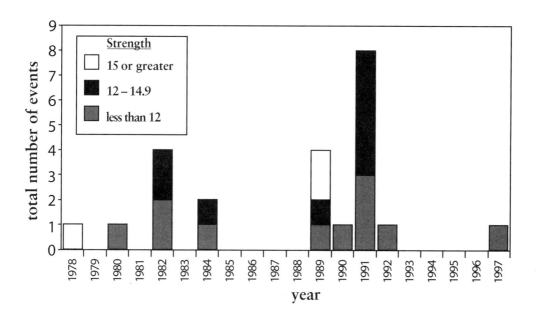

b) There is not really a distinct pattern in the activity of the solar flares.

3. **a)** Students should notice that the strongest solar flares correspond to the periods during which there were the greatest number of sunspots: for example, from 1989 to 1991.

    b) Answers will vary.

Assessment Tools

EarthComm **Notebook Entry-Checklist**
Use this checklist as a quick guide for student self-assessment and/or an opportunity to quickly score student work. Add further criteria specific to your classroom needs or to this particular investigation.

Investigate Notebook Entry-Evaluation Sheet
Point out the criteria listed on this evaluation sheet that are relevant to this particular investigation. Encourage students to internalize the criteria by making them part of your "assessment conversations" as you circulate around the classroom.

Reflecting on the Activity and the Challenge

This is a good opportunity to have your students discuss any communication problems they may have personally experienced because of sunspots. Discuss the possibility of a relationship between current sunspots and weather. For example, ask students whether they have noticed any correlations between temperatures in recent years and the occurrence of a solar maximum.

This is also an excellent opportunity to bring out the notion of cycles in Earth System Science. The analysis of data reveals patterns and cycles in nature, allows the correlation of processes on the Sun with events on Earth, and enables us to make predictions.

NOTES

Earth System Evolution Astronomy

Geo Words

photosphere: the visible surface of the Sun, lying just above the uppermost layer of the Sun's interior, and just below the chromosphere.

chromosphere: a layer in the Sun's atmosphere, the transition between the outermost layer of the Sun's atmosphere, or corona.

corona: the outermost atmosphere of a star (including the Sun), millions of kilometers in extent, and consisting of highly rarefied gas heated to temperatures of millions of degrees.

Digging Deeper

THE SUN AND ITS EFFECTS

Structure of the Sun

From the Earth's surface the Sun appears as a white, glowing ball of light. Like the Earth, the Sun has a layered structure, as shown in *Figure 1*. Its central region (the core) is where nuclear fusion occurs. The core is the source of all the energy the Sun emits. That energy travels out from the core, through a radiative layer and a convection zone above that. Finally, it reaches the outer layers: the **photosphere**, which is the Sun's visible surface, the **chromosphere**, which produces much of the Sun's ultraviolet radiation, and the superheated uppermost layer of the Sun's atmosphere, called the **corona**.

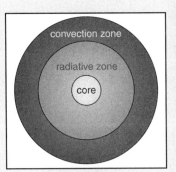

Figure 1 The layered structure of the Sun.

The Sun is the Earth's main external energy source. Of all the incoming energy from the Sun, about half is absorbed by the Earth's surface (see *Figure 2*). The rest is either:

• absorbed by the atmosphere, or
• reflected or scattered back into space by the Earth or clouds.

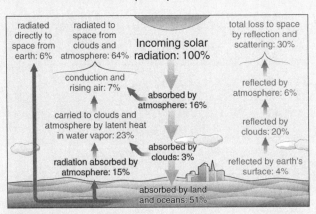

Figure 2 Schematic of Earth's solar energy budget.

Chapter 1

Digging Deeper

Assessment Opportunity

Use a brief quiz (or a class discussion) to assess your students' understanding of the main ideas in the reading and the activity. A few sample questions are provided below:

Question: Label the diagram of the Sun.

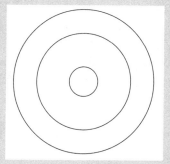

Answer: Student diagrams should be labeled like the one in *Figure 1*.

Question: Use an illustration to explain what happens to the energy that is incoming to the Earth from the Sun.
Answer: Student sketches should resemble *Figure 2*.

Question: What causes sunspots?
Answer: Sunspots form when magnetic field lines just below the Sun's surface are twisted and protrude through the solar photosphere.

Teaching Tip

Use **Blackline Master Astronomy 5.1, Layered Structure of the Sun** to make an overhead of the structure of the Sun, shown in *Figure 1* on page E50. Use this overhead to discuss how the structure of the Sun is different from the structure of the Earth. (Note that the figure has been provided as a BLM without labels, for use by students. You may wish to label your overhead permanently, or add labels to the diagram during your discussion with the students.)

Teaching Tip

Use **Blackline Master Astronomy 5.2, Earth's Solar Energy Budget** to make an overhead transparency of *Figure 2*. Incorporate this overhead into a discussion on the Earth's energy budget.

In this diagram of the Earth's solar energy budget, the lighter gray arrows in the center and right side of the diagram (shown as yellow and blue arrows in *Figure 2* on page E50 of the student text) indicate incoming solar radiation. The darker gray arrows on the left side of the diagram (shown as orange arrows in *Figure 2* on page E50 of the student text) indicate energy radiated by the Earth. Incoming solar radiation can be further subdivided into solar radiation that is absorbed by the Earth and its atmosphere (70%, represented by the center column of arrows) and solar radiation that is reflected back to space without being absorbed by the Earth (30%, represented by the right-hand column of arrows). The total energy radiated by the Earth is equal to 70% of the total incoming solar radiation. This is the same percentage of incoming solar radiation that is absorbed by the Earth.

NOTES

_____

_____

_____

_____

_____

_____

_____

_____

_____

_____

_____

_____

_____

_____

_____

_____

_____

_____

_____

_____

_____

Blackline Master Astronomy 5.1
Layered Structure of the Sun

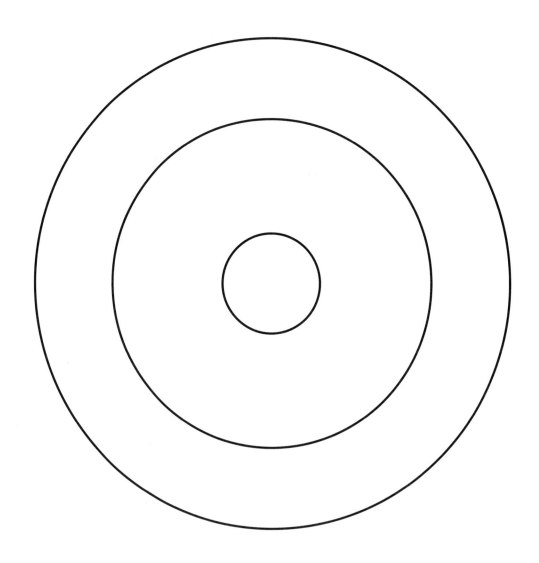

Blackline Master Astronomy 5.2
Earth's Solar Energy Budget

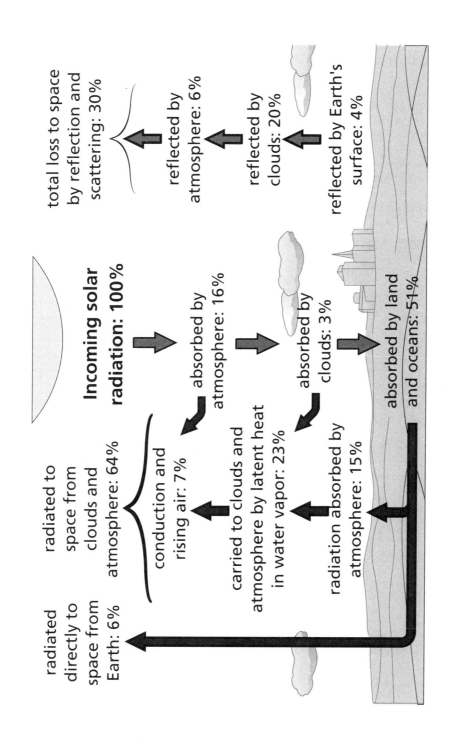

Molecules of dust and gas in the atmosphere interfere with some of the incoming solar radiation by changing its direction. This is called scattering, and it explains the blue color of the sky. The atmosphere scatters shorter visible wavelengths of visible light, in the blue range, more strongly than longer visible wavelengths, in the red and orange range. The blue sky you see on a clear day is the blue light that has been scattered from atmospheric particles that are located away from the line of sight to the Sun. When the Sun is low on the horizon, its light has to travel through a much greater thickness of atmosphere, and even more of the blue part of the spectrum of sunlight is scattered out of your line of sight. The red and orange part of the spectrum remains, so the light you see coming directly from the Sun is of that color. The effect is greatest when there is dust and smoke in the atmosphere, because that increases the scattering. The scattered light that makes the sky appear blue is what makes it possible for you to see in a shaded area.

Figure 3 Dust and smoke in the atmosphere enhance the beauty of sunsets.

Most of the sunlight that passes through the atmosphere reaches the Earth's surface without being absorbed. The Sun heats the atmosphere not directly, but rather by warming the Earth's surface. The Earth's surface in turn warms the air near the ground. As the Earth's surface absorbs solar radiation, it re-radiates the heat energy back out to space as infrared radiation. The wavelength of this infrared radiation is much longer than that of visible light, so you can't see the energy that's re-radiated. You can feel it, however, by standing next to a rock surface or the wall of a building that has been heated by the Sun.

E 51

Teaching Tip

Extremely small particles of dust and smoke in the atmosphere scatter the shorter-wavelength blue light more strongly than the longer-wavelength orange and red light, making the Sun appear orange or red (see *Figure 3* on page E51). The effect is most pronounced when the Sun is low in the sky, because then the Sun's rays travel a longer distance through the atmosphere. The same effect is caused by the molecules of the air, but the presence of particles of dust and smoke enhances the effect.

The coloring of the clouds during a beautiful sunset is caused by the reflection of the reddish or orangish sunlight that is still shining on the clouds. The effect is most spectacular when there are high clouds near the observer but no clouds low on the horizon to block the low-slanting rays of the Sun. The colorful sunlight then shines on the undersides of the high clouds.

Teaching Tip

Figure 4 on page E52 shows the relationship between the predicted sunspot activity (smooth solid curve) and the actual number of sunspots (jagged solid curve). The dashed lines above and below the predicted activity curve represent the 95% confidence interval for the predicted activity. As can be seen, most of the actual data fall within this error envelope. Sunspot activity goes through cycles of greater and lesser numbers of sunspots. Predicting the behavior of a sunspot cycle is fairly reliable once the cycle is underway (about 3 years after the minimum in sunspot number occurs), but predictions made before or near the occurrence of the minimum in sunspot number are less dependable. Further refinement of these early predictions continues and is important for the planning of satellite orbits and space missions, which often requires knowledge of solar activity levels years in advance.

Use **Blackline Master Astronomy 5.3, Cycle 23 Sunspot Number Prediction** of *Figure 4* on page E52 to make an overhead transparency. This overhead could be incorporated into a discussion on sunspots. This overhead also provides an opportunity to discuss the uncertainty of predictions and the significance of the 95% confidence interval, which is shown on this figure by the two dotted curves.

Earth System Evolution Astronomy

Geo Words

albedo: the reflective property of a non-luminous object. A perfect mirror would have an albedo of 100% while a black hole would have an albedo of 0%.

The reflectivity of a surface is referred to as its **albedo**. Albedo is expressed as a percentage of radiation that is reflected. The average albedo of the Earth, including its atmosphere, as would be seen from space, is about 0.3. That means that 30% of the light is reflected. Most of this 30% is due to the high reflectivity of clouds, although the air itself scatters about 6% and the Earth's surface (mainly deserts and oceans) reflects another 4%. (See *Figure 2* on page E50.) The albedo of particular surfaces on Earth varies. Thick clouds have albedo of about 0.8, and freshly fallen snow has an even higher albedo. The albedo of a dark soil, on the other hand, is as low as 0.1, meaning that only 10% of the light is reflected. You know from your own experience that light-colored clothing stays much cooler in the Sun than dark-colored clothing. You can think of your clothing as having an albedo, too!

The Earth's Energy Budget

The amount of energy received by the Earth and delivered back into space is the Earth's energy budget. Like a monetary budget, the energy resides in various kinds of places, and moves from place to place in various ways and by various amounts. The energy budget for a given location changes from day to day and from season to season. It can even change on geologic time scales. Daily changes in solar energy are the most familiar. It is usually cooler in the morning, warmer at midday, and cooler again at night. Visible light follows the same cycle, as day moves from dawn to dusk and back to dawn again. But overall, the system is in balance. The Earth gains energy from the Sun and loses energy to space, but the amount of energy entering the Earth system is equal to the amount of energy flowing out, on a long-term average. This flow of energy is the source of energy for almost all forms of life on Earth. Plants capture solar energy by photosynthesis, to build plant tissue. Animals feed on the plants (or on one another). Solar energy creates the weather, drives the movement of the oceans, and powers the water cycle. All of Earth's systems depend on the input of energy from the Sun. The Sun also supplies most of the energy for human civilization, either directly, as with solar power and wind power, or indirectly, in the form of fossil fuels.

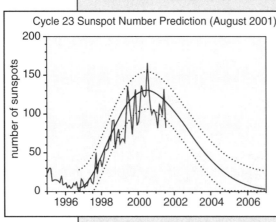

Figure 4 The jagged line represents the actual number of sunspots; the smooth dark line is the predicted number of sunspots.

EarthComm

Blackline Master Astronomy 5.3
Cycle 23 Sunspot Number Prediction

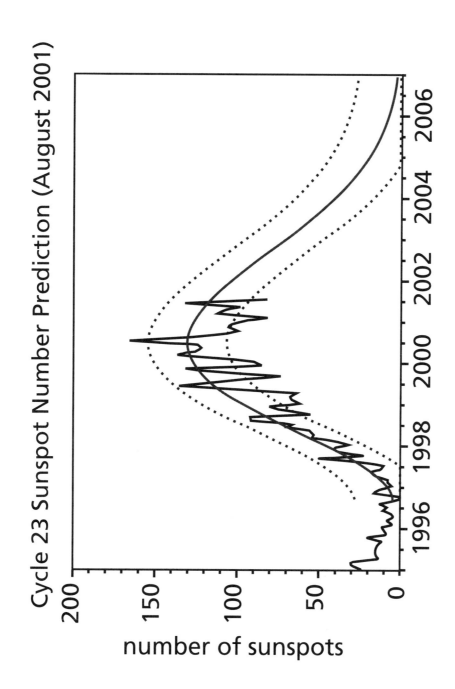

Harmful Solar Radiation

Just as there are benefits to receiving energy from the Sun, there are dangers. The ill effects of sunlight are caused by ultraviolet (UV) radiation, which causes skin damage. The gas called ozone (a molecule made up of three oxygen atoms) found in the upper atmosphere shields the Earth from much of the Sun's harmful UV rays. The source of the ozone in the upper atmosphere is different from the ozone that is produced (often by cars) in polluted cities. The latter is a health hazard and in no way protects you. Scientists have recently noted decreasing levels of ozone in the upper atmosphere. Less ozone means that more UV radiation reaches Earth, increasing the danger of Sun damage. There is general agreement about the cause of the ozone depletion. Scientists agree that future levels of ozone will depend upon a combination of natural and man-made factors, including the phase-out, now under way, of chlorofluorocarbons and other ozone-depleting chemicals.

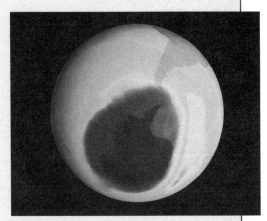

Figure 5 Depletion in the ozone layer over Antarctica. Rather than actually being a hole, the ozone hole is a large area of the stratosphere with extremely low concentrations of ozone.

Sunspots and Solar Flares

Sunspots are small dark areas on the Sun's visible surface. They can be as small as the Earth or as large as Uranus or Neptune. They are formed when magnetic field lines just below the Sun's surface are twisted and poke through the solar photosphere. They look dark because they are about 1500 K cooler than the surrounding surface of the Sun. Sunspots are highly magnetic. This magnetism may cause the cooler temperatures by suppressing the circulation of heat in the region of the sunspot.

Sunspots last for a few hours to a few months. They appear to move across the surface of the Sun over a period of days. Actually, the sunspots move because the Sun is rotating. The number of sunspots varies from year to year and tends to peak in 11-year cycles along with the number of dangerously strong solar flares. Both can affect systems here on Earth. During a solar

E 53

Teaching Tip

The ozone hole, as shown in *Figure 5* on page E53, is a well-defined, large-scale destruction of the ozone layer over Antarctica that occurs each Antarctic spring. The use of the word "hole" is a bit misleading. The hole is in reality a significant thinning or reduction in ozone concentrations, which results in the destruction of up to 70% of the ozone normally present in the upper atmosphere over Antarctica. As of September 2000, the "hole" covered approximately 28.3 million km^2, an area over three times larger than the United States. In the past few years, the size of the ozone hole has stabilized, presumably owing to the great decrease in emission of human-made ozone-destroying refrigerant gases by international treaty.

Earth System Evolution Astronomy

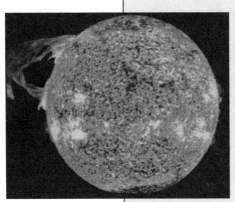

Figure 6 A solar flare.

Geo Words

plasma: a state of matter wherein all atoms are ionized; a mixture of free electrons and free atomic nuclei.

ionosphere: the part of the Earth's atmosphere above about 50 km where the atoms are significantly ionized and affect the propagation of radio waves.

ion: an atom with one or more electrons removed (or added), giving it a positive (or negative) charge.

flare like the one shown in *Figure 6*, enormous quantities of ultraviolet, x-ray, and radio waves blast out from the Sun. In addition, protons and electrons stream from flares at 800 km/hr. These high-radiation events can be devastating to Earth-orbiting satellites and astronauts, as well as systems on the ground. In 1989 a major solar flare created electric currents that caused a surge of power that knocked out a power grid in Canada, leaving hundreds of thousands of people without power. More recently, in 1997, radiation from a flare affected an Earth-orbiting satellite that carried telecommunications traffic. For at least a day people whose beeper messages went through that satellite had no service.

The flow of charged particles (also called a **plasma**) from the Sun is called the solar wind. It flows out from the solar corona in all directions and is responsible for "space weather"—the environment outside our planet. Like severe storms in our atmosphere, space weather can cause problems for Earth systems. Strong outbursts in this ongoing stream of charged particles can disrupt radio signals by disturbing the upper layers of the atmosphere. The sounds of your favorite short-wave radio station or the signals sent by a ham radio operator travel as radio waves (a form of electromagnetic radiation). These signals travel around the Earth by bouncing off the **ionosphere**, a layer of the atmosphere 80 to 400 km above the Earth's surface. The ionosphere forms when incoming solar radiation blasts electrons out of the upper-atmosphere gases, leaving a layer of electrons and of charged atoms, called **ions**. The ionosphere acts like a mirror, reflecting a part of the radio waves (AM radio waves in the 1000 kHz range) back to Earth.

Solar flares intensify the solar wind, which makes the ionosphere thicken and strengthen. When this happens, radio signals from Earth are trapped inside the ionosphere. This causes a lot of interference. As discussed above, solar activity can also be a problem for satellite operations. Astronauts orbiting the Earth and people aboard high-flying aircraft—particularly those who fly polar routes, where exposure to radiation may be greatest—also have cause to worry about space weather. To provide up-to-date information about current solar activity, the United States government operates a Space Environment Center Web site called "Space Weather Now."

At least one effect of space weather is quite wonderful. When the solar wind encounters the Earth's magnetic field, it excites gases in the Earth's atmosphere, causing them to glow. The charged particles from the solar wind end up in an oval-shaped area around the Earth's magnetic poles. The result

E 54

Teaching Tip

Solar flares, like the one pictured in *Figure 6* on page E54, extend out to the layer of the Sun called the corona. This is the outermost atmosphere of the Sun. The energy released during these sudden, rapid, and intense events can be millions of times greater than the energy released by a volcanic explosion on Earth.

Activity 5 The Sun and Its Effects on Your Community

is a beautiful display called an **aurora**, seen in *Figure 7*. People who live in northern areas see auroras more often than those who live near the Equator do. During periods of heavy solar activity, however, an aurora can be seen as far south as Texas and New Mexico. Auroras are often called the northern lights (aurora borealis) or southern lights (aurora australis). From the ground, they often appear as green or red glows, or shimmering curtains of white, red, and green lights in the sky.

Collecting Data about the Sun

How do astronomers collect data about the Sun? From the ground, they use solar telescopes—instruments outfitted with special sensors to detect the different kinds of solar activity. There are dozens of solar telescope sites around the world. They include the Sacramento Peak Solar Observatory in New Mexico, the McMath Solar telescope in Arizona, and the Mount Wilson solar observatory in California. From space, they study the Sun using orbiting spacecraft like the Yohkoh satellite (*Yohkoh* is the Japanese word for "sunbeam"). Other missions include the Transition Region and Coronal Explorer (TRACE), the Ulysses Solar-Polar mission, the Solar and Heliospheric Observatory, the GOES satellites, and many others. These spacecraft are equipped with detectors sensitive to x-rays, radio waves, and other wavelengths of radiation coming from the Sun. In this way, scientists keep very close track of solar activity and use that information to keep the public informed of any upcoming dangers.

Some scientists theorize that sunspot cycles affect weather on Earth. They think that during times of high sunspot activity, the climate is warmer. During times of no or low sunspot activity, the climate is colder. A sharp decrease in sunspots occurred from 1645 to 1715. This period of lower solar activity, first noted by G. Sporer and later studied by E.W. Maunder, is called the Maunder Minimum. It coincided with cooler temperatures on Earth, part of a period now known as the "Little Ice Age." Similar solar minimums occurred between 1420–1530, 1280–1340, and 1010-1050 (the Oort minimum). These periods preceded the discovery of sunspots, so no correlation between sunspots and temperature is available. Solar astronomers number the solar cycles from one minimum to the next starting with number one, the 1755–1766 cycle. Cycle 23 peaked (was at a maximum) in the year 2000. (See *Figure 4* on page E52.) There is still much debate about the connection between sunspot cycles and climate.

Figure 7 The aurora borealis or northern lights light up the sky in the Northern Hemisphere.

Geo Words

aurora: the bright emission of atoms and molecules near the Earth's poles caused by charged particles entering the upper atmosphere.

Check Your Understanding

1. How do solar flares interfere with communication and power systems?

2. In your own words, explain what is meant by the term "solar wind." How does the Sun contribute to "space weather?"

3. Describe the Earth's energy budget.

EarthComm

Teaching Tip

Figure 7 on page E55 shows the aurora borealis as viewed from space. This beautiful display is not restricted to the Earth. Images from the Galileo space probe have shown an aurora on Jupiter. Have any of your students ever seen the aurora?

Check Your Understanding

1. Solar flares produce more intense solar wind—a flow of charged particles. The solar wind thickens and strengthens the ionosphere. Both radio waves from Earth and those traveling to Earth become trapped within the ionosphere and cause interference (static); solar flares sometimes even black out radio communications.

2. The solar wind is a flow of charged particles from the Sun. The outward flow of these charged particles from the Sun is responsible for "space weather."

3. The balance between radiant energy received by the Earth and radiant energy delivered back into space constitutes the Earth's energy budget. The energy budget for a given location changes from day to day and from season to season. Over the entire surface of the Earth and averaged over a period of years, however, the energy budget is very nearly in balance.

Assessment Tool

Check Your Understanding Notebook Entry-Evaluation Sheet
Use this sheet to evaluate the extent to which students understand the key concepts explored in **Activity 5** and explained in the **Digging Deeper** reading section, and to evaluate the students' clarity of expression.

Earth System Evolution Astronomy

Understanding and Applying What You Have Learned

1. Study the graph that you made showing sunspot activity. You have already determined that sunspot activity occurs in cycles. Using graph paper, construct a new graph that predicts a continuation of the cycle from 2001 to 2015. Indicate which years you think would see increased solar-flare activity and more dangerous "space weather."

2. The latest sunspot maximum occurred in 2001. Using the data from your sunspot-activity data table, predict the next sunspot minimum.

3. Make lists of the possible consequences of solar flares to the following members of your community: an air traffic controller, a radio station manager, and the captain of a ship at sea. Can you think of other members of your community that would be affected by solar activity?

4. You have read that Earth's albedo is about 0.30.

   a) In your own words, describe what this means.
   b) Is the Earth's albedo constant? Why or why not?
   c) How does changing a planet's albedo change a planet's temperature? Why does this occur?
   d) If Earth's albedo was higher, but Earth was farther from the Sun, could the Earth have the same temperature? Why or why not?

Preparing for the Chapter Challenge

You have been asked to help people in your community to understand how events from outside the Earth affect their daily lives. Write a short paper in which you address the following questions:

1. How has the Sun affected your community in the past?

2. How has the Sun affected you personally?

3. How might the Sun affect your community in the future?

4. What are some of the benefits attained from a study of the Sun?

5. What are some of the problems caused by sunspots and solar flares?

6. Explain how auroras are caused. Explain also why they can or cannot be viewed in your community.

7. Compare the chances of dangerous effects from the Sun with the chances of an impact event affecting the Earth.

E 56

Understanding and Applying What You Have Learned

1. Predicted extrapolation of the graph should show a continued increase in the number of sunspots for another two years, then a gradual decrease to a solar minimum at about the year 2006, whereupon the number will increase again. Students should predict the years of the maximum and minimum, but you should ask them to consider whether they have enough data to predict the actual highs and lows with confidence.

Sunspot Activity from 1899 to 1998

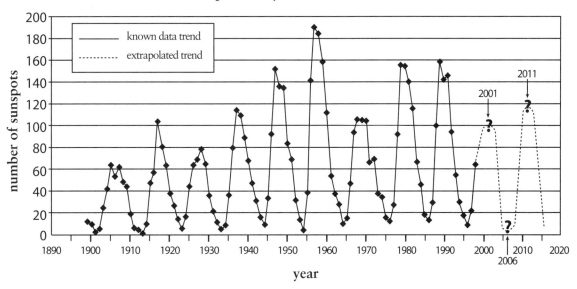

2. The next sunspot minimum should occur around the year 2006 or 2007, 10 to 11 years after the last minimum (in 1996).

3. Students should be able to predict that increased solar activity would interfere with communications, so that an air traffic controller might have difficulty communicating with pilots and an air tragedy might occur. The radio station manager would probably receive a lot of complaints about listeners' reception. And the captain of a ship would have difficulty communicating with the harbor at his or her destination. Accept any reasonable answers for others who would be affected.

4. a) An albedo of 0.30 means that 30% of the light that reaches the Earth is reflected back into space.

    b) No, the Earth's albedo is not constant because features on the surface of the Earth change. For example, during periods of glaciation, increased ice cover means that more solar radiation is reflected back to space.

c) Changing a planet's albedo will affect the planet's temperature. This is because a higher albedo means that more light is reflected and less light is absorbed. This translates to lower amounts of absorbed energy and lower temperatures.

d) No, the Earth would not have the same temperature. The two effects would *supplement* each other rather than *counteract* each other, so the Earth would receive less solar radiation. This would almost certainly result in lower average global temperature. If, on the other hand, the Earth's albedo was lower and the Earth was farther from the Sun, then the global average temperature might come out to be about the same, depending upon the relative magnitudes of the changes.

Preparing for the Chapter Challenge

Students should reread the **Chapter Challenge** and consider how they can use what they learned in **Activity 5** to help people in their community understand the benefits and hazards of sunspots and solar cycles. Answers to the questions will vary depending upon the community in which you live.

For assessment of this section (see **Assessment Rubric for Chapter Challenge on Astronomy** on pages 12 and 13), students should include a discussion on:
- how solar activities influence the Earth
- the hazardous and beneficial effects of solar radiation
- the Sun's composition and structure and that of other stars
- why extraterrestrial influences on the community are a natural part of Earth system evolution

NOTES

Inquiring Further

1. **Viewing sunspots**

   If you have a telescope, you can view sunspots by projecting an image of the Sun onto white cardboard. Never look directly at the Sun, with or without a telescope. Stand with your back to the Sun, and set up a telescope so that the large (front) end is pointing toward the Sun and the other end is pointing toward a piece of white cardboard. You should see a projection of the Sun on the cardboard, including sunspots. If you map the positions of the sunspots daily, you should be able to observe the rotation of the Sun over a couple of weeks. Use the *EarthComm* web site to locate good science sites on the Internet that show daily images of solar activity. Search them out and compare your observations of sunspots to what you see from the large observatories.

    Work with an adult during this activity. Do not look at the bright image for long periods of time.

2. **Aurorae**

   Have people in your community ever seen the northern lights? Even if your community is not very far north, do some research to see if the auroras have ever been spotted from your community.

3. **Solar radiation and airplanes**

   Periods of sunspot maximum increase the dosage of radiation that astronauts and people traveling in airplanes receive. Do some research on how much radiation astronauts receive during sunspot minima and maxima. How much radiation do airplane passengers receive? How do the amounts compare to the solar radiation you receive at the Earth's surface? How do scientists balance safety with the issue of the extra weight that would be added to aircraft, spacecraft, or space suits to provide protection?

4. **The hole in the ozone layer**

   People who live near the South Pole of the Earth are at risk for increased ultraviolet exposure from the Sun. This is due to a thinning in the atmosphere called the ozone hole. Research this ozone hole. Is there a northern ozone hole? Could these ozone holes grow? If so, could your community be endangered in the future?

5. **History of science**

   Research the life of British physicist Edward Victor Appleton, who was awarded the Nobel Prize in physics in 1947 for his work on the ionosphere. Other important figures in the discovery of the properties of the upper atmosphere include Oliver Heaviside, Arthur Edwin Kennelly, F. Sherwood Rowland, Paul Crutzen, and Mario Molina.

E 57

Inquiring Further

1. Viewing sunspots

If you have access to a telescope, this activity provides an opportunity for students to safely see that sunspots do exist. Alternatively, students can visit the *EarthComm* web site to view daily images of sunspot activity.

2. Aurorae

Encourage students to connect the cultural and historical aspects associated with the viewing of aurora with the science behind these events. During times of intense auroral activity, the aurora can sometimes be readily seen down well into the mid-latitudes; so it is possible that your students may have seen the aurora even if they have not travelled to high altitudes.

3. Solar radiation and airplanes

Astronauts receive some protection from solar radiation as long as they remain within the spacecraft, but that protection is not complete. Little is known about the extent to which they are being harmed from solar and cosmic radiation. There is data that indicates that exposure to these harmful forms of radiation can cause health problems. For example, radiation exposure can produce a number of significant changes in various elements of the blood, making an individual more susceptible to disease. Also, ionizing radiations of the kind found in space can produce significant damage to the lens of the eye. Efforts are underway to find usable forms of radiation shielding for spacecraft, but a suitable solution has not yet been found.

The radiation emitted from solar flares can give passengers in airplanes and astronauts in spacecraft a dose of radiation equivalent to a medical x-ray. However, astronauts are at a greater risk when they step outside their spacecraft. Spacewalks, like those that take place during work on the International Space Station, can expose astronauts to radiation levels equivalent to several hundred chest x-rays—enough to cause acute radiation sickness. Efforts are currently underway to improve predictions of space weather, to make sure that astronauts are not out of the spacecraft during extremely strong solar flares.

4. The hole in the ozone layer

The Southern Hemisphere ozone hole is a well-defined, large-scale destruction of the ozone layer over Antarctica that occurs each Antarctic spring. The use of the word "hole" is a bit misleading, because the hole is really a significant thinning, or reduction in ozone concentrations, which results in the destruction of up to 70% of the ozone normally found over Antarctica. As of September 2000, the "hole" covered approximately 28.3 million km^2, an area over three times larger than the United States.

The loss of ozone has not been restricted to the Antarctic: there is evidence of an ozone decrease over the heavily populated northern mid-latitudes (30°N to 60°N).

Unlike the sudden and nearly total loss of ozone over Antarctica at certain altitudes, however, the loss of ozone in mid-latitudes is much less and much slower—only a few percent per year.

In the Arctic regions, atmospheric circulation brings warmer air to the area. The relative warmth of the Arctic is the main reason why a similar ozone hole doesn't form over the North Pole. Students can learn more about the causes and consequences of the ozone hole, and what is being done to limit ozone depletion, by visiting the *EarthComm* web site.

5. History of science
Encourage investigation into the history of science. Important work by scientists in the first half of the 20th century was instrumental to understanding and improving radio communication, and led to the discovery of the ionosphere.

Edward Victor Appleton was born in Bradford, England, on September 6, 1892. Appleton devoted much of his life to researching scientific problems in atmospheric physics, using mainly radio techniques. In 1924, Appleton began a series of experiments in which he shot waves upward to see whether they were reflected. Indeed, the waves were reflected, proving the existence of the reflective layer in the upper atmosphere now called the ionosphere. Appleton was also able to measure the time taken by the waves to travel to the upper atmosphere and back, which allowed him to determine that the layer was 60 miles above ground. The method Appleton used was what is now called frequency-modulation radar. The ionosphere was thus the first "object" detected by radio location. This led to development of radio research and to radar, a military invention of great importance in World War II.

NOTES

ACTIVITY 6 — THE ELECTROMAGNETIC SPECTRUM AND YOUR COMMUNITY

Background Information

Human beings can't help being prejudiced toward visible light because our eyes evolved to be sensitive to just this narrow part of all the wavelengths emitted by the Sun. Our ears are oriented to yet another narrow wavelength range. Creatures that have eyes (and ears) oriented to other wavelengths flourish on this planet. Because we know that these other electromagnetic waves exist, we put them to work to allow us to "see" things our eyes can't perceive.

We have built and deployed thousands of instruments sensitive to, or using, parts of the electromagnetic spectrum other than the part that corresponds to visible light. We cook with microwaves and peer inside the human body with x-ray technology. We use radio waves to communicate with each other. Astronomers routinely study the universe in x-ray and gamma-ray wavelengths, as well as the radio, ultraviolet, and infrared ranges. Atop volcanoes around the world, volcanologists use infrared technology to gauge the rising temperatures of subsurface rivers of lava and the cooling rate of newly deposited molten rock.

Electomagnetic radiation surrounds us in all its forms. Can it pose a threat to us as well? If so, what is the nature of this threat, and how can we harness light's power to serve us as well? In **Activity 6**, your students will be observing the wavelengths of radiation that our Sun emits in order to assess where electromagnetic radiation stands in the hierarchy of threats and benefits that face their community. In the process, they will begin to explore the larger universe through their growing familiarity with the electromagnetic spectrum. They will learn that the full story of their community, the Earth, and the universe in which it resides can be told only using the electromagnetic spectrum as a tool of exploration.

Understanding the Electromagnetic Spectrum
As your students work through the investigations in this unit, they will move along some historic scientific and cultural pathways. **Digging Deeper** traces the growing understanding of the electromagnetic spectrum that began with Isaac Newton's first attempts to split a beam of light through a prism. The reading sums up its look at the electromagnetic spectrum by describing today's orbiting satellites that see the universe through x-ray and ultraviolet eyes. It also notes the ground-based dishes that probe the universe looking for the radio waves emitted from distant objects. Students are also welcomed into the complex world of the spectroscopist, partly in preparation for **Activity 7**, when they will explore the uses of the spectrum to classify stars.

For spectroscopists, the word spectrum has more than one meaning. The first meaning is the one your students have been studying—the complete range of radiation from gamma rays to radio. Theoretically, there is no end to the electromagnetic spectrum: as long as something has energy, it emits radiation. The second usage of spectrum refers to data gathered about the actual wavelengths of light being studied by a researcher. A "picture" of a spectrum looks somewhat similar to the spectrum your students will examine in the investigation.

How Scientists Use the Electromagnetic Spectrum

In many cases, the wavelengths of interest to scientists are tiny portions of the full range of radiation emitted or absorbed by an object. For example, astronomers used the Goddard High Resolution Spectrograph (GHRS) on the Hubble Space Telescope to focus on a narrow range of wavelengths emitted by the star Chi Lupi. These are emissions from atomic hydrogen in the atmosphere of the star. The specific range began at 193.71 nm and ended at 193.91 nm. What were the astronomers looking for in such a small range of the star's spectrum? The GHRS was sensitive to tiny variations in this wavelength of ultraviolet light from Chi Lupi. Those tiny variations told the scientists which chemical elements existed in the star's atmosphere during the time they were observing. As it turns out, the star's atmosphere contained traces of arsenic, chromium, iron, germanium, manganese, nickel, platinum, ruthenium, and zirconium!

How did the astronomers know what they were looking at in this stellar spectrum? Laboratory spectroscopists study the spectrum of each element as it burns under tightly controlled lab conditions. This procedure allows them to construct a chemical fingerprint for the elements. When astronomers compared the spectrum of Chi Lupi to laboratory-standard spectra for elements that emit light in that narrow wavelength range, they knew immediately what they had. This, in turn, helped them formulate a model for the star's atmospheric mixing and rotation rate.

Let's explore the spectrum your students are studying as it is portrayed in *Figure 5* of **Activity 6**. This is actually a snapshot of the wavelengths of light between 152.0 and 158.0 nm emitted from a star called Melnick 42. There are two interesting things going on in this little slice of data. First, the largest peak to the right of center is the spectral signature of ionized carbon flowing away from Melnick 42. On the left side of the spectrum, there is a huge dip in intensity. This indicates that something is absorbing light, and that it is traveling in our direction at about 2900 km/s. The astronomers studying this star know that it is a 100-solar-mass star. It is losing its mass in all directions through a very strong stellar wind; basically, it is ejecting shells of material, and those shells are absorbing light as they move away rapidly.

The Tools of Spectroscopy

Scientists use the tools of spectroscopy to study every kind of object in the universe, including other planets, the Sun, stars, and galaxies. At Jupiter and Saturn, for example, auroral displays emit ultraviolet light that has been studied by the Hubble Space Telescope. Researchers using infrared detectors routinely study the surface of Mars during dust storms to determine changes in the planet's atmospheric and surface temperatures. Infrared studies of distant dust clouds can reveal the presence of newly forming stars hidden within protected starbirth nurseries. Energetic explosions of stars are tracked for months using ultraviolet detectors, and long-term observations of galactic cores are revealing the telltale high-energy signatures of black holes. Detectors on board the Cosmic Background Explorer satellite detected the fading radiation left over from the Big Bang, and instruments on the

Compton Gamma-Ray Observatory studied radiation from some of the most energetic sources in the universe.

Electromagnetic Radiation—Beneficial or Harmful?

Considering the many uses of electromagnetic radiation here on Earth, it would seem that the benefits far outweigh the harm we might face from it. It's true that in the range of dangers we face, harmful radiation from a distant star is not as big a threat as the dangers posed by solar flares at the height of a sunspot cycle. It's important to put these things into perspective—to construct a "threat" continuum to help us sort out which dangerous events are more likely to occur than others.

The Dangers of Electromagnetic Radiation

By itself, the electromagnetic spectrum is not a tool for harm or danger. It is a diagnostic tool, a way of classifying wavelengths of light and helping us understand the activities that generate them.

However, danger from radiation is very real. As your students learned in **Activity 5**, the Sun is our greatest provider of heat and light, but also of harmful radiation. Our atmosphere shields out the worst effects, and we take precautions when we travel to areas where we might be exposed to higher levels of ultraviolet radiation—for example, near the South Pole or on tops of high mountains.

The chances of being irradiated at high altitude does not stop at the mountains or affect only astronauts in orbit around the Earth. High-flying aircraft routinely move through regions where Earth's magnetosphere and atmosphere cannot protect from high-energy radiation. Most flight crews who fly polar routes, for example, are required to wear dosimeters to track how much radiation they have been exposed to during their routine flights. Most passengers are not exposed to enough radiation to pose a threat, but what if you or your students ended up regularly flying polar routes as part of a business? Would it make sense to carry a dosimeter if you flew a polar route once a week? Several times a month? Several times a week?

Your students may have heard or read about the radiation dangers of cellular telephones, or the hazards that might threaten those who live near high-power transmission lines. Pregnant women are routinely shielded during x-rays, and technicians who perform magnetic resonance imaging and CAT scans are required to protect themselves from constant exposure to any possible radiation from their instruments.

Assessing the Danger

What are your students to make of this information? Their first line of defense is to learn as much as they can about the instruments involved—the wavelength ranges in which they emit most strongly, the dangers they pose, and the benefits they provide. Complete knowledge will allow them to assess dangers in a logical way.

In some cases, researchers are still gathering data about the long-term effects of radiation on living organisms. In other cases—as in the claim that cellular phones emit harmful radiation—stories are tricky to verify and not enough information is available to make a wise decision about using such technology.

Science does not yet have all the answers about the benefits and hazards of the electromagnetic radiation that our technologies use, but it has made great strides in understanding how and why it is emitted. In turn, we have learned much about how to harness that radiation for our own use.

More Information – on the Web
Visit the *EarthComm* web site www.agiweb.org/earthcomm/ to access a variety of links to web sites that will help you to deepen your understanding of content and prepare you to teach this activity. Many of the sites also contain images that you can download.

Goals and Assessment

Clarify that the goals indicate what students should understand and be able to do as a result of the activity. Make sure students understand that Chapter Assessments are based upon these goals.

Goal	Location in Activity	Assessment Opportunity
Explain electromagnetic radiation and the electromagnetic spectrum in terms of wavelength, speed, and energy.	**Investigate** Part B **Digging Deeper; Check Your Understanding** Questions 1 – 2	Responses to questions are reasonable, and closely match those given in Teacher's Edition.
Investigate the different instruments astronomers use to detect different wavelengths in the electromagnetic spectrum.	**Investigate** Part A **Investigate** Part C **Digging Deeper; Check Your Understanding** Questions 4 – 5	Observations of spectrum are correct. Research on mission or instrument is thorough and accurate. Questions are answered correctly.
Understand that the atoms of each of the chemical elements have a unique spectral fingerprint.	**Digging Deeper**	
Explain how electromagnetic radiation reveals the temperature and chemical makeup of objects like stars.	**Digging Deeper**	
Understand that some forms of electromagnetic radiation are essential and beneficial to us on Earth, while others are harmful.	**Digging Deeper; Check Your Understanding** Question 3	Observations of spectrum are correct. Research on mission or instrument is thorough and accurate. Questions are answered correctly.

NOTES

Earth System Evolution Astronomy

Activity 6

The Electromagnetic Spectrum and Your Community

Goals

In this activity you will:

- Explain electromagnetic radiation and the electromagnetic spectrum in terms of wavelength, speed, and energy.

- Investigate the different instruments astronomers use to detect different wavelengths in the electromagnetic spectrum.

- Understand that the atoms of each of the chemical elements have a unique spectral fingerprint.

- Explain how electromagnetic radiation reveals the temperature and chemical makeup of objects like stars.

- Understand that some forms of electromagnetic radiation are essential and beneficial to us on Earth, and others are harmful.

Think about It

Look at the spectrum as your teacher displays it on the overhead projector. Record in your *EarthComm* notebook the colors in the order in which they appear. Draw a picture to accompany your notes.

- What does a prism reveal about visible light?
- The Sun produces light energy that allows you to see. What other kinds of energy come from the Sun? Can you see them? Why or why not?

What do you think? Record your ideas in your *EarthComm* notebook. Be prepared to discuss your responses with your small group and the class.

Activity Overview

Students use a spectroscope to observe the visible part of the electromagnetic spectrum under natural sunlight, fluorescent light, and incandescent light. They create plots of the electromagnetic spectrum to get a sense of what part of the spectrum is visible to us, and to understand the proportion of the other types of electromagnetic radiation in the spectrum. Students then research specific missions and instruments to learn how astronomers use electromagnetic radiation to study objects and events in space. The **Digging Deeper** reading section reviews the elements of the electromagnetic spectrum and explores how astronomers use the spectrum to understand celestial objects.

Preparation and Materials Needed

Part A

Students will need spectroscopes to complete this part of the investigation. Kits to make spectroscopes are available commercially, as are ready-made spectroscopes. There is a large range of prices and sophistication. This activity requires only the simplest.

If it is not possible to obtain commercially made spectroscopes, you can construct usable instruments using the following instructions.

How to Make a Spectroscope

You can either make the spectroscopes before class or have your students make them in class.

   a) Cover both ends of a cardboard tube with aluminum foil.

   b) Using scissors, make a thin (0.5 mm) slit in the foil at one end.

   c) Make a square opening, 1 cm by 1 cm, in the foil at the other end.

   d) Carefully cut a piece of diffraction grating into a 2 cm by 2 cm square. Diffraction grating is fragile and should be handled only by the edges.

   e) Use small pieces of an overhead transparency to make a sandwich around the grating.

   f) Tape the "sandwich" over the 1 cm by 1 cm hole in the foil.

You may want to bring in a small table lamp if the lab does not have an incandescent bulb. Most rooms have fluorescent lights. If the physics teacher has a hydrogen, helium, or sodium light you may borrow, use this light to let the students see another kind of spectrum. A small piece of the spectrum will actually show up as narrow bands of light.

Part B

You will need to provide students with copies of local or state highway maps. The maps are used for students to plot the distance of the ultraviolet band.

Part C

Students need to do Internet research to complete this part of the investigation. If you do not have access to the Internet for your class, you can provide students with written information on each of the missions and instruments, or you may find it necessary to have students complete this section outside of class.

Materials

Part A
- Spectroscope (or materials to make one: cardboard tube, aluminum foil, scissors, piece of diffraction grating – 2 cm x 2 cm, two small pieces of overhead transparency, tape)
- Fluorescent light
- Incandescent light

Part B
- Tape
- Metric ruler
- Colored pencils: red, orange, yellow, green, blue, violet
- Calculator
- Paper

Part C
- Internet access*

Think about It

Student Conceptions

Most students will likely have seen the effects of a prism separating the different parts of the visible light spectrum. If in no other circumstance, students have probably seen a rainbow. It is likely, however, that fewer students will realize that each color of the rainbow is one part of the visible light spectrum, and that the prism acts to separate these different parts of the spectrum from one another. Even fewer will have any understanding of why such a separation occurs or is even possible. Most students will at some level realize that the Sun gives off energy other than visible light, but they may not realize that the infrared radiation that they do not see warms their bodies. Even fewer will realize that the Sun gives off other kinds of radiation that they are familiar with, like radio wave and microwave energy.

*The *EarthComm* web site provides suggestions for obtaining resources on space science missions.

Answer for the Teacher Only

You can project the visible spectrum by using an overhead projector and a sheet of holographic diffraction grating (a grating produced by accurate holographic techniques). Lay two pieces of black construction paper on the surface of the overhead projector so that you form a vertical slit 2 mm wide. Turn on the projector and hold the holographic diffraction grating over the upper lens of the projector. Rotate the grating until the best spectrum is produced, then tape the grating in place over the lens. Adjust the size of the spectrum by moving the projector toward or away from the screen. Allow time for discussion in groups about the sequence of colors. The spectrum will always be in the same order. In this investigation students will come to understand that each of the different colors of the spectrum has a different wavelength of radiation, and that all of the visible-light spectrum is only a very small part of the electromagnetic spectrum. Students will also come to understand that many of the types of radiation that they are familiar with in their everyday lives, like radio waves and microwaves, are also part of this continuum of electromagnetic radiation, and that the Sun emits a wide range of different wavelengths of radiation.

Assessment Tool

Think about It Evaluation Sheet
Use this evaluation sheet to help students understand and internalize the basic expectations for the warm-up activity.

Investigate

Part A: Observing Part of the Electromagnetic Spectrum

1. Obtain a spectroscope, similar to the one shown in the illustration. Hold the end with the diffraction grating to your eye. Direct it toward a part of the sky away from the Sun. (**CAUTION**: never look directly at the Sun; doing so even briefly can damage your eyes permanently.) Look for a spectrum along the side of the spectroscope. Rotate the spectroscope until you see the colors going from left to right rather than up and down.

   a) In your notebook, write down the order of the colors you observed.

   b) Move the spectroscope to the right and left. Record your observations.

2. Look through the spectroscope at a fluorescent light.

   a) In your notebook, write down the order of the colors you observed.

3. Look through the spectroscope at an incandescent bulb.

   a) In your notebook, write down the order of the colors you observed.

    Do not look directly at a light with the unaided eye. Use the spectroscope as instructed.

4. Use your observations to answer the following questions:

   a) How did the colors and the order of the colors differ between reflected sunlight, fluorescent light, and the incandescent light? Describe any differences that you noticed.

   b) What if you could use your spectroscope to look at the light from other stars? What do you think it would look like?

Part B: Scaling the Electromagnetic Spectrum

1. Tape four sheets of photocopy paper end to end to make one sheet 112 cm long. Turn the taped sheets over so that the tape is on the bottom.

2. Draw a vertical line 2 cm from the left edge of the paper. Draw two horizontal lines from that line, one about 8 cm from the top of the page, and one about 10 cm below the first line.

3. On the top line, plot the frequencies of the electromagnetic spectrum on a logarithmic scale. To do this, mark off 24 1-cm intervals starting at the left vertical line. Label the marks from 1 to 24 (each number represents increasing powers of 10, from 10^1 to 10^{24}).

4. Use the information from the table of frequency ranges ($\log_{10}$) to divide your scale into the individual bands of electromagnetic radiation. For the visible band, use the entire band, not the individual colors.

E 59

Investigate

Part A: Observing Part of the Electromagnetic Spectrum

> **Teaching Tip**
>
> **IMPORTANT:** Students should never look directly at the Sun. Sources of sunlight include light reflected from car bumpers, windows, a piece of white paper, or aluminum foil. Make certain that students understand that they should never look directly at the Sun. Radiation from the Sun can be dangerous to their eyes, and too much exposure to solar ultraviolet radiation can burn the eye and may, among other things, increase the risk of cataracts.

1. a) From left to right the colors observed are: red, orange, yellow, green, blue, and violet.

    b) Moving the spectroscope from left to right has no apparent affect on the order of the colors or the width of individual color bands. The spectrum does not move with the spectroscope. However, as the spectroscope is moved beyond the source of reflected sunlight, the spectrum fades.

2. a) The order of colors is the same as in the reflected sunlight.

3. a) The order of colors is the same as in the reflected sunlight.

4. a) The order of colors did not differ in any of the three spectra. However, in the florescent light spectra, there were three narrow color bands (in the yellow, green, and violet parts of the spectrum) that were much brighter than the rest of the spectrum. (Note that depending on the type of fluorescent light bulb, this may vary.)

    b) Student answers may vary, but they might suggest that the order of the colors might be the same but that certain color bands might be brighter.

Part B: Scaling the Electromagnetic Spectrum

3. Subdividing each 1-cm interval into 10 equal increments would make it easier to plot the values. Alternatively, students could use a metric ruler when plotting the values. They will otherwise need to interpolate the values by eye. For example, to plot the value 12.47, the students need to place the point 47% of the way from the "12" mark to the "13" mark. It would be much easier and more accurate to do this is if there were marks for 12.1, 12.2, etc.

Earth System Evolution Astronomy

Frequency Range Table			
EMR Bands	**Frequency Range (hertz)**	**Log_{10} Frequency Range (hertz)**	**10^{14} Conversions**
Radio and Microwave	Near 0 to 3.0×10^{12}	0 to 12.47	.
Infrared	3.0×10^{12} to 4.6×10^{14}	12.47 to 14.66	.
Visible	4.6×10^{14} to 7.5×10^{14}	14.66 to 14.88	4.6×10^{14} to 7.5×10^{14}
Red	4.6×10^{14} to 5.1×10^{14}	14.66 to 14.71	4.6×10^{14} to 5.1×10^{14}
Orange	5.1×10^{14} to 5.6×10^{14}	14.71 to 14.75	5.1×10^{14} to 5.6×10^{14}
Yellow	5.6×10^{14} to 6.1×10^{14}	14.75 to 14.79	5.6×10^{14} to 6.1×10^{14}
Green	6.1×10^{14} to 6.5×10^{14}	14.79 to 14.81	6.1×10^{14} to 6.5×10^{14}
Blue	6.5×10^{14} to 7.0×10^{14}	14.81 to 14.85	6.5×10^{14} to 7.0×10^{14}
Violet	7.0×10^{14} to 7.5×10^{14}	14.85 to 14.88	7.0×10^{14} to 7.5×10^{14}
Ultraviolet	7.5×10^{14} to 6.0×10^{16}	14.88 to 16.78	.
X-ray	6.0×10^{16} to 1.0×10^{20}	16.78 to 20	.
Gamma Ray	1.0×10^{20} to...	20 to ...	.

5. To construct a linear scale, you will need to convert the range of frequencies that each band of radiation covers for the logarithmic scale. This will allow you to compare the width of the bands of radiation relative to each other. Convert the frequency numbers for all bands (except visible) to 10^{14} and record them in the table.

Example: 10^{17} is 1000 times greater than 10^{14}, so $2.5 \times 10^{17} = 2500 \times 10^{14}$.

6. On the lower horizontal line, mark off ten 10-cm intervals from the vertical line. Starting with the first interval, label each mark with a whole number times 10^{14}, from 1×10^{14} to 10×10^{14}. Label the bottom of your model "Frequency in hertz." Plot some of the 10^{14} frequencies you calculated on the bottom line of your constructed model. Plot the individual colors of the visible spectrum and color them.

a) Compare the logarithmic and linear scales. Describe the differences.

7. Look at the range of ultraviolet radiation.

a) How high do the ultraviolet frequencies extend (in hertz)?

b) Using the same linear scale that you constructed in **Step 6** (10 cm = 1×10^{14} Hz), calculate the width (in centimeters) of the ultraviolet electromagnetic radiation band.

c) Using this same scale what do you think you would need to measure the distance from the beginning of the ultraviolet band of the electromagnetic radiation to the end of the ultraviolet band of the electromagnetic radiation?

d) Using your calculations above and the linear scale you created in **Steps 5** and **6,** how much wider is

6. **a)** Here, as in so many fields of science and technology, using a logarithmic scale makes it possible to represent, on a convenient scale, numerical values that range over many powers of 10. As you can see from the exercise, it would be impossible to plot such wide-ranging values on a linear scale.

   Each of the scales used has its drawbacks. In the case of the logarithmic scale, the relative sizes of the bands are greatly distorted from their real dimensions. The entire frequency range of radio and microwave radiation (the largest category on the logarithmic scale created in this investigation) is about 3 trillion Hz (from near 0 to 3.0×10^{12}). The range of visible light, however, spans 290 trillion Hz (2.9×10^{14}), even though it appears to be the smallest range on the logarithmic spectrum drawn!

   The linear scale in this exercise is also less than ideal. Although the correct relative proportions of the radiation bands are accurately portrayed, only a limited region of the spectrum can be meaningfully represented. Further, the smallest increment used (1×10^{14}) is too big to provide any detail to the radio and microwave portion of the spectrum. At the scale drawn, the region of this part of the spectrum (from near 0 to 3.0×10^{12}) only measures 3 mm!

7. **a)** The ultraviolet band extends from 7.5×10^{14} Hz to 6.0×10^{16} Hz.

   b) On the logarithmic scale, the UV is approximately 1.9 cm long. The ultraviolet (UV) band spans 5.925×10^{16} Hz (6×10^{16} to 7.5×10^{14}). This equates to 592.5×10^{14}. Using the 10 cm = 1×10^{14} cm linear scale, this band would be 5925 cm long!

   c) Answers will vary, but should be something in excess of about 60 m long.

   d) The visible band on this linear scale was 29 cm long. From the calculation above, the UV band is 204.3 times as wide as the visible band. On the logarithmic scale, the UV band was 8.6 times as wide as the visible band (the UV band was 1.9 cm wide and the visible band was 0.22 cm wide).

Assessment Tools

EarthComm Notebook Entry-Checklist
Use this checklist as a quick guide for student self-assessment and/or an opportunity to quickly score student work. Add further criteria specific to your classroom needs or to this particular investigation.

Investigate Notebook Entry-Evaluation Sheet
Point out the criteria listed on this evaluation sheet that are relevant to this particular investigation. Encourage students to internalize the criteria by making them part of your "assessment conversations" as you circulate around the classroom.

the ultraviolet band than the entire visible band? How does this compare to the relative widths of these two bands on the log scale you created in **Steps 1-4**?

8. X-rays are the next band of radiation.

a) Using the same linear scale ($10 \text{ cm} = 1 \times 10^{14} \text{ Hz}$) calculate the distance from the end of the ultraviolet band to the end of the x-ray band. Obtain a map from the Internet or use a local or state highway map to plot the distance.

b) Based on your results for the width of the x-ray band, what would be your estimate for the width of the gamma-ray band of radiation? What would you need to measure the distance?

Part C: Using Electromagnetic Radiation in Astronomy

1. Astronomers use electromagnetic radiation to study objects and events within our solar system and beyond to distant galaxies. In this part of the activity, you will be asked to research a space science mission and find out how astronomers are using the electromagnetic spectrum in the mission and then report to the rest of the class. The *EarthComm* web site will direct you to links for missions that are either in development, currently operating, or operated in the past. A small sampling is provided in the table on the next page. Many missions contain multiple instruments (it is very expensive to send

instruments into space, so scientists combine several or more studies into one mission), so you should focus upon one instrument and aspect of the mission and get to know it well. The **Digging Deeper** reading section of this activity might help you begin your work.

Questions that you should try to answer in your research include:

• What is the purpose or key question of the mission?

• How does the mission contribute to our understanding of the origin and evolution of the universe or the nature of planets within our solar system?

• Who and/or how many scientists and countries are involved in the mission?

• What instrument within the mission have you selected?

• What wavelength range of electromagnetic radiation does the instrument work at?

• What is the detector and how does it work?

• What does the instrument look like?

• How are the data processed and rendered? Images? Graphs?

• Any other questions that you and your teacher agree upon.

2. When you have completed your research, provide a brief report to the class.

8. a) The x-ray band spans approximately 9.994×10^{19} Hz. Converting so this number is expressed in terms of 10^{14} gives 999,400 (i.e., $999,400 \times 10^{14}$). Using the same 10 cm = 1×10^{14} Hz linear scale yields a total length of the x-ray band of 9,994,000 cm, or 99.94 km.

    b) Answers will vary. Students are not given an upper limit to the gamma-ray frequency range, so they should recognize that the value is quite large.

Part C: Using Electromagnetic Radiation in Astronomy

1. Students should select which mission they will investigate and attempt to answer all of the relevant questions about each instrument. Information about each mission/instrument is provided below.

Hubble–NICMOS Instrument

NICMOS was installed on board the Hubble Space Telescope in February 1997. NICMOS contains three cameras designed for simultaneous operation, and it can detect light with wavelengths between 0.8 and 2.5 μm, much longer than the human-eye limit. This allows infrared imaging and spectroscopic observations of astronomical targets. Since its installation, NICMOS has made observations of newly forming stars and regions containing the farthest and faintest galaxies ever imaged. The instrument is operated by the NICMOS Science Team, composed of scientists from:

- University of Arizona
- Space Telescope European Coordinating Facility (ST-ECF)
- Space Telescope Science Institute (ST-Sci)
- Ball Aerospace
- NASA Goddard Space Flight Center

Cassini Huygens Mission to Saturn and Titan

The Cassini spacecraft was launched in October 1997. Cassini will study Titan's atmosphere and surface. The Cassini spacecraft, including the orbiter and the Huygens probe, weighed about 5600 kg (12,346 lb.) at launch. The Cassini orbiter instruments are capable of a great increase in resolution and coverage over data obtained during the Voyager mission. They will also study variations in the Saturn system over a four-year period instead of snapshots during relatively brief flybys. The Cassini mission objectives are split into five main areas:

- Saturn's atmosphere and interior
- the rings
- the magnetosphere
- Titan
- the icy satellites

The orbiter will conduct long-term, detailed, close-up studies of the rings, satellites, and planet. It will also perform measurements of Saturn's magnetosphere. The Huygens probe will study the clouds, atmosphere, and surface of Saturn's moon, Titan. The probe will distribute data via radio waves, and will begin data collection in November 2004. The Jet Propulsion Laboratory (JPL) near Pasadena, California, manages the Cassini Project.

SIRTF (Space InfraRed Telescope Facility)

SIRTF is scheduled for launch in July 2002. The SIRTF observatory consists of a 0.85-m telescope and three cryogenically cooled science instruments capable of performing imaging and spectroscopy in the 3–180 μm wavelength range. With its great sensitivity, large-format infrared detector arrays, high observing efficiency, and long cryogenic life, SIRTF offers capabilities over previous and existing facilities. The SIRTF will concentrate on four areas of research:

- the formation of planets and stars
- the origin of energetic galaxies and quasars
- the distribution of matter and galaxies
- the formation and evolution of galaxies

Several groups are working on the project, including:

- Jet Propulsion Laboratory SIRTF Science Center of the California Institute of Technology
- Ball Aerospace and Technologies Corporation
- Lockheed Martin Space System Company
- Smithsonian Astrophysical Observatory
- NASA-Goddard Space Flight Center, Cornell University
- University of Arizona

HETE-2 High Energy Transient Explorer

The High Energy Transient Explorer is a small scientific satellite designed to detect and localize gamma-ray bursts. The primary goal of HETE-2 is to determine the origin and nature of cosmic gamma-ray bursts (GRBs). This is accomplished through the simultaneous, broad-band observation in the soft x-ray, medium x-ray, and gamma-ray energy ranges, and the precise localization and identification of the sources of cosmic gamma-ray bursts. HETE-2 carries three science instruments:

- a set of wide-field gamma-ray spectrometers (FREGATE)
- a wide-field x-ray monitor (WXM)
- a set of soft x-ray cameras (SXC)

The HETE program is an international collaboration including:

- Center for Space Research at the Massachusetts Institute of Technology
- Institute for Chemistry and Physics (RIKEN)
- Los Alamos National Laboratory (LANL)
- Centre d'Etude Spatiale des Rayonnements (CESR)
- University of Chicago
- University of California, Berkeley
- University of California, Santa Cruz
- Centre Nationale d'Etudes Spatiales (CNES)
- Ecole Nationale Supérieure de l'Aéronautique et de l'Espace (Sup'Aéro)
- Consiglio Nazionale delle Ricerche (CNR)
- Instituto Nacional de Pesquisas Espaciais (INPE)
- Tata Institute of Fundamental Research (TIFR)

Chapter 1

Earth System Evolution Astronomy

Descriptions of Selected Missions	
Mission/Instrument	**Description**
Hubble – NICMOS Instrument	Hubble's Near Infrared Camera and Multi-Object Spectrometer (NICMOS) can see objects in deepest space—objects whose light takes billions of years to reach Earth. Many secrets about the birth of stars, solar systems, and galaxies are revealed in infrared light, which can penetrate the interstellar gas and dust that block visible light.
Cassini Huygens Mission to Saturn and Titan	The Ultraviolet Imaging Spectrograph (UVIS) is a set of detectors designed to measure ultraviolet light reflected or emitted from atmospheres, rings, and surfaces over wavelengths from 55.8 to 190 nm (nanometers) to determine their compositions, distribution, aerosol content, and temperatures.
SIRTF	The Space InfraRed Telescope Facility (SIRTF) is a space-borne, cryogenically cooled infrared observatory capable of studying objects ranging from our solar system to the distant reaches of the Universe. SIRTF is the final element in NASA's Great Observatories Program, and an important scientific and technical cornerstone of the new Astronomical Search for Origins Program.
HETE-2 High Energy Transient Explorer	The High Energy Transient Explorer is a small scientific satellite designed to detect and localize gamma-ray bursts (GRB's). The primary goals of the HETE mission are the multi-wavelength observation of gamma-ray bursts and the prompt distribution of precise GRB coordinates to the astronomical community for immediate follow-up observations. The HETE science payload consists of one gamma-ray and two x-ray detectors.
Chandra X-Ray Observatory	NASA's Chandra X-ray Observatory, which was launched and deployed by Space Shuttle Columbia in July of 1999, is the most sophisticated x-ray observatory built to date. Chandra is designed to observe x-rays from high-energy regions of the universe, such as the remnants of exploded stars.

Reflecting on the Activity and the Challenge

The spectroscope helped you to see that visible light is made up of different color components. Visible light is only one of the components of radiation you receive from the Sun. In the second part of the activity, you explored models for describing the range of frequencies of energy within electromagnetic radiation.

Finally, you researched a space mission to learn how astronomers are using electromagnetic radiation to understand the evolution of the Earth system. Radiation from the Sun and other objects in the universe is something you will need to explain to your fellow citizens in your **Chapter Challenge** brochure.

Reflecting on the Activity and the Challenge

Remind the students that because the Sun's light is our source of energy, any change in the Sun might gravely affect Earth's community. Use this opportunity to suggest looking up the Big Bang or the "red shift." This is also a good time to review recent technological advances like the Hubble Telescope, the Keck telescope in Hawaii, and the new x-ray telescope launched by NASA.

Digging Deeper

ELECTROMAGNETIC RADIATION

The Nature of Electromagnetic Radiation

In 1666, Isaac Newton found that he could split light into a spectrum of colors. As he passed a beam of sunlight through a glass prism, a spectrum of colors appeared from red to violet. Newton deduced that visible light was in fact a mixture of different kinds of light. About 10 years later, Christiaan Huygens proposed the idea that light travels in the form of tiny waves. It's known that light with shorter wavelengths is bent (refracted) more than light with longer wavelengths when it passes through a boundary between two different substances. Violet light is refracted the most, because it has the shortest wavelength of the entire range of visible light. This work marked the beginning of **spectroscopy**—the science of studying the properties of light. As you will learn, many years of research in spectroscopy has answered many questions about matter, energy, time, and space.

In your study of the Sun you learned that the Sun radiates energy over a very wide range of wavelengths. Earth's atmosphere shields you from some of the most dangerous forms of electromagnetic radiation. You are familiar with the wavelengths of light that do get through and harm you—mostly in the form of sunburn-causing ultraviolet radiation. Now you can take what you learned and apply the principles of spectroscopy to other objects in the universe.

In the **Investigate** section, you used a **spectroscope** to study the Sun's light by separating it into its various colors. Each color has a characteristic wavelength. This range of colors, from red to violet, is called the **visible spectrum**. The visible spectrum is a small part of the entire spectrum of **electromagnetic radiation** given off by the Sun, other stars, and galaxies.

Electromagnetic radiation is in the form of electromagnetic waves that transfer energy as they travel through space. Electromagnetic waves (like ripples that expand after you toss a stone into a pond) travel at the speed of light (300,000 m/s). That's eight laps around the Earth in one second. Although it's not easy to appreciate from everyday life, it turns out that electromagnetic radiation has properties of both particles and waves. The colors of the visible spectrum are best described as waves, but the same energy that produces an electric current in a solar cell is best described as a particle.

Geo Words

spectroscopy: the science that studies the way light interacts with matter.

spectroscope: an instrument consisting of, at a minimum, a slit and grating (or prism) which produces a spectrum for visual observation.

visible spectrum: part of the electromagnetic spectrum that is detectable by human eyes. The wavelengths range from 350 to 780 nm (a nanometer is a billionth of a meter).

electromagnetic radiation: the energy propagated through space by oscillating electric and magnetic fields. It travels at 3×10^8 m/s in a vacuum and includes (in order of increasing energy) radio, infrared, visible light (optical), ultraviolet, x-rays, and gamma rays.

Digging Deeper

Assign the reading for homework, along with the questions in **Check Your Understanding** if desired.

Assessment Opportunity

Reword or restructure the questions in **Check Your Understanding** for a brief quiz. Use the quiz (or a class discussion of the questions in the textbook) to assess your students' understanding of the main ideas in the reading and the activity.

Assessment Tool

Check Your Understanding Notebook Entry-Evaluation Sheet
Use this sheet to evaluate the extent to which students understand the key concepts explored in **Activity 6** and explained in **Digging Deeper**, and to evaluate the students' clarity of expression.

Teaching Tip

Use **Blackline Master Astronomy 6.1, The Electromagnetic Spectrum** of *Figure 1* on page E64 to make an overhead transparency. Incorporate this overhead into a discussion on the electromagnetic spectrum. Take this opportunity to review the main points of the investigation and to make sure that students have an understanding of the spectrum and its uses.

Earth System Evolution Astronomy

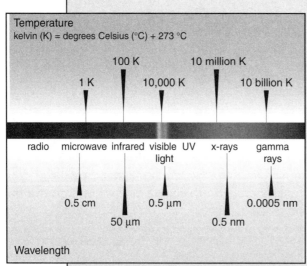

Figure 1 The electromagnetic spectrum. Wavelengths decrease from left to right, and energy increases from left to right. The diagram shows that a relationship exists between the temperature of an object and the peak wavelength of electromagnetic radiation it emits.

Figure 1 summarizes the spectrum of energy that travels throughout the universe. Scientists divide the spectrum into regions by the wavelength of the waves. Long radio waves have wavelengths from several centimeters to thousands of kilometers, whereas gamma rays are shorter than the width of an atom.

Humans can see only wavelengths between 0.4 and 0.7 µm, which is where the visible spectrum falls. A micrometer (µm) is a millionth of a meter. This means that much of the electromagnetic radiation emitted by the Sun is invisible to human eyes. You are probably familiar, however, with some of the kinds of radiation besides visible light. For example, **ultraviolet** radiation gives you sunburn. **Infrared** radiation you detect as heat. Doctors use x-rays to help diagnose broken bones or other physical problems. Law-enforcement officers use radar to measure the speed of a motor vehicle, and at home you may use microwaves to cook food.

Astronomy and the Electromagnetic Spectrum

Humans have traveled to the Moon and sent probes deeper into our solar system, but how do they learn about distant objects in the universe? They use a variety of instruments to collect electromagnetic radiation from these distant objects. Each tool is designed for a specific part of the spectrum. Visible light reveals the temperature of stars. Visible light is what you see when you look at the stars through telescopes, binoculars, or your unaided eyes. All other forms of light are invisible to the human eye, but they can be detected.

Radio telescopes like the Very Large Array (VLA) and Very Large Baseline Array (VLBA) in New Mexico are sensitive to wavelengths in the radio range. Radio telescopes produce images of celestial bodies by recording the different amounts of radio emission coming from an area of the sky

Geo Words

ultraviolet: electromagnetic radiation at wavelengths shorter than the violet end of visible light; with wavelengths ranging from 5 to 400 nm.

infrared: electromagnetic radiation with wavelengths between about 0.7 to 1000 µm. Infrared waves are not visible to the human eye.

radio telescope: an instrument used to observe longer wavelengths of radiation (radio waves), with large dishes to collect and concentrate the radiation onto antennae.

Blackline Master Astronomy 6.1
The Electromagnetic Spectrum

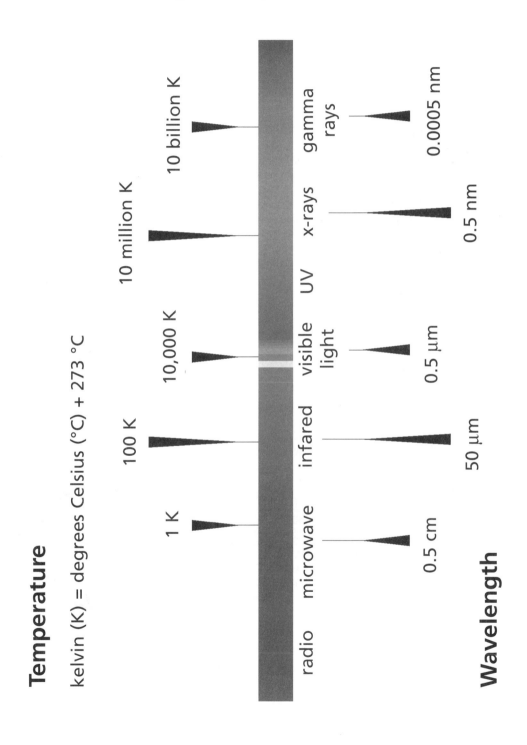

Temperature

kelvin (K) = degrees Celsius (°C) + 273 °C

1 K	100 K	10,000 K	10 million K	10 billion K	

Wavelength

radio	microwave	infared	visible light	UV	x-rays	gamma rays
	0.5 cm	50 μm	0.5 μm		0.5 nm	0.0005 nm

observed. Astronomers process the information with computers to produce an image. The VLBA has 27 large dish antennas that work together as a single instrument. By using recorders and precise atomic clocks installed at each antenna, the signals from all antennas are combined after the observation is completed.

Geo Words

x-ray telescope: an instrument used to detect stellar and interstellar x-ray emission. Because the Earth's atmosphere absorbs x-rays, x-ray telescopes are placed high above the Earth's surface.

The galaxy M81 is a spiral galaxy about 11 million light years from Earth and is about 50,000 light years across. The spiral structure is clearly shown in *Figure 2*, which shows the relative intensity of emission from neutral atomic hydrogen gas. In this pseudocolor image, red indicates strong radio emission and blue weaker emission.

Figure 2 The galaxy M81.

The orbiting Chandra **x-ray telescope** routinely detects the highly energetic radiation streaming from objects like supernova explosions, active galaxies, and black holes. The Hubble Space Telescope is outfitted with a special infrared instrument sensitive to radiation being produced by star-forming nebulae and cool stars. It also has detectors sensitive to ultraviolet light being emitted by hot young stars and supernova explosions.

A wide array of solar telescopes both on Earth and in space study every wavelength of radiation from our nearest star in minute detail. The tools of astronomy expand scientists' vision into realms that human eyes can never see, to help them understand the ongoing processes and evolution of the universe.

Figure 3 Astronauts working on the Hubble Space Telescope high above the Earth's atmosphere.

Teaching Tip

The galaxy M81, shown in *Figure 2* on page E65, has a total visual brightness of approximately magnitude 6.8. (See **Background Information** for **Activity 7** for details regarding magnitude.) Under exceptionally good conditions, M81 has been spotted in the Northern Hemisphere sky with the naked eye. M81 is the brightest dominant galaxy of a nearby group called the M81 group. In 1993, a supernova that occurred in galaxy M81 was discovered by an amateur astronomer.

Teaching Tip

The Hubble Space Telescope, pictured in *Figure 3* on page E65, was deployed on April 25, 1990 from the space shuttle *Discovery*. It circles the Earth at approximately 5 miles per second, completing an orbit of the Earth once every 97 minutes. The Hubble Space Telescope is powered by two 40-foot-long solar arrays that generate a total of 2400 watts of energy. Its location outside of the obscuring effects of the Earth's atmosphere makes the Hubble Space Telescope a unique astronomical tool. Since its deployment, it has provided remarkable new views of the universe that have revolutionized astronomers' thinking about many astronomical mysteries.

Chapter I

Earth System Evolution Astronomy

Geo Words

peak wavelength: the wavelength of electromagnetic radiation with the most electromagnetic energy emitted by any object.

Using Electromagnetic Radiation to Understand Celestial Objects

The wavelength of light with the most energy produced by any object, including the Sun, is called its **peak wavelength**. Objects that are hot and are radiating visible light usually look the color of their peak wavelength. People are not hot enough to emit visible light, but they do emit infrared radiation that can be detected with infrared cameras. The Sun has its peak wavelength in the yellow region of the visible spectrum. Hotter objects produce their peaks toward the blue direction. Very hot objects can have their peaks in the ultraviolet, x-ray, or even gamma-ray range of wavelength. A gas under high pressure radiates as well as a hot solid object. Star colors thus reflect temperature. Reddish stars are a "cool" 3000 to 4000 K (kelvins are celsius degrees above absolute zero, which is at minus 273°C). Bluish stars are hot (over 20,000 K).

One of the most important tools in astronomy is the spectrum—a chart of the entire range of wavelengths of light from an object. Astronomers often refer to this chart as the spectrum of the star. These spectra come in two forms: one resembles a bar code with bright and dark lines (see *Figure 4*), and the other is a graph with horizontal and vertical axes (see *Figure 5*). Think of these spectra as "fingerprints" that reveal many kinds of things about an object: its chemical composition, its temperature and pressure, and its motion toward or away from us.

Figure 4 One of the forms of the spectrum of a star. The data encoded here tell astronomers that this star is bright in some elements and dim in others.

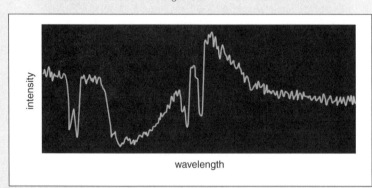

Figure 5 The spectrum of a star can also be represented as a graph with horizontal and vertical axes.

E 66

Teaching Tip

Figures 4 and 5 on page E66 show two different representations of the same stellar spectrum. Note that the dark absorption lines in *Figure 4* correspond to the low intensity wavelengths in *Figure 5*. How would the bright color bands seen during the investigation of fluorescent light spectra be represented on a graph like the one shown in *Figure 5*?

Activity 6 The Electromagnetic Spectrum and Your Community

Each chemical element in the universe has its own unique spectrum. If you know what the spectrum of hydrogen is, you can look for its fingerprint in a star. If you suspect that a star may have a lot of the elements helium or calcium, for example, you can compare the spectrum of the star with the known spectra of helium or calcium. If you see bright lines in the stellar spectrum that match the patterns of bright lines in the helium or calcium spectra, then you have identified those elements in the star. This kind of spectrum is known as an **emission spectrum**. If you look at the star and see dark lines where you would expect to see an element—especially hydrogen—it is likely that something between you and the star is absorbing the element. This kind of spectrum is known as an **absorption spectrum**.

The positions of lines in a spectrum reveal the motion of the star toward or away from Earth, as well as the speed of that motion. You have experienced the effect yourself, when a car, a truck, or a train passes by you with its horn blowing. The pitch of the sound increases as the object approaches you, and decreases as the object passes by and moves away from you. That is because when the object is coming toward you, its speed adds to the speed of the sound, making the wavelength of the sound seem higher to your ear. The reverse happens when the object is moving away from you. The same principle applies to the spectrum from a distant object in space, which might be moving either toward Earth or away from Earth.

Geo Words

emission spectrum: a spectrum containing bright lines or a set of discrete wavelengths produced by an element. Each element has its own unique emission spectrum.

absorption spectrum: a continuous spectrum interrupted by absorption lines or a continuous spectrum having a number of discrete wavelengths missing or reduced in intensity.

Check Your Understanding

1. What are the colors of the spectrum of visible sunlight, from longest wavelength to shortest? Are there breaks between these colors, or do they grade continuously from one to the next? Why?

2. Which wavelengths of light can be more harmful to you than others? Why?

3. What tools do astronomers use to detect different wavelengths of light?

4. How can the speed of a distant object in space be measured?

Understanding and Applying What You Have Learned

1. Think back to **Activity 5** where you learned about the radiation from the Sun and how much of it passes through the Earth's atmosphere.

   a) Which parts of the electromagnetic spectrum that come from the Sun does the atmosphere absorb the most?

   b) Which gases are responsible for blocking each part of the electromagnetic spectrum?

   c) What would a spectrum of the Sun look like if you were searching for those elements from the ground? Would it be an emission spectrum or an absorption spectrum? Why?

EarthComm

Check Your Understanding

1. Red, orange, yellow, green, blue, and violet are the colors of the visible-light spectrum, listed from longest wavelength to shortest wavelength. This is not mentioned in the **Digging Deeper** reading section, but it is given in the table of frequency ranges on page E60. Note that in the visible spectrum, the color indigo is sometimes listed between blue and violet. The colors grade continuously into one another, without breaks.

2. Ultraviolet light is harmful to humans because it can cause skin damage and skin cancer. The reason has to do with how the wavelengths of the ultraviolet radiation interact with the cells of the skin. The longer wavelengths of visible light do not have such an effect.

3. Astronomers use radio telescopes, x-ray telescopes, and solar telescopes to detect different wavelengths of light.

4. The speed of a distant object toward or away from the Earth can be measured using the principle of what is called the red shift, which is described in the final paragraph of the **Digging Deeper** reading section. What is measured is the shift in frequency from the frequency of a known line in the spectrum of electromagnetic radiation emitted from the object. From that shift, the speed of the object toward or away from the Earth can be calculated.

Understanding and Applying What You Have Learned

1. The Earth is a relatively cool body. As such, it naturally radiates with a peak wavelength somewhere in the infrared part of the spectrum (see *Figure 1* on page E64). Someone on a distant planet studying the Earth might choose to investigate infrared wavelengths because the peak wavelength of a radiating body radiates more intensely than other parts of the electromagnetic spectrum. Also, we (humans) generate a lot of radio and microwave radiation from the technological devices that we use. Accordingly, those would be two other interesting types of radiation to study. One can imagine that the Earth would emanate a strange energy spectrum compared to other planetary bodies that have a similar temperature!

2. a) In the electromagnetic spectrum, energy increases as wavelength decreases (for a given intensity of radiation).

   b) Supernova explosions or black holes must be extremely high-energy phenomena. As such, their peak wavelength is probably towards the short wavelength/high energy range of the electromagnetic spectrum. Therefore, x-rays might be a good part of the electromagnetic spectrum to study.

3. The Sun looks yellow to us because its temperature is such that its peak wavelength of radiation is in the yellow part of visible-light radiation. However, just because yellow is the Sun's peak wavelength doesn't mean that it doesn't emit other radiation as well. The Sun also radiates both infrared radiation (which we feel as heat) and higher-energy ultraviolet radiation (which burns our skin).

Earth System Evolution Astronomy

Preparing for the Chapter Challenge

Recall that your challenge is to educate people about the hazards from outer space and to explain some of the benefits from living in our solar system. Electromagnetic radiation has both beneficial and harmful effects on life on Earth. Use what you have learned in this activity to develop your brochure.

1. Make a list of some of the positive effects of electromagnetic radiation on your community. Explain each item on the list.

2. Make a list of some of the negative effects of electromagnetic radiation on your community. Explain each item on the list.

3. Make a list of celestial radiation sources and any effects they have on Earth systems. What are the chances of a stellar radiation source affecting Earth?

Inquiring Further

1. **Using radio waves to study distant objects**

   Radio waves from the Sun penetrate the Earth's atmosphere. Scientists detect these waves and study their strength and frequency to understand the processes inside the Sun that generate them. Do some research on how these waves are studied.

2. **Detecting electromagnetic radiation**

   Investigate some of the instruments that astronomers use to detect electromagnetic radiation besides light. Where are you likely to find ultraviolet detectors? Describe radio telescope arrays.

3. **Technologies and the electromagnetic spectrum**

   Research some of the technologies that depend on the use of electromagnetic radiation. These might include microwave ovens, x-ray machines, televisions, and radios. How do they work? How is electromagnetic radiation essential to their operation? What interferes with their operation?

Preparing for the Chapter Challenge

After completing this activity, students should be able to explain that the electromagnetic spectrum is made up of many different wavelengths of radiation, and that visible light is also made of a range of wavelengths. They should understand that nearly all life on Earth depends on energy from the Sun. Students should review their work in this activity and relate it to the **Chapter Challenge**. Positive effects of the Sun's energy on our community are warmth and light (which are essential for photosynthesis and vision, and can be used as an alternative energy source). Negative effects include the harmful effects of ultraviolet radiation (e.g., skin cancer).

For assessment (see **Assessment Rubric for Chapter Challenge on Astronomy** on pages 12 and 13), remind students that their reports and flow chart should reflect:
- the hazardous and beneficial effects of solar radiation
- why extraterrestrial influences on the community are a natural part of Earth system evolution

Inquiring Further

1. Using radio waves to study distant objects

The science of radio astronomy dates back to 1933, when Karl Jansky published his discovery of cosmic radio waves in the *New York Times*. As a staff member of the Bell Telephone Laboratories, Jansky was part of a team investigating the use of "short waves" (wavelengths of about 10 – 20 m) for transatlantic telephone service. While Jansky was investigating the sources of static that might interfere with radio voice transmissions, he discovered radio wave emissions emanating from the heart of the Milky Way Galaxy. Although Jansky went on to other things, several scientists were intrigued by his discovery. Among those intrigued were John Kraus and Grote Reber, who eventually went on to mold the new emerging science of radio astronomy. Today, radio astronomy is an active field and a number of radio telescopes exist. Two such radio telescopes are the Very Large Array (VLA) and the Very Large Baseline Array (VLBA) in New Mexico. Radio telescopes have been used to study a wide range of astronomical phenomenon, including, among others, planetary nebulae and black holes. The *EarthComm* web site provides links that will help students with their research into the study of cosmic radio waves.

2. Detecting electromagnetic radiation

Many types of devices are used to detect different ranges of electromagnetic radiation. Students should be encouraged to choose one range of radiation to investigate. The *EarthComm* web site provides a number of links to help them get started with their research. Detectors of ultraviolet (UV) radiation are used in a wide range of commercial applications in addition to being used by scientists. A few examples of the use of UV detectors follow. UV detectors are integral parts of flame detectors that are used on oil drilling platforms, in airplane hangars, on rocket launch pads, and in other places. UV detectors are also used in the testing of products that are designed to filter out UV radiation, like sunscreen lotions and sunglasses. Incoming solar UV radiation is monitored using UV detectors installed in rockets and orbiting satellites.

Radio telescope arrays: In order to resolve images (i.e., make sharp images), the diameter of a telescope's collecting area must be many times greater than the wavelength of the radiation it detects. Radio waves have wavelengths that are much longer than those of visible light, so even large radio telescope dishes produce blurry images. A single radio dish would have to be many kilometers across to achieve a sharp image at radio wavelengths, and such telescopes are very difficult to build. Astronomers are able to electronically simulate the effect of a very large dish by using the combined signal from many smaller single-dish radio telescopes. This is a radio telescope array. The *EarthComm* web site provides links that will help students to research some of the radio telescope arrays currently in use, and it also provides links that will guide them to information on the underlying principles that allow astronomers to combine the signals of different radio dishes.

3. Technologies and the electromagnetic spectrum

A wide range of modern technologies relies on the use of electromagnetic radiation. Several examples are given in this **Inquiring Further** section, but you may want to suggest that students concentrate their efforts on only a small number of technologies. Direct students to the *EarthComm* web site to help them get started with their research.

NOTES

ACTIVITY 7 — OUR COMMUNITY'S PLACE AMONG THE STARS

Background Information

On any starry night, it appears to observers that we are literally surrounded by stars. This is because our planetary system is located in a relatively populous arm of the Milky Way Galaxy. From our vantage point in the galaxy, we are treated to the sight of many different kinds of stars, clouds of gas and dust called nebulae, clusters of stars, and many, many distant galaxies. Figuring out what's what out there is a kind of classification game—a way to sort things by color, size, brightness, and shape. In **Activity 7**, your students will explore the ways that astronomers classify objects in the universe. Inherent in those classifications are clues to the nature of celestial objects. And once we understand the nature of a star or planet, we can use our knowledge to assess whether or not that object is a threat to our community.

The Classifying of Stars

Observing a starry sky is like hiking through a forest. You see trees of different shapes, sizes, heights, and ages. If it is a mixed forest, some trees may have leaves and others may have needles. Most will be some shade of green or white or even a reddish green. (Of course, if you're visiting during the autumn, leaf shades will form a beautiful mixture of reds, oranges, and yellows.) What if you wanted to know the ages of the trees? You might decide that taller trees are older than shorter ones. That would be a mistake, because trees of different species grow to different heights. You notice short weed-like things sprouting up from the forest floor. Are they weeds or seedling trees? What connection can you make between acorns or nuts and the trees from which they fell? Dead logs and broken branches litter the forest floor—from which trees did they come? How do you explain sap?

To understand what you're really seeing in the forest, you have to know a lot about forest life. It's the same with stars, except that we are not walking through star fields. We are observing them from a distance, and we can never touch them. In our early history, we were limited to wavelengths of light we could detect with our eyes. The invention of multi-wavelength detectors extended our view and allowed us to explore stars and planets and galaxies more fully.

Today, astronomers classify stars in many ways. Some of the most common sorting routines place stars in categories according to their luminosities, masses, and radii. Astronomers use the Sun as a baseline unit. For example, the orange-red star Betelgeuse in the constellation of Orion, the Hunter, is somewhere between 12 and 17 solar masses. It luminosity is varying, so we can say that it varies between 40,000 and 60,000 times the brightness of the Sun. Finally, it is a huge star—about 1260 solar radii.

The Luminosity of Stars

Luminosity can be defined in two ways— intrinsic brightness and apparent brightness. Intrinsic brightness is a measure of the total radiative output of an object. It is independent of distance from the observer. A star's apparent luminosity, on the other hand, describes how bright it appears to us here on

Earth. How bright it appears depends on a combination of its distance from Earth and its radiative output. The first investigation in **Activity 7** allows students to demonstrate for themselves the difference between intrinsic and apparent brightness.

The brightest star in our skies (after the Sun) is called Sirius. Its apparent brightness is listed as magnitude -1.5, meaning that's how bright it appears on an apparent brightness range called the magnitude scale. On that scale, the Sun is -27! The dimmest stars have magnitudes greater than 30. Most of the stars we see in the sky at night range from magnitude -1.5 to magnitude 6.0. To see dimmer stars, we need telescopes and other instruments.

Classifying Stars by Temperature
Stars are also classified by their temperatures, which are related to their brightness. Hot stars are bright, whereas cooler stars are dim. These differences in themselves are not enough to help astronomers understand the age and evolutionary state of a star, but they are a good start. Today, astronomers plot the temperatures and luminosities of stars on a graph called the Hertzsprung-Russell (HR) diagram. To make any inferences about the ages of the stars, however, they need also to understand such things as the mass of the stars and their chemical makeup.

Classifying Stars by Spectral Class
On the HR diagram supplied to the students, the spectral class of each star type is also plotted. Spectral classes or types may be the most important classifications of all. They help astronomers determine a star's age, function, rates of mass loss, and history. The spectral type of a star is based mainly on its temperature and luminosity (which is related to temperature). (This makes sense: the hotter

a star is, the greater its intrinsic luminosity.) The spectral types indicated on the HR diagram appear as a series of letters paired with numbers. They were originally set by Harvard astronomer, Henry Draper, and published in 1890. They have been through a few revisions since then. Today, stars are classified as types O, B, A, F, G, K, and M. The numbers paired with the letters are simply subdivisions to indicate the stage of evolution each type of star inhabits. In recent years, classes labeled L and T have been added to describe very dim, low-mass objects, which may or may not all be stars. With the introduction of a class of carbon stars, astronomers have also added the classification C.

In addition to the temperature ranges supplied for the students on page E74, the Draper classifications also include the principal features of the visible spectrum of each star type.

As we move from O to M stars, we are essentially moving from hot, young, massive stars through progressively cooler and less massive stars, until we get to M stars, which are the coolest of all.

The hottest stars are classified O and B. O-type stars (which are hot, young, massive stars) have spectra that show ionized helium, doubly ionized nitrogen, triply ionized silicon, and weak hydrogen fingerprints. These are typical of very hot, young stars. B-type stars (which are more evolved and cooler than their O siblings) show neutral helium, singly ionized oxygen and magnesium, and stronger hydrogen features than in O stars.

Moving down the temperature line we get to A-type stars, which show strong hydrogen

Chapter 1

features, and singly ionized magnesium, silicon, iron, tin, calcium, and other elements in their atmospheres. F stars continue the cooling trend, allowing more metals to form in their atmospheres, along with ionized calcium, iron, and chromium.

G stars (which include our Sun), have strong traces of calcium along with the metals, and some traces of the molecule CH. Stars cooler than the Sun are labeled K and M. They show a strong presence of metals in their atmospheres and some molecules of titanium oxide (which causes the reddish cast to the M-type stars).

Classifying Stars by Luminosity

The classification of stars doesn't stop there! Luminosity classes have been added to the mix. This sorting scheme uses Roman numerals to separate supergiants, giants, dwarfs, and subdwarfs. A luminous supergiant star is labeled Ia, and less luminous supergiants are called Ib. Bright giants and normal giants are labeled II and III, respectively. Subgiants are labeled IV. The Sun and all the stars on the main sequence of the Hertzsprung-Russell diagram are labeled V. The application of all these different classification schemes winds up giving stars a hash of letters and numbers as their "proper" luminosity and spectral designations. The description of Sirius, for example, ends up looking like this:

Sirius — Alpha Canis Majoris (the brightest star in the constellation Canis Major)
Magnitude: -1.5
Spectral Class: A1V
Distance: 8.6 light-years

Risk from Supernova Explosions

Distance and time have a way of separating us from all but the most cataclysmic events in nearby space. There are few nearby stars whose supernova explosions would affect us so drastically that life would end on the planet. But from most supernova explosions, we would merely experience a rush of radiation that our atmosphere should deflect.

In 1987, astronomers detected a supernova explosion in the Large Magellanic Cloud, a companion galaxy to the Milky Way that lies about 170,000 light-years away. The explosion was first seen visually, and it turned out that some particles generated by the explosion, called neutrinos, had been detected as they passed through Earth. Those were the only effects on Earth from this cataclysmic event. There are more distant stars that will explode someday, but their distance will be our best protection.

Risk from Other Cosmic Events

In the very long term, many other cosmic events could happen. For example, our Sun and its immediate stellar neighborhood is passing through a cloud of material called the Local Interstellar Cloud. When that cloud moves past us, our whole region of the galaxy may be subject to stronger radiation from nearby stars. No one knows how seriously to take the danger from this. According to the scientist on whose measurements this work is based (Jeffrey Linsky, at the University of Colorado), we could move out of this cloud at any time in the next few thousand years. Is this something we should be worried about? Only time and more research will tell.

In the far distant future, it is clear that our galaxy and the Andromeda Galaxy will collide with each other. Other galactic collisions have torn apart galaxies, but it isn't immediately clear how life on a small planet in an outer spiral arm will be affected. It won't happen for millions of years, and thus isn't a threat to us.

As your students work through their threat assessments, they will find that imminent danger from the stars is highly unlikely. This is particularly true as they compare such chances with the likelihood of such events as hurricanes, earthquakes, solar flares, accidental releases of radiation from power plants, and impacts from small Earth-orbit-crossing objects.

More Information – on the Web

Visit the *EarthComm* web site www.agiweb.org/earthcomm/ to access a variety of links to web sites that will help you to deepen your understanding of content and prepare you to teach this activity. Many of the sites also contain images that you can download.

Chapter I

Goals and Assessment

Clarify that the goals indicate what students should understand and be able to do as a result of the activity. Make sure students understand that Chapter Assessments are based upon these goals.

Goal	Location in Activity	Assessment Opportunity
Understand the place of our solar system in the Milky Way Galaxy.	**Digging Deeper; Understanding and Applying What You Have Learned** Question 1	Responses reflect a strong level of understanding.
Study stellar structure and the stellar evolution (the life histories of stars).	**Investigate** Part B **Digging Deeper; Check Your Understanding** Questions 1 – 3 **Understanding and Applying What You Have Learned** Questions 1 – 3, and 5	Stars are correctly plotted on HR diagrams. Responses to questions are accurate.
Understand the relationship between the brightness of an object (its luminosity) and its magnitude.	**Investigate** Part A **Digging Deeper; Understanding and Applying What You Have Learned** Questions 2 and 6	Observations are correct; responses reflect a strong level of understanding.
Estimate the chances of another star affecting the Earth in some way.	**Digging Deeper; Understanding and Applying What You Have Learned** Question 5	Responses reflect a strong level of understanding.

NOTES

Chapter 1

Activity 7   Our Community's Place Among the Stars

Goals

In this activity you will:

- Understand the place of our solar system in the Milky Way galaxy.

- Study stellar structure and the stellar evolution (the life histories of stars).

- Understand the relationship between the brightness of an object (its luminosity) and its magnitude.

- Estimate the chances of another star affecting the Earth in some way.

Think about It

When you look at the nighttime sky, you are looking across vast distances of space.

- As you stargaze, what do you notice about the stars?
- Do some stars appear brighter than others? Larger or smaller? What about their colors?

What do you think? Record your impressions and sketch some of the stars in your *EarthComm* notebook. Be prepared to discuss your thoughts with your small group and the class.

Activity Overview

Students observe how the brightness of three light bulbs of different wattages varies with distance. This helps them understand the relationship between brightness and distance, and the challenge presented to astronomers when observing objects in the sky. Students then plot data on a Hertzsprung-Russell (HR) diagram to classify the stars and understand the concept of intrinsic brightness. **Digging Deeper** reviews the classification of stars and explores the lives of stars—from birth to death.

Preparation and Materials Needed

Part A

This part of the investigation can be completed as a class. You will need to assemble the materials. Otherwise, there is little advance preparation.

Part B

You will need to make copies of the HR diagrams and *Table 1* for the students. Use **Blackline Masters** provided in this Teacher's Edition.

Materials

Part A

- Three lamps with 40-W, 60-W, and 100-W bulbs with frosted glass envelopes
- Meter stick
- Graph paper
- Light meter (optional)

Part B

- Copy of blank HR diagram (**Blackline Master Astronomy 7.1, HR Diagram**)
- Copy of HR diagram with star categories (**Blackline Master Astronomy 7.2, HR Diagram**)
- Copy of Table 1 (**Blackline Master Astronomy 7.3, Selected Properties of Fourteen Stars.**)

Think about It

Student Conceptions

Assign this section as a homework exercise before you complete **Activity 7** in class. You may want to give students a couple of days, in case it is cloudy or overcast, making stargazing difficult. Students will likely notice the different brightnesses and sizes of stars in the sky. They are also likely to notice that the stars do in fact appear to have different colors.

Answer for the Teacher Only

Obviously, some stars have a greater apparent brightness than others. There are two reasons for this: stars vary greatly in their intrinsic brightness (how much light they emit), and they also vary greatly in their distance from Earth. Stars also vary in their color, although that effect is subtle.

In contrast to planets, the images of the stars we see are pinpoints of light with no lateral dimension, because the stars are so distant from Earth. What our eyes perceive as the "width" of a star is an optical effect.

Also, a star usually appears to twinkle (that is, the image dances slightly as we look at it). That effect has to do with slight distortions of the path of the starlight as it passes through the Earth's atmosphere. Irregular motions of the atmosphere, together with temperature differences from place to place that give the air a slightly different index of refraction, cause the path of the starlight to vary slightly and irregularly with time. This results in what we perceive as twinkling.

Assessment Tool

Think about It Evaluation Sheet
Use this evaluation sheet to help students understand and internalize the basic expectations for the warm-up activity.

NOTES

Blackline Master Astronomy 7.1
HR Diagram

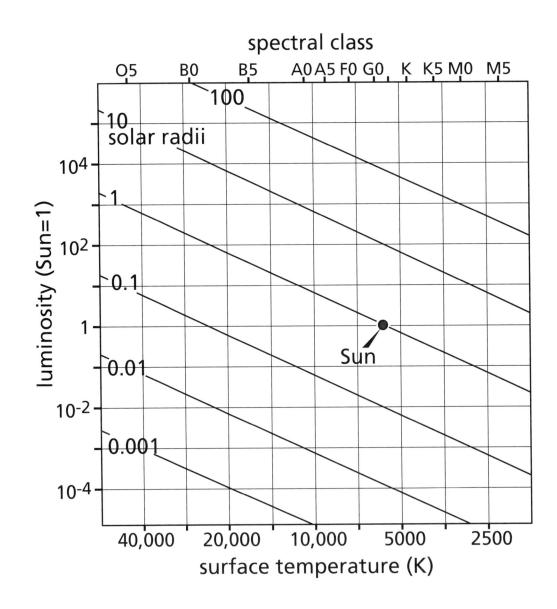

Blackline Master Astronomy 7.2
HR Diagram

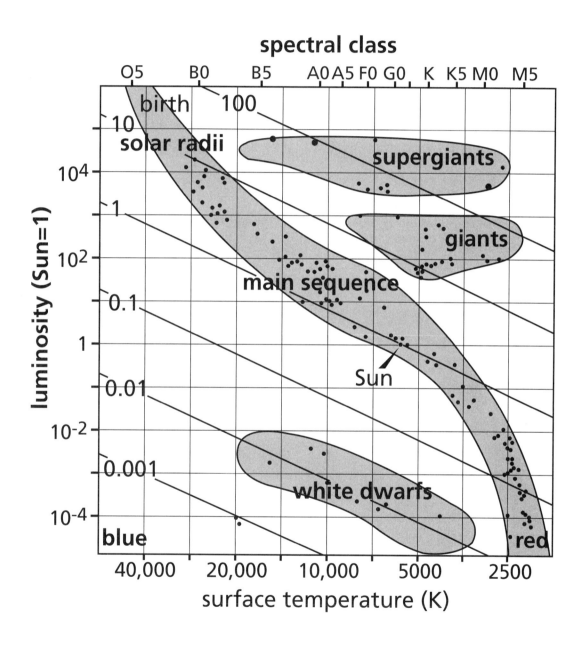

Blackline Master Astronomy 7.3
Selected Properties of Fourteen Stars

Star	Surface Temperature (K)	Luminosity (Relative to Sun)	Distance (Light-Years)	Mass (Solar Masses)	Diameter (Solar Diameters)	Color	Type of Star
Sirius A	9100	22.6	8.6	2.3	2.03	Blue	
Arcturus	4300	115	36.7	4.5	31.5	Red	
Vega	10300	50.8	25.3	3.07	3.1	Blue	
Capella	5300	75.8	42.2	3	10.8	Red	
Rigel	11000	38,679	733	20	62	Blue	
Procyon A	6500	7.5	11.4	1.78	1.4	Yellow	
Betelgeuse	2300	4520–14,968 (variable)	427	20	662	Red	
Altair	7800	11.3	65.1	2	1.6	Yellow	
Aldebaran	4300	156–171 (variable)	65	25	51.5	Red	
Spica	25300	2121	262	10.9	7.3	Blue	
Pollux	4500	31	33	4	8	Red	
Deneb	10500	66,500	1600	25	116	Yellow	
Procyon B	8700	0.0006	11.2	0.65	0.02	White	
Sirius B	24000	0.00255	13.2	0.98	0.022	Blue-white	

Note: Mass, diameter, and luminosity are given in solar units. For example, Sirius A has 2.3 solar masses, has a diameter 2.03 that of the Sun, and has luminosity 22.6 times brighter than the Sun.
1 solar mass = 2 x 10^{30} kg = 330,000 Earth masses; 1 solar diameter = 700,000 km = 110 Earth diameters.

NOTES

Earth System Evolution Astronomy

Investigate

Part A: Brightness versus Distance from the Source

1. Set a series of lamps with 40-, 60-, and 100-W bulbs (of the same size and all with frosted glass envelopes) up at one end of a room (at least 10 m away). Use the other end of the room for your observing site. Turn all the lamps on. Close all of the shades in the room.

   a) Can you tell the differences in brightness between the lamps?

2. Move the lamp with the 40-W bulb forward 5 m toward you.

   a) Does the light look brighter than the 60-W lamp?

   b) Does it look brighter than the 100-W lamp?

3. Shift the positions of the lamps so that the 40-W lamp and the 100-W lamp are in the back of the room and the 60-W lamp is halfway between you and the other lamps.

   a) How do the brightnesses compare?

4. Using a light meter, test one bulb at a time. If you do not have a light meter, you will have to construct a qualitative scale for brightness.

   a) Record the brightness of each bulb at different distances.

5. Graph the brightness versus the distance from the source for each bulb (wattage).

   a) Plot distance on the horizontal axis of the graph and brightness on the vertical axis. Leave room on the graph so that you can extrapolate the graph beyond the data you have collected. Plot the data for each bulb and connect the points with lines.

   b) Extrapolate the data by extending the lines on the graph using dashes.

6. Use your graph to answer the following questions:

   a) Explain the general relationship between wattage and brightness (as measured by your light meter).

   b) What is the general relationship between distance and brightness?

   c) Do all bulbs follow the same pattern? Why or why not?

   d) Draw a light horizontal line across your graph so that it crosses several of the lines you have graphed.

   e) Does a low-wattage bulb ever have the same brightness as a high-wattage bulb? Describe one or two such cases in your data.

   f) The easiest way to determine the absolute brightness of objects of different brightness and distance is to move all objects to the same distance. How do you think astronomers handle this problem when trying to determine the brightness and distances to stars?

7. When you have completed this activity, spend some time outside stargazing. Think about the relationship between brightness and distance as it applies to stars.

   a) Write your thoughts down in your *EarthComm* notebook.

 Do not stare at the light bulbs for extended periods of time.

Investigate

Part A: Brightness versus Distance from the Source

1. If a 10-m space is not available, then the distances in this investigation should be scaled to meet the requirements of the space that is available.

   a) Students should be able to tell the difference between the brightness of the lamps. The difference in brightness between the 40-W and 100-W lamps should be clear. Although the 60-W lamp should be distinct from the other two, students may have more difficulty seeing the difference if the lamps are close together.

2. a) The nearer 40-W lamp should look brighter than the 60-W lamp that is twice as far away.

   b) Although the difference is not as great compared to the 60-W lamp, the nearer 40-W lamp should look brighter than the 100-W lamp that is twice as far away.

3. a) The 60-W lamp should appear brighter than the 40-W lamp that is twice as far away. The difference between the 100-W lamp and the 60-W lamp is less distinct. They will likely appear to be roughly similar in brightness.

4. a) The following data were generated using a light meter over distances up to 2 m away from the light source. Although looking at unshaded light sources for any length of time can make it difficult to develop and gauge a qualitative scale of relative brightness, a qualitative scale can nonetheless be used as a substitute for a light meter.

Distance (m)	Light meter reading for 40-W lamp foot candles (Fc)	Light meter reading for 60-W lamp foot candles (Fc)	Light meter reading for 100-W lamp foot candles (Fc)
0.5	32	68	125
1	7.5	17	35
2	1.9	4.3	8.4
4	0.5	1.1	2.1

Note that the decrease in brightness does not vary as a linear function of distance. It is important to note, however, that the sensitivity of the light meter used will affect the results obtained. Error in the measurements will be greatest at the closest distances to the light source. This is because relative to measurements taken farther away, small changes in distance affect a large change in brightness. The data reported here are qualitative, and should be used only as an example. Your results will vary. In some cases, depending on the light meter used, the decrease in brightness with increasing distance from the source will quickly become too small for the light meter to measure accurately. The important point of this exercise is to illustrate that the relationship between light intensity and distance is not linear. (Actually, light intensity falls off as the square of the distance from the source.)

5.

Change in Brightness with Distance from Source

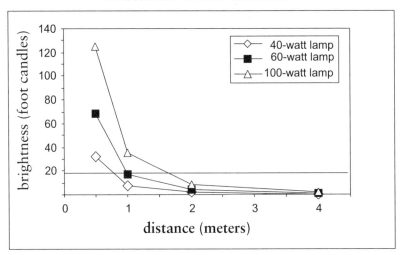

6. a) At the same distance, a higher-wattage lamp is brighter than a lower-wattage lamp.

   b) Brightness decreases with increasing distance.

   c) Each of the bulbs showed the same general pattern, with the decrease in brightness getting smaller at greater distances (i.e., the brightness decreased rapidly at first as one moved farther from the source, then the rate of decrease in brightness with distance lessened with increasing distance.)

   d) See above graph.

   e) Extrapolating from the graph produced in **Step 5**, at 1.5 m from the light source the 100-watt lamp had approximately the same brightness as did the 60-watt lamp when measured 1 m from the light source.

   f) Because both the distance to a star and its absolute brightness affect the brightness that we observe from Earth, scientists must determine at least one of these two parameters independently to interpret the significance of the brightness of a star as it is seen from Earth. Details of the method that astronomers use to determine the distance to distant stars are given in this Teachers Edition in the **Background Information to Activity 1**. In brief, scientists use the same basic principles of triangulation to measure the distance to stars that surveyors on Earth use to measure the distance to much closer objects.

7. Answers will vary, but students should realize that the observations that they made are applicable to stars, and that the distance of a star from the Earth must be considered when judging the significance of the relative brightness of that star as they observe it in the night sky.

Assessment Tools

EarthComm Notebook Entry-Checklist

Use this checklist as a quick guide for student self-assessment and/or an opportunity to quickly score student work. Add further criteria specific to your classroom needs or to this particular investigation.

Investigate Notebook Entry-Evaluation Sheet

Point out the criteria listed on this evaluation sheet that are relevant to this particular investigation. Encourage students to internalize the criteria by making them part of your "assessment conversations" as you circulate around the classroom.

Part B: Luminosity and Temperature of Stars

1. An important synthesis of understanding in the study of stars is the Hertzsprung-Russell (HR) diagram. Obtain a copy of the figure below. Examine the figure and answer the following questions:

   a) What does the vertical axis represent?

   b) What does the horizontal axis represent?

   c) The yellow dot on the figure is the Sun. What is its temperature and luminosity?

   d) Put four more dots on the diagram labeled A through D to show the locations of stars that are:

      A. hot and bright
      B. hot and dim
      C. cool and dim
      D. cool and bright

2. Obtain a copy of the *Table 1* and the HR diagram that shows the locations of main sequence stars, supergiants, red giants, and white dwarfs.

   a) Using the luminosity of the stars, and their surface temperatures, plot the locations of stars shown in *Table 1* on a second HR diagram.

3. Classify each of the stars into one of the following four categories, and record the name in your copy of the table:

   — Main sequence
   — Red giants
   — Supergiants
   — White dwarfs

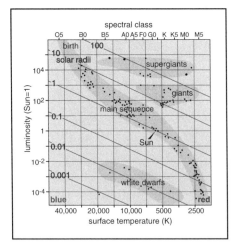

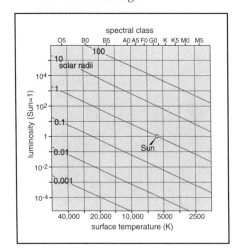

Part B: Luminosity and Temperature of Stars

1. a) The vertical axis represents the luminosity of a star, also called its intrinsic brightness. The luminosity of a star represents the total output of light by the star. It is independent of the distance of the star from the Earth.

    b) The horizontal axis represents the surface temperature of a star, in kelvins.

    c) The Sun has a luminosity of 1 and a surface temperature of about 5500 K.

    d)

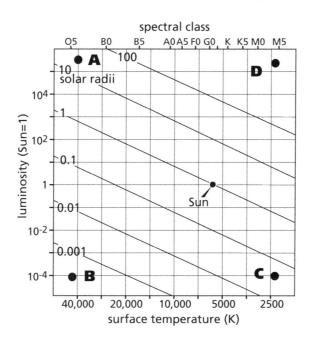

2. a) Note that the positions of the points plotted on this diagram should be taken as approximate.

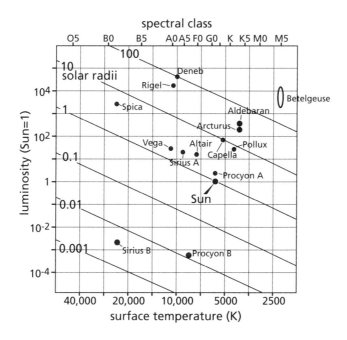

Teaching Tip

Note that the scales on both axes of the Hertzsprung-Russell diagram are not linear. You may want to discuss strategies for best approximating where to plot each of the stars on this diagram. One useful tip is to use the stars' diameters as a guide. The diameters shown in *Table 1* and the contours of the radii in the H-R diagram are both ratioed to that of our Sun, so their numerical values are identical and no conversion is necessary to plot the diameters on the H-R diagram.

3. Note that in this classification exercise, the terms "red giants" and "giants" are being used interchangeably. Red giants fall within the "Giants" field as it is shown in the figure on page E71 of the Student Edition.

Star	Type of star
Sirius A	Main sequence
Arcturus	Giant
Vega	Main sequence
Capella	Main sequence
Rigel	Supergiant
Procyon A	Main sequence
Betelgeuse	Giant/Supergiant
Altair	Main sequence
Aldebaran	Giant
Spica	Main sequence
Pollux	Giant
Deneb	Supergiant
Procyon B	White dwarf
Sirius B	White dwarf

NOTES

Earth System Evolution Astronomy

Table 1 Selected Properties of Fourteen Stars							
Star	Surface Temperature (K)	Luminosity (Relative to Sun)	Distance (Light Years)	Mass (Solar Masses)	Diameter (Solar Diameters)	Color	Type of Star
Sirius A	9100	22.6	8.6	2.3	2.03	Blue	
Arcturus	4300	115	36.7	4.5	31.5	Red	
Vega	10300	50.8	25.3	3.07	3.1	Blue	
Capella	5300	75.8	42.2	3	10.8	Red	
Rigel	11000	38,679	733	20	62	Blue	
Procyon A	6500	7.5	11.4	1.78	1.4	Yellow	
Betelgeuse	2300	4520–14,968 (variable)	427	20	662	Red	
Altair	7800	11.3	65.1	2	1.6	Yellow	
Aldebaran	4300	156–171 (variable)	65	25	51.5	Red	
Spica	25300	2121	262	10.9	7.3	Blue	
Pollux	4500	31	33	4	8	Red	
Deneb	10500	66,500	1600	25	116	Yellow	
Procyon B	8700	0.0006	11.2	0.65	0.02	White	
Sirius B	24000	0.00255	13.2	0.98	0.008	Blue-white	

Note: Mass, diameter, and luminosity are given in solar units. For example, Sirius A has 2.3 solar masses, has a diameter 2.03 that of the Sun, and has luminosity 22.6 times brighter than the Sun.
1 solar mass = 2×10^{30} kg = 330,000 Earth masses; 1 solar diameter = 700,000 km = 110 Earth diameters.

Reflecting on the Activity and the Challenge

Measuring the apparent differences in brightness of the light bulbs at different distances helps you to see that distance and brightness are important factors in helping you understand the objects in our universe. When you look at the stars at night, you are seeing stars at different distances and brightnesses. In your **Chapter Challenge** you will be telling people about the effects of distant objects on the Earth. When you assess danger from space, it is important to understand that stars in and of themselves don't pose a danger unless they are both relatively nearby and doing something that could affect Earth. The spectral characteristics of stars help you to understand their temperature, size, and other characteristics. In turn, that helps you to understand if a given star is or could be a threat to Earth. The light from distant stars can also be used to understand our own star, and our own solar system's makeup and evolution.

EarthComm

Reflecting on the Activity and the Challenge

This exercise not only reinforces the relative distances to our neighboring stars (solidifying the notion that they most likely do not, in and of themselves, pose a direct threat to the Earth) but also provides an opportunity to discuss stellar evolution and to place the Sun into the context of the life cycle of a star. This will help students to further evaluate the potential effects that the Sun can have on the Earth.

Digging Deeper

EARTH'S STELLAR NEIGHBORS

Classifying Stars

You already know that our solar system is part of the Milky Way galaxy. Our stellar neighborhood is about two-thirds of the way out on a spiral arm that stretches from the core of the galaxy. The galaxy contains hundreds of billions of stars. Astronomers use a magnitude scale to describe the brightness of objects they see in the sky. A star's brightness decreases with the square of the distance. Thus, a star twice as far from the Earth as an identical star would be one-fourth as bright as the closer star. The first magnitude scales were quite simple—the brightest stars were described as first magnitude, the next brightest stars were second magnitude, and so on down to magnitude 6, which described stars barely visible to the naked eye. The smaller the number, the brighter the star; the larger the number, the dimmer the star.

Today, scientists use a more precise system of magnitudes to describe brightness. The brightest star in the sky is called Sirius A, and its magnitude is −1.4. Of course, the Sun is brighter at −27 and the Moon is −12.6! The dimmest naked-eye stars are still sixth magnitude. To see anything dimmer than that, you have to magnify your view with binoculars or telescopes. The best ground-based telescopes can detect objects as faint as 25th magnitude. To get a better view of very faint, very distant objects, you have to get above the Earth's atmosphere. The Hubble Space Telescope, for example can detect things as dim as 30th magnitude!

Figure 1 This NASA Hubble Space Telescope near-infrared image of newborn binary stars reveals a long thin nebula pointing toward a faint companion object which could be the first extrasolar planet to be imaged directly.

Digging Deeper

Assign the reading for homework, along with the questions in **Check Your Understanding** if desired.

Assessment Opportunity

Use a quiz to assess student understanding of the concepts presented in this activity. Some sample questions are listed below.

Question: How can a supergiant star be distinguished from a white dwarf?
Answer: The Hertzsprung-Russell diagram can be used to classify different kinds of stars. A supergiant star has a much higher luminosity than a white dwarf star of a similar surface temperature (or spectral class).

Question: What are two possible reasons that one star observed in the night sky might appear brighter than another?
Answer: The brightness of a star, as it is observed from Earth, can be affected by either the intrinsic brightness of the star (its luminosity) or its distance from the Earth. If one star appears brighter than another to an observer on Earth, the brighter star may be closer to the Earth, it may have a greater intrinsic brightness, or its apparent greater relative brightness may be due to some combination of the effects of its distance and intrinsic brightness.

Question: How are neutron stars and supernovae related to one another?
Answer: A neutron star is the remnant of a massive star that has experienced a supernova explosion.

Teaching Tip

The image shown in *Figure 1* on page E73 was taken from NASA's Hubble Space Telescope using the Near Infrared Camera and Multi-Object Spectrometer (NICMOS). The bright objects near the center of image are binary protostars, which illuminate an extended cloud of gas and dust from which the stars formed. So much dust surrounds these protostars that they are virtually invisible at optical wavelengths. However, near-infrared light penetrates the overlying dust, revealing the newborn stars within.

At lower left, there is a point of light many times fainter than the binary protostars. Theoretical calculations indicate that this companion object is much too dim to be an ordinary star. Instead, it has been suggested that a hot protoplanet several times the mass of Jupiter is more consistent with the observed brightness.

Earth System Evolution Astronomy

Perhaps you have seen a star described as a G-type star or an O-type star. These are stellar classifications that depend on the color and temperature of the stars. They also help astronomers understand where a given star is in its evolutionary history. To get such information, astronomers study stars with spectrographs to determine their temperature and chemical makeup. As you can see in the table below, there are seven main categories of stars:

Stellar Classification	Temperature (kelvins)
O	25,000 K and higher
B	11,000–25,000 K
A	7500–11,000 K
F	6000–7500 K
G	5000–6000 K
K	3500–5000 K
M	less than 3500 K

The Lives of Stars

Astronomers use the term **luminosity** for the total rate at which a star emits radiation energy. Unlike apparent brightness (how bright the star appears to be) luminosity is an intrinsic property. It doesn't depend on how far away the star is. In the early 1900s Ejnar Hertzsprung and Henry Norris Russell independently made the discovery that the luminosity of a star was related to its surface temperature. In the second part of this activity, you worked with a graph that shows this relationship. It is called the Hertzsprung-Russell (HR) diagram in honor of the astronomers who discovered this relationship. The HR diagram alone does not tell you how stars change. By analogy, if you were to plot the IQ versus the weight of everyone in your school, you would probably find a very poor relationship between these two variables. Your graph would resemble a scatter plot more than it would a line. However, if you plotted the height versus weight for the same people, you are more likely to find a strong relationship (data would be distributed along a trend or line). The graph doesn't tell you why this relationship exists — that's up to you to determine. Similarly, the HR diagram shows that stars don't just appear randomly on a plot of luminosity versus temperature, but fall into classes of luminosity (red giants, white dwarfs, and so on).

The life cycle of a star begins with its formation in a cloud of gas and dust called a **molecular cloud**. The material in the cloud begins to clump

Geo Words

luminosity: the total amount of energy radiated by an object every second.

molecular cloud: a large, cold cloud made up mostly of molecular hydrogen and helium, but with some other gases, too, like carbon monoxide. It is in these clouds that new stars are born.

Blackline Master Astronomy 7.4

Comparison of the Relative Diameters of Some Different Stars

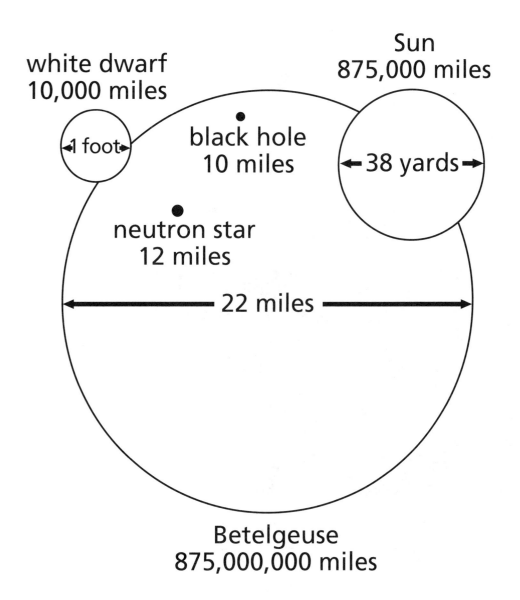

white dwarf
10,000 miles

1 foot

Sun
875,000 miles

38 yards

black hole
10 miles

neutron star
12 miles

22 miles

Betelgeuse
875,000,000 miles

Note that images are not drawn to scale. Numerical diameters are scaled to 10,000 miles equals 1 foot.

together, mixing and swirling. Eventually the core begins to heat as more material is drawn in by gravitational attraction. When the temperature in the center of the cloud reaches 15 million kelvins, the stellar fusion reaction starts up and a star is born. Such stars are called main-sequence stars. Many stars spend 90% of their lifetimes on the main sequence.

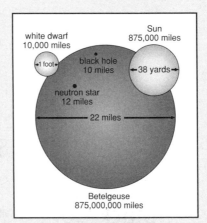

Figure 2 Scaling stars to 10,000 miles to one foot reveals the relative sizes of various stars.

Newborn stars are like baby chickens pecking their way out of a shell. As these infant stars grow, they bathe the cloud surrounding them in strong ultraviolet radiation. This vaporizes the cloud, creating beautiful sculpted shapes in the cloud. In the photograph in *Figure 3*, the Hubble Space Telescope studied a region of starbirth called NGC 603. Notice the cluster of bright white stars in the center "cavern" of the cloud of gas and dust. Their ultraviolet light has carved out a shell of gas and dust around the stellar newborns.

Figure 3 The starforming region NGC 603 in the galaxy M33.

Figure 4 The Orion Nebula is an example of a molecular cloud, from which new stars are born.

Teaching Tip

Use **Blackline Master Astronomy 7.4, Comparison of the Relative Diameters of Some Different Stars** to make an overhead of *Figure 2* on page E75. Use this overhead to discuss with students the different sizes of various stars. Does the relative size of the Sun surprise them?

Teaching Tip

Figure 3 on page E75 of the Student Edition shows a Hubble Space Telescope image of the vast nebula called NGC 604. NGC 604 lies in the neighboring spiral galaxy M33, which is located about 2.7 million light-years from Earth. This is the site where new stars are forming in a spiral arm of the galaxy, and although such nebulae are common, this one is particularly large being nearly 1500 light-years across. At the heart of NGC 604 are more than 200 hot stars that are much more massive than our Sun.

Teaching Tip

Figure 4 on page E75 shows an image of the Orion Nebula. This image is based on a composite of 81 images obtained using the European Southern Observatory's Very Large Telescope (VLT) at the Paranal Observatory. The Orion Nebula is located within the Orion constellation. This beautiful nebula is bright enough to be seen with the naked eye, and the use of a small telescope reveals it to be a complex of gas and dust tens of light-years wide. At its core several massive and hot stars known as the Trapezium stars illuminate the nebula. At the heart of this nebula there are in fact about 1000 very young stars that are only about one million years old.

Earth System Evolution Astronomy

How long a star lives depends on its mass (masses of selected stars are shown in *Table 1* in the **Investigate** section of the activity). Stars like our Sun will live about 10 billion years. Smaller, cooler stars might go on twice that long, slowly burning their fuel. Massive supergiant stars consume their mass much more quickly, living a star's life only a few tens of millions of years. Very hot stars also go through their fuel very quickly, existing perhaps only a few hundred thousand years. The time a star spends on the main sequence can be determined using the following formula:

Time on main sequence = $\dfrac{1}{M^{2.5}} \times$ **10 billion years**

where *M* is the mass of the star in units of solar masses.

Even though high-mass stars have more mass, they burn it much more quickly and end up having very short lives.

In the end, however, stars of all types must die. Throughout its life a star loses mass in the form of a stellar wind. In the case of the Sun this is called the solar wind. As a star ages, it loses more and more mass. Stars about the size of the Sun and smaller end their days as tiny, shrunken remnants of their former selves, surrounded by beautiful shells of gas and dust. These are called planetary nebulae. In about five billion years the Sun will start to resemble one of these ghostly nebulae, ending its days surrounded by the shell of its former self.

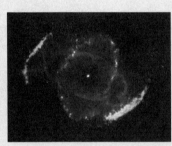
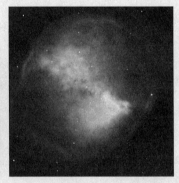

Figure 5A, 5B 5C Three examples of the deaths of stars about the size of the Sun. **A:** The Butterfly Nebula. **B:** The Cat's-Eye Nebula. In both cases at least the dying star lies embedded in a cloud of material exhaled by the star, as it grew older. **C:** The Dumbbell Nebula. European Southern Observatory.

E 76

Teaching Tip

Figure 5 on page E76 shows pictures of the Butterfly (NGC6302), Cat's-Eye (NGC 6543), and Dumbbell (NGC6853) nebulae. These are planetary nebulae associated with the deaths of stars. The term planetary nebula is a bit of a misnomer because these phenomena have nothing to do with planetary formation, which is thought to occur much earlier in the life cycles of stars. Instead, these nebulae have been traditionally referred to as planetary nebulae because of their resemblance to a planet in a small telescope.

Massive stars (supergiants tens of times more massive than the Sun) also lose mass as they age, but at some point their cores collapse catastrophically. The end of a supergiant's life is a cataclysmic explosion called a **supernova**. In an instant of time, most of the star's mass is hurled out into space, leaving behind a tiny remnant called a **neutron star**. If the star is massive enough, the force of the explosion can be so strong that the remnant is imploded into a **stellar black hole**—a place where the gravity is so strong that not even light can escape.

The material that is shed from dying stars (whether they end their days as slowly fading dwarf stars, or planetary nebulae, or supernovae) makes its way into the space between the stars. There it mixes and waits for a slow gravitational contraction down to a new episode of starbirth and ultimately star death. Because humans evolved on a planet that was born from a recycled cloud of stellar mass, they are very much star "stuff"—part of a long cycle of life, death, and rebirth.

Astronomers search the universe to study the mechanics of star formation. Star nurseries and star graveyards are scattered through all the galaxies. In some cases, starbirth is triggered when one galaxy collides with (actually passes through) another. The clouds of gas and dust get the push they need to start the process.

Scientists also search for examples of planetary nebulae. They want to understand when and how these events occur. Not only are these nebulae interesting, but also they show scientists what the fate of our solar system will be billions of years from now.

What would happen if there were a supernova explosion in our stellar neighborhood sometime in the future? Depending on how close it was, you could be bombarded with strong radiation and shock waves from the explosion. The chances of this happening are extremely small—although some astronomers think that a supernova some five billion years ago may have provided the gravitational kick that started our own proto–solar nebula on the road to stardom and planetary formation.

Figure 6 The Crab Nebula is the remnant of a supernova explosion first observed in the year AD 1054.

Check Your Understanding

1. How do astronomers classify stars?

2. Write a brief outline of how stars are born.

3. What determines the way a star dies?

Teaching Tip

Figure 6 (on page E77) shows a three-color composite image of the Crab Nebula (also known as Messier 1). The Crab Nebula is the remnant of a supernova explosion at a distance of about 6000 light-years, observed almost 1000 years ago, in AD 1054. Near its center there is a neutron star.

Check Your Understanding

1. Astronomers classify stars by their brightness, which is determined by using a magnitude scale. Astronomers also use color and temperature in their classification of stars.

2. A star is born when the material in a cloud of gas and dust begins to clump together because of gravitational attraction. Eventually, the core material heats up until it reaches a temperature of 15 million K. A stellar fusion reaction starts, and a star is born.

3. How a star dies is determined by the size of the star. Stars of masses similar to the Sun die by gradually shedding mass through stellar wind. Larger stars die in cataclysmic explosions called "supernovae."

Assessment Tool

Check Your Understanding Notebook Entry-Evaluation Sheet
Use this sheet to evaluate the extent to which students understand the key concepts explored in **Activity 7** and explained in the **Digging Deeper** reading section, and to evaluate the students' clarity of expression.

Earth System Evolution Astronomy

Understanding and Applying What You Have Learned

1. Using an astronomy computer program or a guidebook to the stars, make a list of the 10 nearest stars, and their distances, magnitudes, and spectral classes. What do their classes tell you about them?

2. What is mass loss and how does it figure in the death of a star? Is the Sun undergoing mass loss?

3. What happens to the material left over from the death of a star?

4. Two identical stars have different apparent brightnesses. One star is 10 light years away, and the other is 30 light-years away from us. Which star is brighter, and by how much?

5. Refer to *Table 1* to answer the questions below:

   a) Calculate how long the Sun will spend on the main sequence.
   b) Calculate how long Spica will spend on the main sequence.
   c) Relate your results to the statement that the more massive the star, the shorter they live.

6. Explain the relationships between temperature, luminosity, mass, and lifetime of stars.

Preparing for the Chapter Challenge

You are about to complete your **Chapter Challenge**. In the beginning you were directed to learn as much as you could about how extraterrestrial objects and events could affect the Earth and your community. In order to do this you have explored the stars and planets, looking at all the possibilities. By now you have a good idea about how frequently certain kinds of events occur that affect Earth. The Sun is a constant source of energy and radiation.

In this final activity you learned our solar system's place in the galaxy, and you read about how stars are born and die. Because the birth of our solar system led directly to our planet, and the evolution of life here, it's important to know something about stars and how they come into existence.

You now know that the solar system is populated with comets and asteroids, some of which pose a threat to Earth over long periods of time.

Understanding and Applying What You Have Learned

1.

Star	Distance (ly)	Magnitude	Spectral Class
Sun	0.000016	-26.9	G2
Proxima Centauri	4.2	11.3	M5e
Alpha Centauri A	4.3	0.33	G0
Alpha Centauri B	4.3	1.70	K5
Barnard's Star	5.96	9.5	M5
Wolf 359	7.6	13.5	M6e
Lalande 21185	8.11	7.5	M2
Alpha Sirius	8.7	-1.47	A0
Beta Sirius	8.7	8.3	white dwarf
A Luyten 726-8	8.93	12.5	M6e

The spectral class of a star tells astronomers the surface temperature of the star. It also tells something about what the star was like when it was born and how it evolves toward its eventual death.

2. Mass loss refers to the fact that a star loses mass in the form of stellar wind as it ages. The older a star gets, the more mass it loses. Yes, the Sun is undergoing mass loss.

3. The material that is shed from dying stars makes its way into the space between the stars, where it mixes and waits for a slow gravitational contraction down to a new episode of starbirth, and ultimately star death.

4. Provided that each star has the same intrinsic brightness, the star that is 10 light-years away appears brighter to an Earth observer. The star that is 30 light-years away is three times as far away as the star that is 10 light-years away. Because the brightness of a star decreases as the square of its distance from the Earth, the brightness of the star that is 30 light-years away is only one-ninth that of the brightness of the star that is 10 light-years away.

5. Students should use the equation given in the **Digging Deeper** reading section to complete the calculations.

   a) The Sun will remain on the main sequence about 10 billion years. Time on main sequence = $[1/(1 \text{ solar mass})^{2.5}] \times 10$ billion years.

   b) Spica will remain on the main sequence about 2.55×10^7 years (25.5 million years). Time on main sequence = $[1/(10.9 \text{ solar mass})^{2.5}) \times 10$ billion years.

   c) Spica has a much larger mass than the Sun, so it will remain on the main sequence for a much shorter period of time.

6. The Hertzsprung-Russell diagram, shown in **Part B** of **Investigate**, gives the relationship between the radius, the surface temperature, and the luminosity of stars. From this diagram it can be seen that at a given surface temperature, intrinsic luminosity increases as size (radius) of the star increases. Similarly, for a given radius, luminosity increases as surface temperature increases. The equation in the **Digging Deeper** reading section gives the lifetime as a function of the size of the star. This equation shows that as the mass of the star increases, the time the star spends on the main sequence (which is related to the stars' lifetime) decreases. In other words, there is an inverse correlation between a stars' mass and its expected life span.

Preparing for the Chapter Challenge

Students should now have the understanding that they need to complete the **Chapter Challenge**. In this activity students were introduced to our solar system's relative place in our galaxy. In preparation for the completion of the **Chapter Challenge**, you may wish to have students write a short paper that discusses our solar system's place in the galaxy. As part of this, students can consider the Sun's place in the overall scheme of different star types and the lifecycles of stars, and how this may be beneficial or detrimental to the future of life on Earth.

NOTES

The evolution of the Earth's orbit and its gravitational relationship with the Moon make changes to the Earth's climate, length of year, and length of day. The solar system is part of a galaxy of other stars, with the nearest star being only 4.21 light-years away. The Sun itself is going through a ten-billion-year-long period of evolution and will end as a planetary nebula some five billion years in the future. Finally, our Milky Way galaxy is wheeling toward a meeting with another galaxy in the very, very distant future. Your challenge now that you know and understand these things is to explain them to your fellow citizens and help them understand the risks and benefits of life on this planet, in this solar system, and in this galaxy.

Inquiring Further

1. **Evolution of the Milky Way galaxy**

   The Milky Way galaxy formed some 10 billion years ago, when the universe itself was only a fraction of its current age. Research the formation of our galaxy and find out how its ongoing evolution influenced the formation of our solar system.

2. **Starburst knots in other galaxies**

   Other galaxies show signs of star birth and star death. You read about a starbirth region called NGC 604 in the **Digging Deeper** reading section of this activity. Astronomers have found evidence of colliding galaxies elsewhere in the universe. In nearly every case, such collisions have spurred the formation of new stars. In the very distant future the Milky Way will collide with another galaxy, and it's likely that starburst knots will be formed. Look for examples of starbirth nurseries and starburst knots in other galaxies and write a short report on your findings. How do you think such a collision would affect Earth (assuming that anyone is around to experience it)?

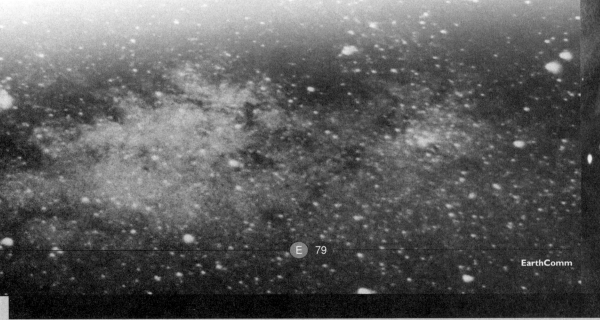

E 79

EarthComm

Inquiring Further

1. Evolution of the Milky Way Galaxy

Astronomers believe that the galaxy formed out of a large, fairly spherical cloud of cold gas, rotating slowly in space. At some point in time, the cloud began to collapse in on itself, or condense, in the same way that the clouds which formed individual stars condensed. Initially, some stars may have formed as the gas cloud began to fragment around the edges, with each fragment condensing further to form a star or group of stars.

The cloud continued to collapse, with more and more stars being formed as it did so. Because the cloud was rotating, the spherical shape began to flatten out into a disc, and the stars, which were formed at this time, filled the disc regions. As the formation of new stars continued, some of those that had been created earlier had time to evolve to the end of their active lifetimes, and these stars began to shed their atmospheres or explode in huge supernova events. In the process, these older citizens of the still-young galaxy enriched the gas in the cloud with the new, heavier elements that they had formed, and the new stars being created in the disc regions contained the heavier elements.

Astronomers call these younger, enriched stars "population 1 stars," and the older stars "population 2 stars." The Milky Way Galaxy of today is a very different place than the cold gas from which it formed over 10 billion years ago. It is no longer a spherical mass of hydrogen. Today, astronomers with radio telescopes have charted the clouds of gas and have found that the Milky Way is a spiral galaxy.

2. Starburst knots in other galaxies

The *EarthComm* web site provides references to guide your students to information regarding some starburst nurseries and starburst knots that exist in other galaxies. You may wish to investigate this information and assign specific examples for your students to research.

Earth Science at Work

ATMOSPHERE: *Astronaut*
There is no atmosphere in space; therefore, astronauts must have pressurized atmosphere in their spacecraft cabins. Protective suits protect them when they perform extra-vehicular activities.

BIOSPHERE: *Exobiologist*
"Did life ever get started on Mars?" By learning more about the ancient biosphere and environments of the early Earth, exobiologists hope that they may be able to answer such questions when space missions return with rocks gathered on Mars.

CRYOSPHERE: *Glaciologist*
Ice is abundant on the Earth's surface, in the planetary system, and in interstellar space. Glaciologists study processes at or near the base of glaciers and ice sheets on Earth and other planets.

GEOSPHERE: *Planetary geologist*
By researching Martian volcanism and tectonism, or the geology of the icy satellites of Jupiter, Saturn, and Uranus, planetary geologists hope to develop a better understanding of our place in the universe.

HYDROSPHERE: *Lifeguard*
Surfers and other water-sport enthusiasts rely on lifeguards to inform them of the time of high and low tides. Low tides or high tides can create dangerous situations.

How is each person's work related to the Earth system, and to Astronomy?

 E 80

NOTES

Astronomy and Your Community: End-of-Chapter Assessment

1. The average distance from the Earth to the Sun is
   a) 1 astronomical unit
   b) 10 parsecs
   c) 10,000,000 leagues
   d) 100,000,000 meters

2. You are asked to make a scale model of two objects that are 50,000,000 km apart. If you used a scale of 1m = 10,000,000 km, how far apart will the two objects be on the model?
   a) 5 m
   b) 5 km
   c) 50,000,000 m
   d) 50,000,000 km

3. What is the size of the smallest object that can be represented in the above model, if the smallest part of that model is no less than 1 mm (0.001 m)?
   a) 001 m
   b) .001 km
   c) 10,000 m
   d) 10,000 km

4. According to current theory used to explain the origin of our solar system, the planets in our solar system:
   a) Formed from the same nebular cloud as the Sun.
   b) Were captured into orbit by the Sun's gravity.
   c) Contain about one-tenth the mass of the Sun.
   d) Formed by fusion of hydrogen in their cores.

Refer to the diagram below to answer questions 5 and 6.

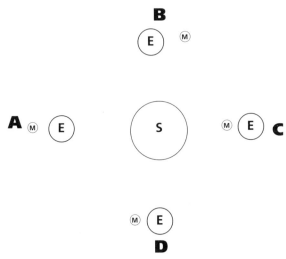

(Note that the diagram is not drawn to scale)

5. Which of the four configurations shown correspond to the new moon phase?
   a) A
   b) B
   c) C
   d) D

6. Which of the following best describes the tidal conditions associated with part B of the diagram?
   a) high tide
   b) low tide
   c) spring tide
   d) neap tide

7. The force that creates the ocean tides is:
   a) The gravitational attraction of the Moon.
   b) The gravitational attraction of the Sun.
   c) The gravitational attraction of the Moon and Sun.
   d) The rotation of the Earth

8. The Earth has fewer visible craters than the Moon because:
   a) Most of the craters are hidden by the oceans.
   b) The moon has shielded the Earth from most impacts.
   c) Craters have been destroyed by geologic processes.
   d) Volcanic activity on the Earth has covered most craters.

9. If the Earth rotated twice as fast on its axis:
   a) The tides would be twice as high as now.
   b) There would be four high tides each day.
   c) Successive high tides would be separated by less time.
   d) There would be twice as many spring tides.

10. As the foci of an ellipse get closer together:
    a) The ellipse becomes more eccentric
    b) The ellipse becomes more like a circle
    c) The volume of the ellipse decreases
    d) The area of the ellipse increases

11. The orbit of the Earth about the Sun is shaped most like:
    a) Pluto's orbit
    b) a Comet's orbit
    c) a circle
    d) a figure eight

12. During the course of the year, the distance between the Earth and Sun currently:
    a) Changes by about 4%.
    b) Changes by about 15%.
    c) Changes by about 30%.
    d) Is constant.

13. The tilt of the Earth's rotation axis:
    a) Varies on an annual cycle.
    b) Varies on a 41,000 year cycle.
    c) Varies between 20° and 30°.
    d) Varies between 0° and 10°.

14. If the tilt of the Earth's rotation axis were less than it is today, then compared to today winter would be
    a) warmer
    b) cooler
    c) the same temperature

15. The size of an impact crater is:
    a) Always bigger than the impactor.
    b) About the same size as the impactor.
    c) Dependent only upon the velocity of the impactor.
    d) Dependent only on the mass of the impactor.

16. To calculate the energy of a meteoric impact one needs to know only the:
    a) Mass and velocity of the meteorite.
    b) Density and velocity of the meteorite.
    c) Diameter and density of the meteorite.
    d) Type of meteorite and its velocity.

17. A future scientist observes that the sunspot cycle has just reached a minimum. The year is 2016. On the basis of this observation, the scientist would then predict the next maximum in sunspot activity to occur in:
    a) 2018
    b) 2021
    c) 2026
    d) 2036

18. One evening, several regional electrical power grids were disrupted. At the same time, cellular telephone users lost service because several communication satellites stopped functioning properly. Which of the solar phenomena listed below is the most likely cause of these disruptions?
    a) solar flares
    b) sunspot activity
    c) Oort minima
    d) coronal turbulence

19. If the entire Earth was suddenly and permanently covered with a layer of glistening, white snow, the albedo of the Earth would _____ making the Earth a much_____ place.
    a) decrease, warmer
    b) increase, warmer
    c) decrease, colder
    d) increase, colder

20. If the Sun stopped producing energy, then all fusion would cease in its
    a) radiative zone
    b) lithosphere
    c) core
    d) corona

21. The order of colors observed in the visible light spectrum is determined by the:
    a) Intensity of the light.
    b) Wavelength of the light.
    c) Quality of the prism.
    d) Position of the observer.

22. An astronomer is studying the electromagnetic radiation from two different astronomical bodies. He finds that the radiation spectrum from the first body has a peak wavelength in the visible light portion of the spectrum, near the blue end. The second body has its peak radiation in the x-ray range of the spectrum. From this information the astronomer correctly concludes that the
    a) second body is hotter than the first
    b) second body is colder than the first
    c) second body is moving toward Earth faster than the first
    d) second body is moving toward Earth slower than the first

23. The gaseous elements ionized in the fluorescent light bulb that you observed caused the light spectrum to:
    a) Have a different order of colors.
    b) Exhibit characteristic bright emission lines.
    c) Be shifted towards shorter wavelengths.
    d) Be dimmer and less distinct.

Refer to the diagram below to answer questions 24-26.

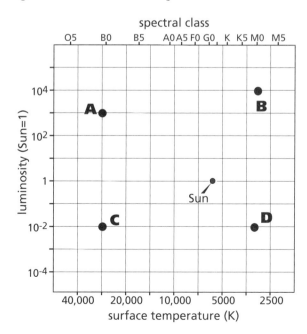

Chapter 1

24. Which of the four stars shown (A, B, C, D) has the largest diameter?
    a) A
    b) B
    c) C
    d) D
25. If all four stars were equidistant from the Earth, which stars would appear to have the same brightness in the night sky?
    a) A and C
    b) C and D
    c) none would appear the same
    d) all would appear the same

26. If star C were 4 times as far away from Earth as Star D, then star C would have an apparent brightness to an Earth observer that was
    a) four times that of star D
    b) the same as star D
    c) one-fourth that of star D
    d) one-sixteenth that of star D

27. Four stars described below all reside on the main sequence. Which will have the shortest expected life span?

Star	Surface Temperature (K)	Luminosity (relative to Sun)	Diameter (solar diameters)	Mass (solar masses)
A	10,300	50.8	3.1	3.07
B	5300	75.8	10.8	3.0
C	2300	7.5	1.4	1.78
D	25300	2121	7.3	10.9

Answer Key

1. a
2. a
3. d
4. a
5. c
6. d
7. c
8. c
9. c
10. b
11. c
12. a
13. b
14. a
15. a
16. a
17. b
18. a
19. d
20. c
21. b
22. a
23. b
24. b
25. b
26. d
27. d

Teacher Review

Use this section to reflect on and review the investigation. Keep in mind that your notes here are likely to be especially helpful when you teach this investigation again. Questions listed here are examples only.

Student Achievement

What evidence do you have that all students have met the science content objectives?

_____

Are there any students who need more help in reaching these objectives? If so, how can you provide this?_____

_____

What evidence do you have that all students have demonstrated their understanding of the inquiry processes?_____

_____

Which of these inquiry objectives do your students need to improve upon in future investigations? _____

_____

What evidence do the journal entries contain about what your students learned from this investigation? _____

_____

Planning

How well did this investigation fit into your class time?_____

_____

What changes can you make to improve your planning next time? _____

_____

Guiding and Facilitating Learning

How well did you focus and support inquiry while interacting with students?

_____

What changes can you make to improve classroom management for the next investigation or the next time you teach this investigation? _____

_____

How successful were you in encouraging all students to participate fully in science learning?

How did you encourage and model the skills values, and attitudes of scientific inquiry?

How did you nurture collaboration among students?

Materials and Resources

What challenges did you encounter obtaining or using materials and/or resources needed for the activity?

What changes can you make to better obtain and better manage materials and resources next time?

Student Evaluation

Describe how you evaluated student progress. What worked well? What needs to be improved?

How will you adapt your evaluation methods for next time?

Describe how you guided students in self-assessment.

Self Evaluation

How would you rate your teaching of this investigation?

What advice would you give to a colleague who is planning to teach this investigation?

NOTES

2

Climate Change
...and Your Community

EARTH SYSTEM EVOLUTION
CHAPTER 2

CLIMATE CHANGE ...
AND YOUR COMMUNITY

Chapter Overview

The **Chapter Challenge for Climate Change and Your Community** is for students to write a series of newspaper articles that explore global climate change. As students move through the chapter, they gain the knowledge needed to write articles concerning:

- how global climate has changed over time
- what causes global climate change
- what global warming means
- how global warming might affect their community

Finally, students are asked to use the knowledge they have acquired to write an editorial piece that states:

- whether or not their community should be concerned about global warming
- what steps the community should take in response to the possibility of global warming

Chapter Goals for Students

- Understand how changes in climate are tied to Earth systems.

- Participate in scientific inquiry and construct logical conclusions based on evidence.

- Recognize the global impacts of a change in climate.

- Appreciate the value of Earth science information in improving quality of life, both globally and within the community.

Chapter Timeline

Chapter 2 takes about three weeks to complete, assuming one 45-minute period per day, five days per week. Adjust this guide to suit your school's schedule and standards. Build flexibility into your schedule by manipulating homework and class activities to meet your students' needs.

A sample outline for presenting the chapter is shown on the following pages. This plan assumes that you assign homework at least three nights a week, and that you assign **Understanding and Applying What You Have Learned** and **Preparing for the Chapter Challenge** as group work to be completed during class. This outline also assumes that **Inquiring Further** sections are reserved as additional, out-of-class activities.

This outline is only a sample, not a suggested or recommended method of working through the chapter. Adjust your daily and weekly plans to meet the needs of your students and your school.

Day	Activity	Homework
1	Getting Started; Activity; Chapter Challenge; Assessment Criteria	
2	Activity 1 – Investigate; Parts A and B	
3	Activity 1 – Investigate; Part C	Digging Deeper; Check Your Understanding
4	Activity 1 – Review; Understanding and Applying; Preparing for the Chapter Challenge	
5	Activity 2 – Investigate; Parts A and B	Digging Deeper; Check Your Understanding
6	Activity 2 – Review; Understanding and Applying; Preparing for the Chapter Challenge	
7	Activity 3 – Investigate; Parts A to C	
8	Activity 3 – Investigate; Parts D and E	Digging Deeper; Check Your Understanding
9	Activity 3 – Review; Understanding and Applying; Preparing for the Chapter Challenge	
10	Activity 4 – Investigate	Digging Deeper; Check Your Understanding
11	Activity 4 – Review; Understanding and Applying; Preparing for the Chapter Challenge	
12	Activity 5 – Investigate; Parts A to C	Digging Deeper; Check Your Understanding

Chapter 2

Day	Activity	Homework
13	**Activity 5 – Investigate; Part C continued**	
14	**Activity 5 – Review; Understanding and Applying; Preparing for the Chapter Challenge**	
15	**Activity 6 – Investigate**	**Digging Deeper; Check Your Understanding**
16	**Complete Chapter Report**	**Finalize Chapter Report**
17	**Present Chapter Report**	

National Science Education Standards

Writing a series of newspaper articles about global warming sets the stage for the **Chapter Challenge**. Students learn how and why global climate has changed over time. They study the effects of climate change on the community. Through a series of activities, students begin to develop the content understandings outlined below.

CONTENT STANDARDS

Unifying Concepts and Processes
- Systems, order and organization
- Evidence, models, and explanation
- Constancy, change, and measurement
- Evolution and equilibrium

Science as Inquiry
- Identify questions and concepts that guide scientific investigations
- Design and conduct scientific investigations
- Use technology and mathematics to improve investigations
- Formulate and revise scientific explanations and models using logic and evidence
- Communicate and defend a scientific argument
- Understand scientific inquiry

Earth and Space Science
- Energy in the Earth system
- Geochemical cycles
- Origin and evolution of the Earth system

Science and Technology
- Communicate the problem, process, and solution
- Understand science and technology

Science in Personal and Social Perspectives
- Personal and community health
- Population growth
- Natural resources
- Environmental quality
- Natural and human-induced hazards
- Science and technology in local, national, and global challenges

History and Nature of Science
- Science as a human endeavor
- Nature of scientific knowledge
- Historical Perspectives

Chapter 2

Key Science Concepts and Skills	
Activities Summaries	**Earth Science Principles**

Activity 1: Present-Day Climate in Your Community
Students examine data on temperature and precipitation to describe the climate of their community. They look at a local topographic map to determine what physical features in their community might influence climate. Students then compare the climate of their community to that of a different community.
	• Climate, local and global
	• Weather
	• Factors affecting climate:
	   • latitude
	   • elevation
	   • geography

Activity 2: Paleoclimates
Students correlate changes in the appearance of tree rings to the Little Ice Age to understand the significance of growth rings in trees as evidence for climate change. They create a model of sediment layers to understand how fossil pollen can be used to understand climate change through time.
	• Paleoclimate
	• Evidence of climate change:
	   • fossil pollen
	   • ice cores
	   • deep-sea sediments
	   • glacial sediments
	   • tree rings

Activity 3: How Do Earth's Orbital Variations Affect Climate? Students complete a series of exercises on paper and then use a globe to understand the causes of seasons on Earth. They also use the globe to understand how the tilt of the Earth's axis causes the seasons. Students then draw a series of ellipses to understand the shape of the Earth's orbit. Finally, they complete an experiment to understand how energy from the Sun varies with distance.	• Axial tilt • Eccentricity • Obliquity • Precession • Milankovitch cycles
Activity 4: How Do Plate Tectonics and Ocean Currents Affect Global Climate? Students build models to understand how the positions of the continents have affected ocean circulation around Antarctica over the past 55 million years. They consider how changes in ocean circulation affect climate.	• Ocean circulation • Ocean currents and global climate • Plate tectonics and global climate • Pangea
Activity 5: How Do Carbon Dioxide Concentrations in the Atmosphere Affect Global Climate? Students examine atmospheric carbon dioxide concentrations over the last century and over the last 160,000 years. They compare the data to global temperatures to find the relationship between the two. Students then design and conduct an experiment to help them understand the greenhouse effect.	• Atmospheric carbon dioxide concentrations • Greenhouse gases and the greenhouse effect • Carbon budget
Activity 6: How Might Global Warming Affect Your Community? Students work in small groups to brainstorm about the effects of increased temperature on the community. They select one outcome of global warming and then, on paper, design an experiment to test their idea.	• Making predictions • Computer modeling • Effects of global warming

Equipment List for Chapter Two:

Materials needed for each group per activity.

Activity 1 Part A

• Topographic map of community*

Activity 1 Part B

• Topographic map of a different community*

Activity 1 Part C

• Variable, depending on experimental design

Activity 2 Part A

• Copies of the photo of tree rings (**Blackline Master Climate Change 2.1, Tree Rings**)

Activity 2 Part B

• Blue, red, green, and yellow modeling clay
• Rolling pin (optional)
• Waxed paper (optional)
• Beaker or small milk carton
• Small copper pipe or empty barrel from a ballpoint pen
• Thin stick (that fits inside pipe)

Activity 3 Parts A and C

• Compass (to draw a circle)
• Ruler
• Protractor
• Blue, red, and black colored pencils

Activity 3 Part B

• Globe
• Small thermometer
• Duct tape
• Black marker
• Lamp with a reflector or halogen bulb
• Meter stick

Activity 3 Part D

• Piece of rope, 4 m long
• Two wooden dowels
• Ruler
• Chalk, at least four different colors

Activity 3 Part E

• Scissors
• Ruler
• Poster board
• Lamp without a shade
• Meter stick
• Chalk

Activity 4

• Copies of the Southern Hemisphere continental configurations: Eocene, Miocene, and present (**Blackline Master Climate Change 4.1 A - C, Continental Configurations**)
• Cake pan, 8 inches round
• Modeling clay
• Water
• Blue-colored ice cubes

Activity 5 Part A

• Graph paper (**Blackline Master Astronomy 2.2**)

*See the *EarthComm* web site for information about how to obtain these resources at http://www.agiweb.org/earthcomm

Chapter 2

Activity 5 Part B

• Copy of graph showing ice core data (use **Blackline Master 5.1 Graph of Data from Ice Core in Antarctica**)

Activity 5 Part C

• Suggested materials for greenhouse-effect experiment: two identical 2-L plastic bottles or beakers or small aquarium tanks or dish tubs with glass "lids," water, clear plastic bags, thermometers, ice cubes, lamps, graph paper

• See the *EarthComm* web site for further suggestions on modeling the greenhouse effect

Activity 6

• Poster board

*See the *EarthComm* web site for information about how to obtain these resources at http://www.agiweb.org/earthcomm

NOTES

CLIMATE CHANGE
...and Your Community

Getting Started

The Earth's climate has changed many times over geologic history.

• What kinds of processes or events might cause the Earth's climate to change?

What do you think? Write down your ideas about these questions in your *EarthComm* notebook. Be prepared to discuss your ideas with your small group and the class.

Scenario

Your local newspaper would like to run a series of articles about global warming. However, the newspaper's science reporter is unavailable. The newspaper has come to your class to ask you and your classmates to write the articles. These feature articles and an editorial will be run in the Science and Environment section of the newspaper. The newspaper editor wants to give the readers of the paper a thorough scientific background to understand the idea of global climate change.

Chapter Challenge

Article 1: How Has Global Climate Changed Over Time?

Many people are not aware that the Earth's climate has changed continually over geologic time. This article should contain information about:

• the meaning of "climate," both regional and global;
• examples of different global climates in the geologic past;
• how geologists find out about past climates, and
• a description of your community's present climate and examples of past climates in your part of the country.

Getting Started

Uncovering students' conceptions about Climate Change and the Earth System

Use **Getting Started** to elicit students' ideas about the main topic. The goal of **Getting Started** is not to seek closure (i.e., the "right" answer) but to provide you, the teacher, with information about the students' starting point and about the diversity of ideas and beliefs in the classroom. By the end of the chapter, students will have developed a more detailed and accurate understanding of how the Earth's climate has changed through time.

This **Getting Started** question is intended to provide you with important information about student's conceptions of what climate means, and what things can cause changes in climate. Most students will have likely heard about climate change or global warming through the mass media, if not from prior coursework. The degree to which the activities of humankind are causing the Earth's climate to change is a strongly debated issue, and students are likely to have heard different opinions. They may have even formed their own opinions. They will probably be aware that many people believe that the burning of gasoline (and other fossil fuels, like oil and coal) is the cause of this potential human-induced global warming, but they may not understand why there is controversy about this subject.

Confusion between ozone depletion and global warming is common, although they are two separate issues. Some high school students do not yet fully understand the difference between climate and weather. Without knowing the difference, they may be more likely to assign greater weight to a single warm summer or winter than is warranted, given the longer timescales at which climate and climate change are evaluated.

Students are less likely to have thought deeply about the long-term history of climate change on the Earth, its causes, or the nature of the evidence used to investigate climate change. Students may know that the Earth has experienced ice ages, but they will likely not understand the causes, nor fully understand the role that sedimentary deposits, pollen studies, tree rings, and other evidence play in measuring the real extent and duration of ice ages. Some students may cite catastrophic events like meteorite impact, volcanic activity, or changes in solar output as causes of climate change.

Ask students to work initially independently or in pairs. After time has been given for students to reflect individually on their response to this question, have them exchange their conceptions with their group. Then hold a class discussion. Students will be more likely to share their conceptions with the class if they have the backing of a group. Avoid labeling answers right or wrong, encouraging all participation. Accept all responses, and encourage clarity of expression and detail.

Article 2:
Causes of Global Climate Change

Some people might not be aware that human production of greenhouse gases is not the only thing that can cause the Earth's climate to change. There are many different factors that may affect how and when the Earth's climate changes. This article should include information about:

• Milankovitch cycles;

• plate tectonics;

• ocean currents, and

• carbon dioxide levels.

Article 3:
What is "Global Warming" and How Might It Affect Our Community?

Although almost everyone has heard the terms "greenhouse gases" and "global warming," there is a lot of confusion about what these terms actually mean. This article should contain information about:

• greenhouse gases;

• how humans have increased the levels of carbon dioxide in the atmosphere;

• why scientists think increased carbon dioxide might lead to global warming;

• possible effects of global warming, focusing on those that would have the greatest impact on your community, and

• why it is difficult to predict climate change.

Editorial

The final piece is not an article but rather an editorial in which the newspaper expresses its opinion about a particular topic. In the editorial, you should state:

• whether your community should be concerned about global warming and why, and

• what steps, if any, your community should take in response to the possibility of global warming.

Assessment Criteria

Think about what you have been asked to do. Scan ahead through the chapter activities to see how they might help you to meet the challenge. Work with your classmates and your teachers to define the criteria for assessing your work. Record all of this information. Make sure that you understand the criteria as well as you can before you begin.
Your teacher may provide you with a sample rubric to help you get started.

E83

Scenario and Chapter Challenge

Read (or have a student read) the **Chapter Challenge** aloud to the class. Allow students to discuss what they have been asked to do. Have students meet in teams to begin brainstorming what they would like to include in their **Chapter Challenge** articles. Ask them to summarize briefly in their own words what they have been asked to do.

Alternatively, lead a class discussion about the challenge and the expectations. Review the titles of the activities in the Table of Contents. To remind students that the content of the activities corresponds to the content expected for the articles they will be writing, ask them to explain how the title of each activity relates to the expectations for the **Chapter Challenge**. Familiarize students with the way activities are structured, pointing out the sections that are common to all activities. Note particularly the section titled **Preparing for the Chapter Challenge**.

Guiding questions for discussion include:
- What do the activities have to do with the expectations of the challenge?
- What have you been asked to do?
- What should a good final **Chapter Challenge** contain?

Assessment Criteria

Review the attributes of high-quality, informative newspaper articles. A sample rubric for assessing the **Chapter Challenge** is shown on the following pages. You can copy and distribute the rubric as is, or you can use it as a baseline for developing scoring guidelines and expectations that suit your needs. For example:
- You may wish to ensure that core concepts and abilities derived from your local or state science frameworks also appear on the rubric.
- You may wish to modify the format of the rubric to make it more consistent with your evaluation system.

However you decide to evaluate the Chapter Report, keep in mind that all expectations should be communicated to students and that the expectations should be outlined at the start of their work. Please review **Assessment Criteria** (pages xxiv to xxv of this Teacher's Edition) for a more detailed explanation of the assessment system developed for the *EarthComm* program.

Chapter 2

Assessment Rubric for Chapter Challenge on Climate Change

Meets the standard of excellence.
 	
5
	Significant information is presented about <u>all</u> of the following:
	• What climate means, at both the regional and global scale.
	• Examples of different global climates in the geologic past.
	• How geologists find out about past climates.
	• A description of your community's present and past climate.
	• What causes climate change, including:
	  • Milankovitch cycles
	  • plate tectonics
	  • ocean currents
	  • atmospheric carbon dioxide concentrations
	• How "greenhouse gases" contribute to global warming.
	• How humans have increased the levels of CO_2 in the atmosphere.
	• Why scientists think increased atmospheric CO_2 might lead to global warming.
	• What the possible effects of global warming may be, with a community focus.
	• What difficulties are encountered in predicting climate change.
	• Whether or not your community should be concerned about global warming.
	• What your community can do in response to global warming.
	<u>All</u> of the information is accurate and appropriate.
	The writing is clear and interesting.

Approaches the standard of excellence.
 	
4
	Significant information is presented about <u>most</u> of the following:
	• What climate means, at both the regional and global scale.
	• Examples of different global climates in the geologic past.
	• How geologists find out about past climates.
	• A description of your community's present and past climate.
	• What causes climate change, including:
	  • Milankovitch cycles
	  • plate tectonics
	  • ocean currents
	  • atmospheric carbon dioxide concentrations
	• How "greenhouse gases" contribute to global warming.
	• How humans have increased the levels of CO_2 in the atmosphere.
	• Why scientists think increased atmospheric CO_2 might lead to global warming.
	• What the possible effects of global warming may be, with a community focus.
	• What difficulties are encountered in predicting climate change.
	• Whether or not your community should be concerned about global warming.
	• What your community can do in response to global warming.
	<u>All</u> of the information is accurate and appropriate.
	The writing is clear and interesting.

Assessment Rubric for Chapter Challenge on Climate Change

Meets an acceptable standard.
 	
3
	<u>*Significant*</u> information is presented about <u>*most*</u> of the following:
	• What climate means, at both the regional and global scale.
	• Examples of different global climates in the geologic past.
	• How geologists find out about past climates.
	• A description of your community's present and past climate.
	• What causes climate change, including:
	   • Milankovitch cycles
	   • plate tectonics
	   • ocean currents
	   • atmospheric carbon dioxide concentrations
	• How "greenhouse gases" contribute to global warming.
	• How humans have increased the levels of CO_2 in the atmosphere.
	• Why scientists think increased atmospheric CO_2 might lead to global warming.
	• What the possible effects of global warming may be, with a community focus.
	• What difficulties are encountered in predicting climate change.
	• Whether or not your community should be concerned about global warming.
	• What your community can do in response to global warming.
	<u>*Most*</u> of the information is accurate and appropriate.
	The writing is clear and interesting.

Below acceptable standard and requires remedial help.
 	
2
	<u>*Limited*</u> information is presented about the following:
	• What climate means, at both the regional and global scale.
	• Examples of different global climates in the geologic past.
	• How geologists find out about past climates.
	• A description of your community's present and past climate.
	• What causes climate change, including:
	   • Milankovitch cycles
	   • plate tectonics
	   • ocean currents
	   • atmospheric carbon dioxide concentrations
	• How "greenhouse gases" contribute to global warming.
	• How humans have increased the levels of CO_2 in the atmosphere.
	• Why scientists think increased atmospheric CO_2 might lead to global warming.
	• What the possible effects of global warming may be, with a community focus.
	• What difficulties are encountered in predicting climate change.
	• Whether or not your community should be concerned about global warming.
	• What your community can do in response to global warming.
	<u>*Most*</u> of the information is accurate and appropriate.
	Generally, the writing does not hold the reader's attention.

Chapter 2

Assessment Rubric for Chapter Challenge on Climate Change

| **Basic level that requires remedial help or demonstrates a lack of effort.**

1 | _Limited_ information is presented about the following:
• What climate means, at both the regional and global scale.
• Examples of different global climates in the geologic past.
• How geologists find out about past climates.
• A description of your community's present and past climate.
• What causes climate change, including:
  • Milankovitch cycles
  • plate tectonics
  • ocean currents
  • atmospheric carbon dioxide concentrations
• How "greenhouse gases" contribute to global warming.
• How humans have increased the levels of CO_2 in the atmosphere.
• Why scientists think increased atmospheric CO_2 might lead to global warming.
• What the possible effects of global warming may be, with a community focus.
• What difficulties are encountered in predicting climate change.
• Whether or not your community should be concerned about global warming.
• What your community can do in response to global warming.
Little of the information is accurate and appropriate.
The writing is difficult to follow. |

NOTES

ACTIVITY I — PRESENT-DAY CLIMATE IN YOUR COMMUNITY

Background Information

Elements of Climate

Climate is more than just the long-term average of temperature and precipitation in a region. Daily, weekly, and yearly range in temperature is also an important determinant of climate. In addition to yearly averages of precipitation, frequency is important: does rain fall often in small amounts, or does it tend to fall less often but in large amounts?

The relative timing of yearly temperature and precipitation is very important for growth of trees and shrubs. In areas with large seasonal variations in precipitation, the wet season might coincide with the warm growing season. Or, the wet season might coincide with the cold season, in which trees and shrubs are dormant. Arid or semiarid regions with the former conditions can support much more woody vegetation, given the same level of annual precipitation, than a region characterized by the latter condition. Several other factors, like wind speed, wind direction, cloud cover (which is correlated with amount of precipitation, but not strongly in some areas), relative humidity, and frequency of occurrence of unusually strong storms like tropical cyclones (hurricanes and typhoons) are also important.

The most sensitive indicator of climate is vegetation. In fact, the connection between vegetation and climate is so strong that climatologists have to a great extent been guided by vegetation type in choosing criteria by which to classify climates.

The Physical Environment

The physical environment is the combination of all the physical factors that make your surroundings what they are. Ultimately, your community owes its uniqueness to the interaction of the physical environment with the people of the community, through time. Regionally, the physical environment is controlled by just a few major factors:

- latitude
- elevation
- nearby geographic features like mountain ranges or water bodies
- characteristic weather patterns

Latitude

Latitude is a measure of distance from the Equator. The latitude of your community can be described as the angle between a line from the center of the Earth to your community on the one hand, and a line from the center of the Earth to the Equator (at the same longitude) on the other hand. Latitude varies from zero at the Equator to 90° at the poles.

The intensity of direct sunlight on a horizontal area of the land surface, given a clear atmosphere above, depends upon the latitude of the area. Areas near the Equator receive, on average, more solar radiation than do areas at higher latitudes, because the Sun shines almost directly overhead for most of the year. At high latitudes, the Sun shines at a low angle, so that a given rate of delivery of solar radiation is spread over a greater area of the land. This means less intensity per unit land area. (Regions near the poles even spend part of the year in 24-hour darkness.) As a result, equatorial regions receive more solar radiation than they lose by long-wave radiation to space, and high-latitude regions receive less solar radiation than they lose by long-wave radiation to space. This difference in heating and cooling is the fundamental cause of all of the Earth's weather. Of course, the low-latitude areas don't keep getting warmer, and the high-latitude areas don't keep getting colder! Rather, the long-term balance of heat is maintained by movement of heat from low latitudes to high latitudes by the winds and ocean currents.

Elevation
The elevation of a given point on the land surface of the Earth is the vertical distance of that point above sea level. The concept of sea level, when applied to inland areas, is not obvious. One way to think about it is to imagine a very long tunnel with impermeable walls that starts from under the ocean somewhere and extends to deep beneath the given point on the land surface, and from there vertically upward as a shaft to the surface. The level of the seawater at the bottom of the shaft is what's meant by "sea level" at that point on the continent. Altitude is a concept that's related to, but different from, elevation: altitude is the term used for the height of a point in the atmosphere above sea level.

The elevation of a local region of a land mass also affects the physical environment of that region. On average, places at high elevations are cooler than places in the same region at lower elevations. In many places at high elevation, glaciers form because summer warmth is not sufficient to melt all the snow that fell during the previous winter. For example, glaciers are found on the mountains in Glacier National Park in Montana, which are located both at high elevation and high latitude.

Geographic Features
Geographic features—like mountain ranges, lakes, and oceans—also affect the physical environment of a region. Mountains have a strong effect on precipitation in nearby areas. The windward side of a mountain range often receives unusually heavy rainfall. Moisture-laden winds drop much of their moisture

when rain or snow pass over the range. Conversely, the leeward side of the range is called a rain shadow. It often receives little rain, because the air has lost so much of its moisture as it crosses the mountain range. The desert on the eastern side of the Sierra Nevada in the western United States has low rainfall for this reason.

Nearby large bodies of water have a strong effect on the physical environment. Because of the extremely high heat capacity of water, proximity to a large water body tends to moderate seasonal variations in air temperature. This is especially true on the western coastal areas of continents in the zone of prevailing westerly winds.

Lake-effect snow is common downwind of the large lakes of the midwestern and eastern United States. Cold winds blow across the lake, accumulating moisture from the still-warm lake water, and then heavy precipitation falls on the cold leeward side of the lake.

Characteristic Weather Patterns
Many of the effects of characteristic weather patterns, like seasonal variation in precipitation, prevailing wind direction, and frequency of severe weather, have already been discussed in the **Elements of Climate** section above. Characteristic weather patterns can play an important role in controlling the climate of a region, and, in some cases, work in conjunction with local geographic features to cause a range of microclimates within a larger region.

Specific Heat Capacity
Not all substances have the same relationship between heat transfer and temperature change. For example, the transfer of a given amount of heat from a volume of water to an equal volume of air would affect the temperature of the air to a greater extent than it would affect the temperature of the

water. That is because it takes a much greater amount of heat to change the temperature of a given volume of water than it does to affect the same temperature change in an equal volume of air. This effect has important consequences that buffer or regulate the climates of some regions.

Heat is a form of energy. It is the thermal energy of motion of the atoms and molecules that constitute matter. Heat is transferred between different bodies of matter in two ways: conduction or radiation.

In conduction, heat is transferred from a hotter body to a colder body in direct contact with one another. That happens because the faster-moving atoms or molecules of the hotter body transfer some of their energy of motion to the slower-moving atoms or molecules of the colder body. Heat energy is thus transferred from the hotter body to the colder body. In the process, the colder body is made hotter and the hotter body is made colder.

Radiation causes heat transfer between bodies that are separated in space. All matter radiates energy in the form of electromagnetic waves. The radiating body loses heat, and its temperature decreases; a body that absorbs the radiation gains heat, and its temperature rises.

Different kinds of matter differ in the relationship between the gain or loss of heat on the one hand, and the increase or decrease of temperature on the other hand, during heat transfer. In a general way, that has to do with the nature of the bonding forces among the constituent atoms and molecules of the given material.

The specific heat capacity of a material is the amount of heat needed per unit mass to raise the temperature of the material by one degree Celsius. Materials with a high specific heat capacity act as good reservoirs of heat, because it takes a lot of heat to change their temperature. Liquid water has an extremely high heat capacity, higher than almost any other substance. The reason has to do with the existence of what are called hydrogen bonds between the positively charged "side" of the water molecule and the negatively charged "side." The structure of ice involves such hydrogen bonds. When ice melts, some but not all of the hydrogen bonds are broken. As heat is added to water, the heat must act not just to increase the thermal motions of the molecules but also to break more hydrogen bonds.

More Information – on the Web
Visit the *EarthComm* web site www.agiweb.org/earthcomm/ to access a variety of links to web sites that will help you deepen your understanding of content and prepare you to teach this activity. Many of the sites also contain images that you can download.

Goals and Assessment

Clarify that the goals indicate what students should understand and be able to do as a result of the activity. Make sure students understand that Chapter Assessments are based upon these goals.

Goal	Location in Activity	Assessment Opportunity
Identify factors of the physical environment.	**Investigate** Parts A and B **Digging Deeper; Understanding and Applying What You Have Learned** Question 2	All relevant physical features are noted. Answers to questions are reasonable, and reflect maps and available data.
Use a topographic map to gather data about elevation, latitude, and physical features.	**Investigate** Parts A and B	Topographic map is accurately interpreted; all relevant features are noted.
Interpret climate data tables.	**Investigate** Parts A and B	Data interpretations are accurate.
Compare and contrast climate information from two different parts of the United States.	**Investigate** Part B **Check Your Understanding** Question 3 **Understanding and Applying What You Have Learned** Question 4	Comparisons are accurate. Answers to questions are reasonable, and reflect maps and available data.
Understand how physical features can influence the climate of an area.	**Investigate** Parts A and B **Digging Deeper; Check Your Understanding** Questions 1 and 4 **Understanding and Applying What You Have Learned** Questions 1 – 3	Climatic descriptions are accurate. Correlation between climate and physical features is reasonable.

Earth System Evolution Climate Change

Activity 1   Present-Day Climate in Your Community

Goals

In this activity you will:

- Identify factors of the physical environment.

- Use a topographic map to gather data about elevation and latitude, and physical features.

- Interpret data from a climate data table.

- Compare and contrast climate information from two different parts of the United States.

- Understand how physical features can influence the climate of an area.

Think about It

A friend e-mails you from Italy to ask what your environment and climate are like. You plan to e-mail her a reply.

- How would you describe the physical environment of your community?
- How would you describe the climate of your community?

What do you think? Record your ideas about these questions in your *EarthComm* notebook. Be prepared to discuss your responses with your small group and the class.

Activity Overview

Students use data on temperature and precipitation to describe the climate of their community. They look at a local topographic map to determine what physical features in their community might influence climate. Students then compare their community's climate with the climate of a different community. **Digging Deeper** explains the difference between weather and climate and explores the factors that can affect climate, including latitude, elevation, and geographic features. The reading also explains the difference between regional and global climate.

Preparation and Materials Needed

Preparation

You will need to obtain topographic maps of your community and a community different from your own. You will also need to decide on the list of items that you would like to provide for **Part C** of this investigation and gather those materials.

Materials

Part A
- Topographic map of community*

Part B
- Topographic map of a different community*

Part C
- Variable, depending on experimental design

Think about It

Student Conceptions

Students may think of their physical environment as their school, houses, etc. After completing the activity, they should realize that the physical environment consists of natural geographical features. They may not be aware of the difference between climate and weather. Student descriptions of climate will vary but will likely include information about temperature and the amount of rainfall the community receives.

*The *EarthComm* web site provides suggestions for obtaining these resources.

Chapter 2

Answers for the Teacher Only

The climate of an area is actually a subset of the physical environment of a given area. In addition to climate, the broader concept of the physical environment includes such features as the topography (the "lay of the land"), the existence and characteristics of water bodies like streams, rivers, and lakes, and the nature of the subsurface, including ground water and caverns. The bedrock geology of the area and the nature of the soil are also important aspects of the physical environment.

Assessment Tool

Think about It Evaluation Sheet
Use this evaluation sheet to help students understand and internalize the basic expectations for the warm-up activity.

NOTES

_____

_____

_____

_____

_____

_____

_____

_____

_____

_____

_____

_____

_____

_____

_____

_____

_____

_____

_____

_____

Chapter 2

Investigate

Part A: Physical Features and Climate in Your Community

1. Depending on where your community is located and how large it is, you might wish to expand your definition of "community" to include a larger area, like your county or state. For example, your town does not have to be right on the ocean to have its climate influenced by the ocean.

a) Write a "definition" of the area that you will examine as your "community."

2. Use the climate data tables provided on the following pages and topographic maps of your "community" to describe the climate in your community.

Climatic Zones of North America (Mercator Projection)

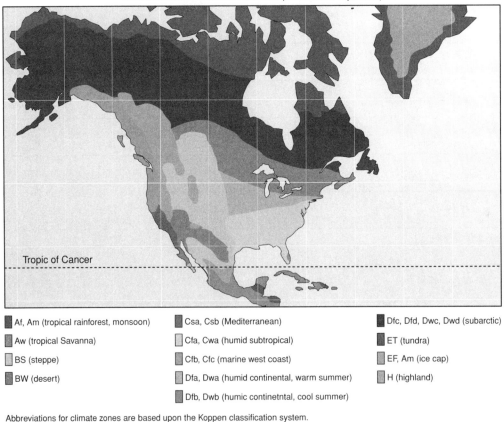

Tropic of Cancer

■ Af, Am (tropical rainforest, monsoon)	■ Csa, Csb (Mediterranean)	■ Dfc, Dfd, Dwc, Dwd (subarctic)
■ Aw (tropical Savanna)	□ Cfa, Cwa (humid subtropical)	■ ET (tundra)
■ BS (steppe)	■ Cfb, Cfc (marine west coast)	■ EF, Am (ice cap)
■ BW (desert)	■ Dfa, Dwa (humid continental, warm summer)	■ H (highland)
	■ Dfb, Dwb (humic continetntal, cool summer)	

Abbreviations for climate zones are based upon the Koppen classification system.

Investigate

Assessment Tools

EarthComm Notebook Entry-Checklist
Use this checklist as a quick guide for student self-assessment and/or an opportunity to quickly score student work. Add further criteria specific to your classroom needs or to this particular investigation.

Investigate Notebook Entry-Evaluation Sheet
Point out the criteria listed on this evaluation sheet that are relevant to this particular investigation. Encourage students to internalize the criteria by making them part of your "assessment conversations" as you circulate around the classroom. For example, while students are working, ask them criteria-driven questions like:
- Is your work thorough and complete?
- Are all of you participating in the activity?
- Do you each have a role to play in solving this problem? And so on.

Part A: Physical Features and Climate in Your Community

1. Answers will vary depending on where your community is located. You may wish to provide some guidance to help students pick an appropriately-sized "community." Alternatively, the topic of what "community" is most appropriate for this investigation can be discussed as a class.

2. Answers will vary depending on where your community is located. If your community is not represented by the given data tables, visit the *EarthComm* web site to collect data from a closer area.

Chapter 2

Earth System Evolution Climate Change

Country and Station	Latitude ′ ″	Longitude ′ ″	Elevation Feet	Record Length Yrs.	\<br\>January Max. °F	\<br\>January Min. °F	\<br\>April Max. °F	\<br\>April Min. °F	\<br\>July Max. °F	\<br\>July Min. °F	\<br\>October Max. °F	\<br\>October Min. °F	Extreme Max. °F	Extreme Min. °F
United States (Conterminous):														
Albuquerque, NM	35 03N	106 37W	5311	30	46	24	69	42	91	66	71	45	105	−17
Asheville, NC	35 26N	82 32W	2140	30	48	28	67	42	84	61	68	45	100	−16
Atlanta, GA	33 39N	84 26W	1010	30	52	37	70	50	87	71	72	52	105	− 8
Austin, TX	30 18N	97 42W	597	30	60	41	78	57	95	74	82	60	109	− 2
Birmingham, AL	33 34N	86 45W	620	30	57	36	76	50	93	71	79	52	106	−10
Bismarck, ND	46 46N	100 45W	1647	30	20	0	55	32	86	58	59	34	114	−45
Boise, ID	43 34N	116 13W	2838	30	36	22	63	37	91	59	65	38	112	−28
Brownsville, TX	25 54N	97 26W	16	30	71	52	82	66	93	76	85	67	106	12
Buffalo, NY	42 56N	78 44W	705	30	31	18	53	34	80	59	60	41	99	−21
Cheyenne, WY	41 09N	104 49W	6126	30	37	14	56	30	85	55	63	32	100	−38
Chicago, IL	41 47N	87 45W	607	30	33	19	57	41	84	67	63	47	105	−27
Des Moines, IA	41 32N	93 39W	938	30	29	11	59	38	87	65	66	43	110	−30
Dodge City, KS	37 46N	99 58W	2582	30	42	20	66	41	93	68	71	46	109	−26
El Paso, TX	31 48N	106 24W	3918	30	56	30	78	49	95	69	79	50	112	− 8
Indianapolis, IN	39 44N	86 17W	792	30	37	21	61	40	86	64	67	44	107	−25
Jacksonville, FL	30 25N	81 39W	20	30	67	45	80	58	92	73	80	62	105	7
Kansas City, MO	39 07N	94 36W	742	30	40	23	66	46	92	71	72	49	113	−23
Las Vegas, NV	36 05N	115 10W	2162	30	54	32	78	51	104	76	80	53	117	8
Los Angeles, CA	33 56N	118 23W	97	30	64	45	67	52	76	62	73	57	110	23
Louisville, KY	38 11N	85 44W	477	30	44	27	66	43	89	67	70	46	107	−20
Miami, FL	25 48N	80 16W	7	30	76	58	83	66	89	75	85	71	100	28
Minneapolis, MN	44 53N	93 13W	834	30	22	2	56	33	84	61	61	37	108	−34
Missoula, MT	46 55N	114 05W	3190	30	28	10	57	31	85	49	58	30	105	−33
Nashville, TN	36 07N	86 41W	590	30	49	31	71	48	91	70	76	49	107	−17
New Orleans, LA	29 59N	90 15W	3	30	64	45	78	58	91	73	80	61	102	7
New York, NY	40 47N	73 58W	132	30	40	27	60	43	85	68	66	50	106	−15
Oklahoma City, OK	35 24N	97 36W	1285	30	46	28	71	49	93	72	74	52	113	−17
Phoenix, AZ	33 26N	112 01W	1117	30	64	35	84	50	105	75	87	55	122	16
Pittsburgh, PA	40 27N	80 00W	747	30	40	25	63	42	85	65	65	45	103	−20
Portland, ME	43 39N	70 19W	47	30	32	12	53	32	80	57	60	37	103	−39
Portland, OR	45 36N	122 36W	21	30	44	33	62	42	79	56	63	45	107	− 3
Reno, NV	39 30N	119 47W	4404	30	45	16	65	31	89	46	69	29	106	−19
Salt Lake City, UT	40 46N	111 58W	4220	30	37	18	63	36	94	60	65	38	107	−30
San Francisco, CA	37 37N	122 23W	8	30	55	42	64	47	72	54	71	51	109	20
Sault Ste. Marie, MI	46 28N	84 22W	721	30	23	8	46	30	76	54	55	38	98	−37
Seattle, WA	47 27N	122 18W	400	30	44	33	58	40	76	54	60	44	100	0
Sheridan, WY	44 46N	106 58W	3964	30	34	9	56	31	87	56	62	33	106	−41
Spokane, WA	47 38N	117 32W	2356	30	31	19	59	36	86	55	60	38	108	−30
Washington, DC	38 51N	77 03W	14	30	44	30	66	46	87	69	68	50	104	−18
Wilmington, NC	34 16N	77 55W	28	30	58	37	74	51	89	71	76	55	104	0

Source: NOAA

NOTES

_____

_____

_____

_____

_____

_____

_____

_____

_____

_____

_____

_____

_____

_____

_____

_____

_____

_____

_____

_____

_____

Chapter 2

Average Precipitation in the United States

Country and Station United States (Conterminous):	Record Length Years	Jan. In.	Feb. In.	Mar. In.	Apr. In.	May In.	Jun. In.	Jul. In.	Aug. In.	Sep. In.	Oct. In.	Nov. In.	Dec. In.	Year In.
Albuquerque, NM	30	0.4	0.4	0.5	0.5	0.8	0.6	1.2	1.3	1.0	0.8	0.4	0.5	8.4
Asheville, NC	30	4.2	4.0	4.8	4.0	3.7	3.5	5.9	4.9	3.6	3.1	2.8	3.6	48.1
Atlanta, GA	30	4.4	4.5	5.4	4.5	3.2	3.8	4.7	3.6	3.3	2.4	3.0	4.4	47.2
Austin, TX	30	2.4	2.6	2.1	3.6	3.7	3.2	2.2	1.9	3.4	2.8	2.1	2.5	32.5
Birmingham, AL	30	5.0	5.3	6.0	4.5	3.4	4.0	5.2	4.9	3.3	3.0	3.5	5.0	53.1
Bismarck, ND	30	0.4	0.4	0.8	1.2	2.0	3.4	2.2	1.7	1.2	0.9	0.6	0.4	15.2
Boise, ID	30	1.3	1.3	1.3	1.2	1.3	0.9	0.2	0.2	0.4	0.8	1.2	1.3	11.4
Brownsville, TX	30	1.4	1.5	1.0	1.6	2.4	3.0	1.7	2.8	5.0	3.5	1.3	1.7	26.9
Buffalo, NY	30	2.8	2.7	3.2	3.0	3.0	2.5	2.6	3.1	3.1	3.0	3.6	3.0	35.6
Cheyenne, WY	30	0.5	0.6	1.2	1.9	2.5	2.1	1.8	1.4	1.1	0.8	0.6	0.5	15.0
Chicago, IL	30	1.9	1.6	2.7	3.0	3.7	4.1	3.4	3.2	2.7	2.8	2.2	1.9	33.2
Des Moines, IA	30	1.3	1.1	2.1	2.5	4.1	4.7	3.1	3.7	2.9	2.1	1.8	1.1	30.5
Dodge City, KS	30	0.6	0.7	1.2	1.8	3.2	3.0	2.3	2.4	1.5	1.4	0.6	0.5	19.2
El Paso, TX	30	0.5	0.4	0.4	0.3	0.4	0.7	1.3	1.2	1.1	0.9	0.3	0.5	8.0
Indianapolis, IN	30	3.1	2.3	3.4	3.7	4.0	4.6	3.5	3.0	3.2	2.6	3.1	2.7	39.2
Jacksonville, FL	30	2.5	2.9	3.5	3.6	3.5	6.3	7.7	6.9	7.6	5.2	1.7	2.2	53.6
Kansas City, MO	30	1.4	1.2	2.5	3.6	4.4	4.6	3.2	3.8	3.3	2.9	1.8	1.5	34.2
Las Vegas, NV	30	0.5	0.4	0.4	0.2	0.1	*	0.5	0.5	0.3	0.2	0.3	0.4	3.8
Los Angeles, CA	30	2.7	2.9	1.8	1.1	0.1	0.1	*	*	0.2	0.4	1.1	2.4	12.8
Louisville, KY	30	4.1	3.3	4.6	3.8	3.9	4.0	3.4	3.0	2.6	2.3	3.2	3.2	41.4
Miami, FL	30	2.0	1.9	2.3	3.9	6.4	7.4	6.8	7.0	9.5	8.2	2.8	1.7	59.9
Minneapolis, MN	30	0.7	0.8	1.5	1.9	3.2	4.0	3.3	3.2	2.4	1.6	1.4	0.9	24.9
Missoula, MT	30	0.9	0.9	0.7	1.0	1.9	1.9	0.9	0.7	1.0	1.0	0.9	1.1	12.9
Nashville, TN	30	5.5	4.5	5.2	3.7	3.7	3.3	3.7	2.9	2.9	2.3	3.3	4.2	45.2
New Orleans, LA	30	3.8	4.0	5.3	4.6	4.4	4.4	6.7	5.3	5.0	2.8	3.3	4.1	53.7
New York, NY	30	3.3	2.8	4.0	3.4	3.7	3.3	3.7	4.4	3.9	3.1	3.4	3.3	42.3
Oklahoma City, OK	30	1.3	1.4	2.0	3.1	5.2	4.5	2.4	2.5	3.0	2.5	1.6	1.4	30.9
Phoenix, AZ	30	0.7	0.9	0.7	0.3	0.1	0.1	0.8	1.1	0.7	0.5	0.5	0.9	7.3
Pittsburgh, PA	30	2.8	2.3	3.5	3.4	3.8	4.0	3.6	3.5	2.7	2.5	2.3	2.5	36.9
Portland, ME	30	4.4	3.8	4.3	3.7	3.4	3.2	2.9	2.4	3.5	3.2	4.2	3.9	42.9
Portland, OR	30	5.4	4.2	3.8	2.1	2.0	1.7	0.4	0.7	1.6	3.6	5.3	6.4	37.2
Reno, NV	30	1.2	1.0	0.7	0.5	0.5	0.4	0.3	0.2	0.2	0.5	0.6	1.1	7.2
Salt Lake City, UT	30	1.4	1.2	1.6	1.8	1.4	1.0	0.6	0.9	0.5	1.2	1.3	1.2	14.1
San Francisco, CA	30	4.0	3.5	2.7	1.3	0.5	0.1	*	*	0.2	0.7	1.6	4.1	18.7
Sault Ste. Marie, MI	30	2.1	1.5	1.8	2.2	2.8	3.3	2.5	2.9	3.8	2.8	3.3	2.3	31.3
Seattle, WA	30	5.7	4.2	3.8	2.4	1.7	1.6	0.8	1.0	2.1	4.0	5.4	6.3	39.0
Sheridan, WY	30	0.6	0.7	1.4	2.2	2.6	2.6	1.2	0.9	1.2	1.1	0.8	0.6	15.9
Spokane, WA	30	2.4	1.9	1.5	0.9	1.2	1.5	0.4	0.4	0.8	1.6	2.2	2.4	17.2
Washington, DC	30	3.0	2.5	3.2	3.2	4.1	3.2	4.2	4.9	3.8	3.1	2.8	2.8	40.8
Wilmington, NC	30	2.9	3.4	4.0	2.9	3.5	4.3	7.7	6.9	6.3	3.0	3.1	3.4	51.4

Source: NOAA

EarthComm

NOTES

Earth System Evolution Climate Change

Country and Station United States (Conterminous): Data through 1998	Record Length	Average Snowfall (includes ice pellets)											
		Jan.	Feb.	Mar.	Apr.	May	Jun.	Jul.	Aug.	Sep.	Oct.	Nov.	Dec.	Year
	Years	In.	In.	In.	In.	In.	In.	In.	In.	In.	In.	In.	In.	In.
Albuquerque, NM	59	2.5	2.1	1.8	0.6	0	T	T	T	T	0.1	1.2	2.6	10.9
Asheville, NC	34	5	4.3	2.8	0.6	T	T	T	T	0	T	0.7	2	15.4
Atlanta, GA	62	0.9	0.5	0.4	T	0	0	0	0	0	T	0	0.2	2
Austin, TX	57	0.5	0.3	T	T	T	0	0	0	0	0	0.1	T	0.9
Birmingham, AL	55	0.6	0.2	0.3	0.1	T	T	T	0	T	T	T	0.3	1.5
Bismarck, ND	59	7.6	7	8.6	4	0.9	T	T	T	0.2	1.8	7	7	44.1
Boise, ID	59	6.5	3.7	1.6	0.6	0.1	T	T	T	T	0.1	2.3	5.9	20.8
Brownsville, TX	59	T	T	T	0	0	0	0	T	0	0	T	T	T
Buffalo, NY	55	23.7	18	11.9	3.2	0.2	T	T	T	T	0.3	11.2	22.8	91.3
Cheyenne, WY	63	6.6	6.3	11.9	9.2	3.2	0.2	T	T	0.9	3.7	7.1	6.3	55.4
Chicago, IL	39	10.7	8.2	6.6	1.6	0.1	T	T	T	T	0.4	1.9	8.1	37.6
Des Moines, IA	57	8.3	7.2	6	1.8	0	T	T	0	T	0.3	3.1	6.7	33.4
Dodge City, KS	56	4.3	3.9	5	0.8	T	T	T	T	0	0.3	2.1	3.6	20
El Paso, TX	57	1.3	0.8	0.4	0.3	T	T	T	0	T	0	0.9	1.6	5.3
Indianapolis, IN	67	6.6	5.6	3.4	0.5	0	T	0	T	0	0.2	1.9	5.1	23.3
Jacksonville, FL	57	T	0	0	T	0	T	T	0	0	0	0	0	T
Kansas City, MO	64	5.7	4.4	3.4	0.8	T	T	T	0	T	0.1	1.2	4.4	20
Las Vegas, NV	48	0.9	0.1	0	T	0	0	0	T	0	T	0.1	0.1	1.2
Los Angeles, CA	62	T	T	T	0	0	0	0	0	0	0	0	T	T
Louisville, KY	51	5.4	4.6	3.3	0.1	T	T	T	0	0	0.1	1	2.1	16.6
Miami, FL	56	0	0	0	0	T	0	0	0	0	0	0	0	T
Minneapolis, MN	60	10.2	8.2	10.6	2.8	0.1	T	T	T	T	0.5	7.9	9.4	49.7
Missoula, MT	54	12.3	7.3	6	2.1	0.7	T	T	T	T	0.8	6.2	11.3	46.7
Nashville, TN	56	3.7	3	1.5	0	0	T	0	T	0	0	0.4	1.4	10
New Orleans, LA	50	0	0.1	T	T	T	0	0	0	0	0	T	0.1	0.2
New York, NY	130	7.5	8.6	5.1	0.9	T	0	T	0	0	0	0.9	5.4	28.4
Oklahoma City, OK	59	3.1	2.4	1.5	T	T	T	T	T	T	T	0.5	1.8	9.3
Phoenix, AZ	61	T	0	T	T	T	0	0	0	0	T	0	T	T
Pittsburgh, PA	46	11.7	9.2	8.7	1.7	0.1	T	T	T	T	0.4	3.5	8.2	43.5
Portland, ME	58	19.6	16.9	12.9	3	0.2	0	0	0	T	0.2	3.3	14.6	70.7
Portland, OR	55	3.2	1.1	0.4	T	0	T	0	T	T	0	0.4	1.4	6.5
Reno, NV	54	5.8	5.2	4.3	1.2	0.8	0	0	0	0	0.3	2.4	4.3	24.3
Salt Lake City, UT	70	13.8	10	9.4	4.9	0.6	T	T	T	0.1	1.3	6.8	11.7	58.6
San Francisco, CA	69	0	T	T	0	0	0	0	0	0	0	0	0	T
Sault Ste. Marie, MI	55	29	18.4	14.7	5.8	0.5	T	T	T	0.1	2.4	15.8	31.1	117.8
Seattle, WA	48	2.9	0.9	0.6	0	T	0	0	0	T	0	0.7	2.2	7.3
Sheridan, WY	58	11	10.4	12.6	9.9	2	0.1	T	0	1.3	4.6	9.2	10.9	72
Spokane, WA	51	15.6	7.5	3.9	0.6	0.1	T	0	0	T	0.4	6.3	14.6	49
Washington, DC	55	5.5	5.4	2.2	T	T	T	T	T	0	0	0.8	2.8	16.7
Wilmington, NC	47	0.4	0.5	0.4	T	T	T	T	0	0	0	T	0.6	1.9

Trace (T) is recorded for less than 0.05 inch of snowfall.

Last updated on 5/25/2000 by NRCC.

Source: The National Climatic Data Center/NOAA

E 88

NOTES

Chapter 2

Frost-Free Days in the United States			
City/State	**Last Frost Date**	**First Frost Date**	**No. of Frost-Free Days per Year**
Albany, NY	May 7	September 29	144 days
Albuquerque, NM	April 16	October 29	196 days
Atlanta, GA	March 13	November 12	243 days
Baltimore, MD	March 26	November 13	231 days
Birmingham, AL	March 29	November 6	221 days
Boise, ID	May 8	October 9	153 days
Boston, MA	April 6	November 10	217 days
Charleston, SC	March 11	November 20	253 days
Charlotte, NC	March 21	November 15	239 days
Cheyenne, WY	May 20	September 27	130 days
Chicago, IL	April 14	November 2	201 days
Columbus, OH	April 26	October 17	173 days
Dallas, TX	March 18	November 12	239 days
Denver, CO	May 3	October 8	157 days
Des Moines, IA	April 19	October 17	180 days
Detroit, MI	April 24	October 22	181 days
Duluth, MN	May 21	September 21	122 days
Fargo, ND	May 13	September 27	137 days
Fayetteville, AR	April 21	October 17	179 days
Helena, MT	May 18	September 18	122 days
Houston, TX	February 4	December 10	309 days
Indianapolis, IN	April 22	October 20	180 days
Jackson, MS	March 17	November 9	236 days
Jacksonville, FL	February 14	December 14	303 days
Las Vegas, NV	March 7	November 21	259 days
Lincoln, NB	March 13	November 13	180 days
Los Angeles, CA	None likely	None likely	365 days
Louisville, KY	April 1	November 7	220 days
Memphis, TN	March 23	November 7	228 days
Miami, FL	None	None	365 days
Milwaukee, WI	May 5	October 9	156 days
New Haven, CT	April 15	October 27	195 days
New Orleans, LA	February 20	December 5	288 days
New York, NY	April 1	November 11	233 days
Phoenix, AZ	February 5	December 15	308 days
Pittsburgh, PA	April 16	November 3	201 days
Portland, ME	May 10	September 30	143 days
Portland, OR	April 3	November 7	217 days
Richmond, VA	April 10	October 26	198 days
Salt Lake City, UT	April 12	November 1	203 days
San Francisco, CA	January 8	January 5	362 days
Seattle, WA	March 24	November 11	232 days
St. Louis, MO	April 3	November 6	217 days
Topeka, KS	April 21	October 14	175 days
Tulsa, OK	March 30	November 4	218 days
Washington, D.C.	April 10	October 31	203 days
Wichita, KS	April 13	October 23	193 days

E 89

EarthComm

NOTES

Earth System Evolution Climate Change

Include the following important climatic factors in your description of the climate in your community:

- Average daily temperatures in the winter and summer.

- Record high and low temperatures in the winter and summer.

- Average monthly precipitation in the winter and summer.

- Average winter snowfall.

- Growing season (number of days between last spring frost and first autumn frost).

3. Inspect a topographic map of your town.

   a) What is the latitude of your town?

   b) What is the elevation of your school?

   c) What is the highest elevation in your town?

   d) What is the lowest elevation in your town?

   e) Which of the following physical features can be found in or fairly near your community: mountains, rivers, valleys, coasts, lakes, hills, plains, or deserts? Specify where they are in relation to your community.

   f) Describe some of the ways that the physical features of your community might influence the climate.

Part B: Physical Features and Climate in a Different Community

1. Select a community that is in a part of the United States that is very different from where you live. For example, if you live in the mountains, pick a community on the plains. If you live near an ocean, pick a community far from a large body of water.

   a) Record the community and the reason you chose that community in your *EarthComm* notebook.

2. Describe the climate in this community.

   a) Include information for the same climatic factors that you used to describe your own community.

3. Inspect a topographic map of this community.

   a) Describe the same physical features that you did for your community.

4. Compare the physical features and climates of the two communities.

   a) In what ways might the physical features influence climate in the two places?

Part C: Heating and Cooling of Land versus Water

1. How do the rates at which rock and soil heat and cool compare to the rate at which water heats and cools? How might this affect climate in your community?

   a) Write down your ideas about these two questions.

   b) Develop a hypothesis about the rate at which rock or soil heat and cool compared to the rate at which water heats and cools.

2. Using materials provided by your teacher, design an experiment to investigate the rates of cooling and heating of soil or rock and water. Note the variable that you are

E 90

3. All of the answers for **Question 3** will have to be determined from the topographic map itself.

   a) The latitude is found by use of marks on one or the other of the vertical boundaries of the map. This might be confusing for the students, because various other numerical values, representing special geographic coordinate systems, are also shown along the boundaries of the map.

   b) - d) Elevations are given by the numbers marked beside the contour lines. Usually, every fifth contour is labeled. Students can find the elevation of the school by identifying which two contour lines the school lies between and then making an interpolation. In some areas with prominent hills or mountains, the elevation of the highest point is given; in other areas, the students will have to identify the relatively high areas and then use the pattern of the contour lines to estimate the elevation of the highest point. The lowest point is usually along a river valley where the river flows off the given map, although in some areas, the lowest elevation might be just a low point on the land surface. The steepness of slopes is revealed by the spacing of the contour lines: areas of steep slopes show closely spaced contour lines, and areas with gentle slopes show widely spaced contour lines. As the students are working, be prepared to help them locate the highest and lowest elevations and other features on the map.

   e) Answers will vary depending on the community.

   f) Answers will vary depending on the community. Many possibilities exist, some of which are discussed in this Teacher's Edition **Background Information** section for this activity. One possibility is that nearby mountains may affect the rain patterns making it arid if the community is on the leeward side of the mountains or wet if the community is on the windward side of the mountains. Another possibility is that a nearby ocean might affect temperatures, lessening seasonal temperature variation.

Part B: Physical Features and Climate in a Different Community

1. a) Students should select a community that is different from their own. You may wish to give them a few choices, in order to help in standardizing grading.

2. a) Answers will vary depending upon the community that has been selected.

3. Answers will vary depending upon the community that has been selected.

4. a) Answers will vary depending upon the variation between the two communities.

Part C: Heating and Cooling of Land versus Water

1. a) Students are unlikely to have thought about this very much, so their responses are likely to be based on personal experience. Accordingly, student responses

will likely vary significantly. This question is intended to introduce the concept of specific heat capacity discussed earlier in this Teacher's Edition. Specifically, this part of the investigation is intended to lead students to discover that the temperature of water is not as easily changed as that of air, rock, or soil. Generally speaking, the expected results of students' experiments will be that the rock and soil heat and cool quicker than the water. This follows because water has a much higher heat capacity than either rock or soil. Note that the heat capacity of rock and soil, and the associated rate of heating and cooling, can be quite variable depending on rock type, soil moisture, etc., but both will be less than that of water.

Teaching Tip

All other things being equal, both the thermal conductivity and the heat capacity of a material determine the rate of temperature change in a dynamic situation (i.e., when thermal conditions are changing). Thermal conductivity is a property of materials that expresses the heat flux, or the rate at which heat flows through the material, given a certain temperature gradient. The thermal conductivity is generally lower for water than either for soil or solid rock, further reinforcing the observation that the water heats and cools more slowly than these geologic materials.

   b) Hypotheses will vary, but should be both well considered and testable. Encourage students to write out their hypothesis in the form of a formal hypothesis statement (i.e., "We hypothesize that...").

2. a) Answers will vary according to experimental design, but should include an analysis of the variables, controls, and safety concerns in addition to the basic experimental design. The variable being tested is the nature of the material (water, soil, or rock). Some important variables to control in this experiment include:
- nature of heat source
- distance of material being tested to that heat source
- ambient temperature around material being tested
- volume of material being tested and the container used to hold it
- time of exposure to the heat source
- convection of air into and out of the container

NOTES

manipulating (the independent variable), the variable that you are measuring (the dependent variable), and the controls within your experiment.

a) Record your design, variables, controls, and any safety concerns in your notebook.

3. When your teacher has approved your design, conduct your experiment.

4. In your *EarthComm* notebook, record your answers to the following questions:

a) Which material heated up faster?

b) Which material cooled more quickly?

c) How did your results compare with your hypothesis?

d) How does this investigation relate to differences in climate between places near a body of water, versus places far from water?

Have your teacher approve your design before you begin your experiment. Do not touch any heat source. Report any broken thermometers to your teacher. Clean up any spills immediately.

Reflecting on the Activity and the Challenge

In this activity, you learned about the physical features and climate of your community. You also compared the physical features and climate of your community to that of another community in the United States. This helped you begin to see ways in which physical features influence climate. This will help you explain the meaning of the term "climate" and describe the climate of your community in your newspaper article.

Digging Deeper

WEATHER AND CLIMATE

Factors Affecting Climate

Weather refers to the state of the atmosphere at a place, from day to day and from week to week. The weather on a particular day might be cold or hot, clear or rainy. **Climate** refers to the typical or average weather at a place, on a long-term average. For example, Alaska has a cold climate, but southern Florida has a tropical climate. Minnesota's climate is hot in the summer and cold in the winter. Western Oregon has very rainy winters. Each of these regions has a definite climate, but weather that varies from day to day, often unpredictably.

The climate of a particular place on Earth is influenced by several important factors: latitude, elevation, and nearby geographic features.

Geo Words

weather: the state of the atmosphere at a specific time and place.

climate: the general pattern of weather conditions for a region over a long period of time (at least 30 years).

EarthComm

Teaching Tip

Some possible materials that can be used to design the experiment for **Part C** of this **Investigation** include:
- soil and/or crushed rock
- water
- clear plastic cups
- a heat lamp or other heat source (sunny window sill, etc.)
- plastic wrap (to control convection of air into and out of the cup)
- alcohol thermometers
- ruler (to quantify distance from heat source)

3. Be sure that students' experiments have been approved before they are conducted.

4. a) Students should find that the rock heated more quickly. The water should be the slowest to heat.

    b) Students should find that the rock cooled more quickly. The water should be the slowest to cool.

    c) Answers will vary depending on students' original hypotheses.

    d) Students should note that because water is slow to heat or cool (much slower than the air or the ground), the temperature of places near large bodies of water should be affected by their presence. Specifically, large bodies of water tend to buffer seasonal temperature changes in places near them by keeping summers cooler and winters warmer than they would be otherwise.

Teaching Tip

In general, students should find that the temperature of the rock is most easily changed, and that the temperature of the water is the most difficult to change. The soil sample should fall somewhere between the two. The specific heat capacity of soil varies significantly depending on the nature of the soil (how sandy it is, etc.) and its moisture content. Regardless, the soil's temperature should be more easily changed than that of the water.

Reflecting on the Activity and the Challenge

This activity helped students to understand what is meant by the term climate. Take this opportunity to review with students what they have learned in the investigation. Ask them to write their own definition of climate based on their findings. Also discuss with students the importance of the relationship between latitude, elevation, and climate. Physical features can modify the climate, but the most important determining factors are elevation and latitude. Helping students understand their environment and

Chapter 2

location in relation to climate will help them understand and discuss any climate changes. A consideration of these factors will be important in focusing on the **Chapter Challenge**.

Digging Deeper

Assign the reading for homework, along with the questions in **Check Your Understanding** if desired.

Assessment Opportunity

Reword or restructure the questions in **Check Your Understanding** for a brief quiz. Use the quiz (or a class discussion of the questions in the textbook) to assess your students' understanding of the main ideas in the reading and the activity.

Assessment Tool

Check Your Understanding Notebook Entry-Evaluation Sheet
Use this sheet to evaluate the extent to which students understand the key concepts explored in **Activity 1** and explained in **Digging Deeper**, and to evaluate the students' clarity of expression.

NOTES

Earth System Evolution Climate Change

Geo Words

latitude: a north-south measurement of position on the Earth. It is defined by the angle measured from the Earth's equatorial plane.

elevation: the height of the land surface relative to sea level.

glacier: a large long-lasting accumulation of snow and ice that develops on land and flows under its own weight.

windward: the upwind side or side directly influenced to the direction that the wind blows from; opposite of leeward.

leeward: the downwind side of an elevated area like a mountain; opposite of windward.

rain shadow: the reduction of precipitation commonly found on the leeward side of a mountain.

Latitude

Latitude is a measure of the distance of a point on the Earth from the Equator. It is expressed in degrees, from zero degrees at the Equator to 90° at the poles. The amount of solar energy an area receives depends upon its latitude. At low latitudes, near the Equator, the Sun is always nearly overhead in the middle of the day, all year round. Near the poles, the Sun is low in the sky even in summer, and in the winter it is nighttime 24 hours of the day. As a result, regions near the Equator are much warmer than regions near the poles. Assuming a constant elevation, temperatures decrease by an average of about one degree Fahrenheit for every three degrees latitude away from the Equator.

Elevation

Elevation, the height of a point on the Earth's surface above sea level, also affects the physical environment. Places at high elevations are generally cooler than places at low elevations in a given region. On average, temperatures decrease by about 3.6°F for every 1000 ft. (300 m) gain in elevation. In many places at high elevation, **glaciers** form because the summers are not warm enough to melt all of the snow each year. The mountains in Glacier National Park in Montana, which are located at high elevation as well as fairly high latitude, contain glaciers, as shown in *Figure 1*.

Geographic Features

Geographic features, like mountain ranges, lakes, and oceans, affect the climate of a region. As shown in *Figure 2*, mountains can have a dramatic effect on precipitation in nearby areas. The **windward** side of a mountain chain often receives much more rainfall than the leeward side. As wind approaches the mountains, it is forced upwards. When the air rises, it cools, and water vapor condenses into clouds, which produce precipitation. Conversely, the **leeward** side of a mountain range is in what is called a **rain shadow**. It often receives very little rain. That is because the air has already lost much of its

Figure 1 Mountains at high elevation and high latitudes often contain glaciers.

Teaching Tip

Elevation is another key factor that affects the climate of an area. Take a moment to discuss with the students how the elevation of your community affects the local climate. Would students expect to find glaciers like those shown in *Figure 1* in their community today? Why or why not? Do they think that glaciers could ever form in their community?

Chapter 2

moisture on the windward side. When the air descends the leeward slope of the mountain, it warms up as the greater air pressure compresses it. That causes clouds to evaporate and the humidity of the air to decrease. The deserts of the southwestern United States have low rainfall because they are in the rain shadow of the Sierra Nevada and other mountain ranges along the Pacific coast.

Geo Words

heat capacity: the quantity of heat energy required to increase the temperature of a material or system; typically referenced as the amount of heat energy required to generate a 1°C rise in the temperature of 1 g of a given material that is at atmospheric pressure and 20°C.

Figure 2 The rain shadow effect. Most North American deserts are influenced by this effect.

Large bodies of water can also affect climate dramatically. The ocean has a moderating effect on nearby communities. Temperatures in coastal communities vary less than inland communities at similar latitude. This is true both on a daily basis and seasonally. The effect is especially strong where the coast faces into the prevailing winds, as on the West Coast of the United States. Kansas City's average temperature is 79°F in July and 26°F in January. San Francisco's average temperature is 64°F in July and 49°F in January. On average, New York City's January temperatures vary only 11°F during a day, whereas Omaha's January temperatures vary 20°F during a day. In each of these two cases the difference between the two cities' climates is too great to be related to their latitude, which is only different by less than about 1.5°. Instead, the differences in climate in these two places are because water has a much higher **heat capacity** than soil and rock. That means that much more heat is needed to raise the temperature of water than to raise the

Teaching Tip

This photo of Death Valley (*Figure 2* on page E93), in interior southern California, is representative of large areas of the semiarid to arid American Southwest. Here, the aridity is largely the result of the presence of high mountain ranges to the west, along the Pacific coast. As moisture-laden storm winds blow in from the Pacific Ocean during coastal storms, so much of the moisture is extracted by precipitation from the rising air as it crosses the mountains that little is left for precipitation over the interior. In general, the lower the elevation of the inland area, the less the rainfall.

Chapter 2

Earth System Evolution Climate Change

Geo Words

lake-effect snow: the snow that is precipitated when an air mass that has gained moisture by moving over a relatively warm water body is cooled as it passes over relatively cold land. This cooling triggers condensation of clouds and precipitation.

global climate: the mean climatic conditions over the surface of the Earth as determined by the averaging of a large number of observations spatially distributed throughout the entire region of the globe.

Little Ice Age: the time period from mid-1300s to the mid-1400s AD. During this period, global temperatures were at their coldest since the beginning of the Holocene.

Check Your Understanding

1. What is the difference between weather and climate?

2. What is the difference between regional climate and global climate?

3. Compare the climate of a city along the Pacific coast with that of a city with a similar latitude but located inland.

4. Explain how there can be snow on the top of a mountain near the Equator.

5. What is the "Little Ice Age"?

temperature of soil or rock. In the same way, water cools much more slowly than soil and rock. Land areas warm up quickly during a sunny day and cool down quickly during clear nights. The ocean and large lakes, on the other hand, change their temperature very little from day to day. Because the ocean absorbs a lot of heat during the day and releases it at night, it prevents daytime temperatures in seaside communities from climbing very high and prevents nighttime temperatures from falling very low (unless the wind is blowing from the land to the ocean!). By the same token, oceans store heat during the summer and release it during the winter, keeping summers cooler and winters warmer than they would be otherwise.

Lake-effect snow is common in late autumn and early winter downwind of the Great Lakes in north–central United States. Cold winds blow across the still-warm lake water, accumulating moisture from the lake as they go. When they reach the cold land, the air is cooled, and the water precipitates out of the clouds as snow. The warm oceans also supply the moisture that feeds major rainstorms, not just along the coast but even far inland in the eastern and central United States.

Global Climate

Climates differ from one region to another, depending on latitude, elevation, and geographical features. However, the entire Earth has a climate, too. This is called **global climate**. It is usually expressed as the year-round average temperature of the entire surface of the Earth, although average rainfall is also an important part of global climate. Today, the average temperature on the surface of the Earth is about 60°F. But the Earth's climate has changed continually over geologic time. During the Mesozoic Era (245–65 million years ago), when the dinosaurs roamed the Earth, global climate was warmer than today. During the Pleistocene Epoch (1.6 million–10,000 years ago), when mastodons and cave people lived, global climate seesawed back and forth between cold glacial intervals and warmer interglacial intervals. During glacial intervals, huge sheets of glacier ice covered much of northern North America. Just a few hundred years ago, the climate was about 3°F cooler than now. The time period from about the mid-1300s to the mid-1800s is called the **Little Ice Age**, because temperatures were generally much colder than today, and glaciers in many parts of the world expanded. Global temperatures have gradually increased since then, as the Earth has been coming out of the Little Ice Age.

Over the next several activities, you will be looking at what causes these changes in global climate, and how human activity may be causing global climate change.

E 94

Check Your Understanding

1. Climate refers to the typical or average weather at a place, on a long-term average. Weather refers to the current state of the atmosphere at a place, from day to day.

2. Regional climate refers to the climate in one local region, which varies with latitude, elevation, and geographical features. Global climate refers to the averaged climate of the entire Earth.

3. The climate of a city along the Pacific coast tends to vary much less than the climate of inland communities because of the moderating effects of the ocean on temperature.

4. There can be snow on top of mountains near the Equator because, on average, temperatures decrease by about 3.6°F for every 300 m of elevation. Therefore, the tops of mountains tend to be much cooler. Also, precipitation is more likely to occur on mountaintops than lower down.

5. The Little Ice Age refers to the time period from about the mid-1300s to the mid-1800s, when temperatures were generally much colder than today, and glaciers in some parts of the world expanded.

Chapter 2

Understanding and Applying What You Have Learned

1. What is the nearest body of water to your community? In what ways does it affect the physical environment of your community?

2. Identify one physical feature in or near your community and explain how it affects the climate. You may need to think regionally. Is there a mountain range, a large lake, or an ocean in your state?

3. How would changing one feature of the physical environment near your community affect the climate? Again, you may need to think on a regional scale. Name a feature and tell how the climate would be different if that feature changed. How would this change life in your community?

4. Can you think of any additional reasons to explain the differences in climate between your community and the other community you looked at in the **Investigate** section?

Preparing for the Chapter Challenge

1. Clip and read several newspaper articles about scientific topics.

2. Using a style of writing appropriate for a newspaper, write a few paragraphs in which you:

   • explain the term climate;
   • describe the climate of your community (including statistics about seasonal temperatures, rainfall, and snowfall);
   • explain what physical factors in your community or state combine to produce this climate;
   • explain the difference between regional climate and global climate, and
   • describe the Earth's global climate.

Inquiring Further

1. **Weather systems and the climate of your community**

   Investigate how weather systems crossing the United States affect the climate in your community. In many places, weather systems travel in fairly regular paths, leading to a somewhat predictable series of weather events.

2. **Jet stream**

   Do some research on the jet stream. How does its position affect the climate in your community?

Understanding and Applying What You Have Learned

1. Answers will vary depending on where your community is located. In general, bodies of water moderate climate because the large heat capacity of water acts to moderate temperature variations.

2. Answers will vary depending on where your community is located. Several examples of the effects of geographic features can be found in this Teacher's Edition in the **Background Information** section for this activity.

3. Answers will vary depending on where your community is located and the feature discussed. Answers should consider not only the resulting change in climate but also how that change would affect life in their community.

4. Answers will vary depending upon the community that has been selected and upon the variation between the two communities.

Preparing for the Chapter Challenge

This section gives students an opportunity to apply what they have learned to the **Chapter Challenge**. They can work on this as a homework assignment or during class time within groups.

The first part of this section is designed to familiarize students with the style of writing found in newspaper articles. You may wish to have each student bring in one or two articles, which you can then copy and distribute to each student for reference.

In the second part of this section, students are asked to write a newspaper article that reviews the findings of **Activity 1**. Aspects of this newspaper article can be used to complete the **Chapter Challenge**.

Criteria for assessment (see **Assessment Rubric for Chapter Challenge on Climate Change** on pages 296 to 298) include:
- a discussion of the meaning of climate, both regional and global
- a description of your community's present climate

Assessment Opportunity

You may want to use a rubric like the one shown on the following page to assess student articles. Review these criteria with your students so that they can be certain to include the appropriate information in their article.

Item	Missing	Incomplete/Inaccurate	Complete/Accurate
Explanation of the term "climate."			
Description of the climate of your community, including statistics about seasonal temperatures and precipitation.			
Explanation of what physical factors in the community influence the climate.			
Explanation of the difference between global and regional climate.			
Description of the Earth's global climate.			

Inquiring Further

1. Weather systems and the climate of your community

Answers will vary depending upon where your community is located. Links that will lead students to information about weather systems that commonly affect the United States can be found on the *EarthComm* web site.

2. Jet stream

The jet stream is a well-defined high-speed flow of air in the upper atmosphere. To be considered a jet stream, the winds must be faster than 57 mph. The jet stream flows at altitudes high above the surface of the Earth. It generally flows from west to east, but it usually follows a curved, undulatory path around the Earth. The jet stream separates cold polar air to its north from warmer air to its south. How the jet stream directly affects your community will vary. Students can visit the *EarthComm* web site to learn more about the jet stream and their community.

NOTES

Chapter 2

ACTIVITY 2 — PALEOCLIMATES

Background Information

Dendroclimatology

Dendroclimatology is the science of using the patterns and relative sizes of annual growth rings of trees as indicators of climate. The trunk and branches of a tree grow in diameter by addition of new cells in what is called the cambium, a thin layer, only a few cells thick, just beneath the bark and at the outer margin of the wood of the tree. New cells are produced in the cambium to make new bark and new wood during the growing season.

Each year's growth of new wood is marked by a distinct ring, as seen in a cross section normal to the axis of the trunk or branch. The ring is seen as a couplet of relatively light-colored wood, called earlywood, in the inner part of the ring, and relatively dark-colored wood, called latewood, in the outer part of the ring. The earlywood is the result of rapid growth in the early part of the growing season, in spring and early summer. The latewood is the result of slower growth in the later part of the growing season, in late summer and early fall. The tubular cells of the earlywood are large and thin-walled, and those of the latewood are smaller but with thicker and darker walls. The transition from the lighter wood of the inner part of the ring to the darker wood of the outer part of the ring is gradual, but the transition from the outer margin on one ring to the inner margin of the next ring is abrupt. It is this asymmetry that makes identification of the boundaries of the rings easy.

Ring width varies as a function of a number of factors. During the life of the tree, the early robust growth results in relatively thick rings in the inner part of the trunk, whereas slower growth later in the life of the tree results in relatively thin rings. When the succession of ring thickness is used for paleoclimate study, this effect is taken into account by comparing the succession among many different trees in a given area. Local effects having to do with shading of one tree by another, or by death of a neighboring tree (thus abruptly increasing the sunlight available to a given tree), also necessitate the use of more than a single tree in paleoclimate studies.

The predominant effects on ring width in trees of a given local area, however, are temperature and precipitation. Cooler temperatures during the growing season cause a slowdown in the metabolic processes in the tree, resulting in less addition of new wood. A related effect is that in a cool year, the growing season is shorter. (Growing season is the time period between breaking of dormancy and start of new growth in the spring, and cessation of growth at the end of the season.) Rate of growth of new wood is also strongly affected by availability of soil moisture to roots.

One of the principal problems in interpreting paleoclimate on the basis of tree rings lies in separation of the temperature effect from the precipitation effect. The problem is most readily surmounted in mountainous regions where a given tree species occupies a definite range in elevations. It is generally believed that in the lowermost part of the elevation range, precipitation exerts the dominant control on ring width, whereas in the uppermost part of the elevation range, where availability of soil moisture is generally less variable from year to year, temperature is the dominant control.

The single most important tree species from which climatic information has been obtained is the bristlecone pine (*Pinus longeava*), which grows in the higher elevations of the dry mountains in interior east-central California in the rain shadow of the Sierra Nevada. Individual bristlecone pines are known to be as old as thousands of years. By intercorrelating older, dead trees, a chronology that extends back for several thousand years has been developed. This chronology, coupled with information on paleoclimate revealed from ring width, provides a qualitative picture of climate that extends far into the past.

Oxygen Isotopes and Paleoclimate

The study of the relative abundance of the different isotopes of an element is becoming a very powerful tool in many scientific disciplines, and the Earth sciences are no exception. The mechanisms that affect and control isotopic ratios are now becoming well understood for a large number of different elements. This knowledge, combined with technological developments that have enabled much more sensitive measurements of these ratios, have led to many advances in a number of scientific disciplines, including the science of paleoclimatology.

The atoms of each of the chemical elements can exist in more than one form, depending upon the number of neutrons in the nucleus of the atom. These atoms of different atomic weights are called isotopes. Different isotopes of a given chemical element have exactly the same chemical behavior, because the number of electrons is the same, but they differ slightly in certain physical processes because of the differing atomic weights. The element

oxygen, for example, has three naturally occurring isotopes, with atomic weights of 16, 17, and 18; these are written conventionally as ^{16}O, ^{17}O, and ^{18}O. About 99.8% of oxygen is ^{16}O, about 0.2% is ^{18}O, and only about 0.04% is ^{17}O.

The relative proportions of ^{18}O and ^{16}O can be determined with high accuracy using a mass spectrometer, a device that sorts atoms by their atomic weight. Mass spectrometers work on the principle that an electrically charged particle that passes near a magnet is deflected by the magnet. A beam of ionized atoms is passed by an electromagnet, and atoms with slightly different atomic weights are deflected at slightly different angles. This separates the different isotopes from one another; the atoms are then caught in collector cups, and the proportions of the various isotopes are measured.

The $^{18}O/^{16}O$ ratio is conventionally expressed in a slightly different way, which you are likely to encounter if you read the literature on paleoclimate. A quantity written as $\delta^{18}O$ (heard as "delta-eighteen-oh") is defined to represent the fractional change in the $^{18}O/^{16}O$ ratio compared to a universally agreed-upon reference standard. It is officially defined by subtracting $^{18}O/^{16}O$ of the standard from $^{18}O/^{16}O$ of the sample, dividing the result by $^{18}O/^{16}O$ of the standard, and then multiplying by one thousand to be expressed as parts per thousand (also called per mil), denoted by the symbol ‰. The per mil representation is analogous to representation as parts per hundred, denoted by the symbol %, which we call percent.

The isotopes of oxygen become separated, to a small but measurable extent, by various natural processes. When water evaporates from a liquid surface, the evaporated water is lighter than the remaining liquid, because the lighter water molecules are faster moving and are more likely to escape across the air–water interface. Correspondingly, precipitated water is heavier than the water vapor that remains in the atmosphere. Similar fractionation occurs during the transition from vapor to ice and from ice to vapor. The net result of these processes is that when water is extracted from the oceans to form continental ice sheets, the glacial ice is relatively light, with $^{18}O/^{16}O$ ratios a few percent lower than in seawater. The $^{18}O/^{16}O$ ratio of average seawater is therefore a good—although qualitative—proxy for the extent of continental glaciation.

Certain metabolic processes in organisms also enrich ^{18}O relative to ^{16}O. This fractionation has a measurable dependence on temperature. This effect is especially relevant to foraminifera, which are unicellular protozoans that live in great numbers in the oceans. Some foraminifera are planktonic, meaning that they live a free-floating existence in the warm near-surface waters of the oceans; others are benthic, meaning that they live on the ocean bottom. Benthic foraminifera are a good measure of total ice volume, because the effects of water temperature are subdued in the always-cold deep waters. Planktonic foraminifera, on the other hand, provide a good signal of water temperature in the surface oceans, although the overall effect of change in the $^{18}O/^{16}O$ ratio due to extraction of ocean water to feed glaciers must be taken into account in interpreting that signal.

The fractionation of oxygen isotopes in evaporation and condensation depends on temperature in a known way, so the isotopic ratio $^{18}O/^{16}O$ has the potential to be a proxy for temperature. In this respect, the isotopes of hydrogen are even more sensitive than those of oxygen. There are two naturally occurring stable isotopes of hydrogen: ^{1}H (usually denoted just as H), and ^{2}H (called deuterium, denoted by D).

At first thought, it does not seem to make sense that hydrogen isotopes would be a better temperature proxy than oxygen isotopes, because a water molecule that contains one deuterium atom weighs only 5.5% more than an ordinary water molecule with two atoms of ^{1}H. In contrast, water with ^{18}O weighs 11% more. The reason is that the energy of vibration of the hydrogen–oxygen bond, which influences the extent of fractionation during evaporation and condensation, depends on the fractional change in the mass of the atom, which is a factor of two for D and only 1.06 for ^{18}O. In consequence, when water evaporates, the vapor has several percent less deuterium than the liquid, which is a far larger effect than for ^{18}O, and this effect varies strongly with water temperature.

Use of deuterium as a temperature proxy involves some troublesome uncertainties, however. This uncertainty is partly because this effect has to be disentangled in some way from the glaciation effect discussed above. Also, the final measured result in, for example, a sample from an air bubble trapped deep in glacier ice, depends on the history of evaporation, condensation, and precipitation as snow, with each step in the process imprinting a temperature signal.

More Information – on the Web
Visit the *EarthComm* web site www.agiweb.org/earthcomm/ to access a variety of links to web sites that will help you deepen your understanding of content and prepare you to teach this activity. Many of the sites also contain images that you can download.

Goals and Assessment

Clarify that the goals indicate what students should understand and be able to do as a result of the activity. Make sure students understand that Chapter Assessments are based upon these goals.

Goal	Location in Activity	Assessment Opportunity
Understand the significance of growth rings in trees as indicators of environmental change.	**Investigate** Part A **Digging Deeper; Check Your Understanding** Question 1	Change in tree-ring appearance is correctly indicated. Connection between tree-ring change and appearance of Little Ice Age is made correctly.
Understand the significance of ice cores from glaciers as indicators of environmental change.	**Digging Deeper; Check Your Understanding** Question 3	Response to question is correct, and closely matches response given in Teacher's Edition.
Investigate and understand the significance of geologic sediments as indicators of environmental change.	**Investigate** Part B **Digging Deeper; Check Your Understanding** Question 2 **Understanding and Applying What You Have Learned** Question 2	Sketch of sediment core is useful; measurements of thickness are correct. Responses to questions are reasonable, and are based on model and reading.
Examine the significance of glacial sediments and landforms as evidence for climate change.	**Digging Deeper; Understanding and Applying What You Have Learned** Question 1	Responses are reasonable for your community.
Investigate and understand the significance of fossil pollen as evidence for climate change.	**Investigate** Part B **Digging Deeper; Understanding and Applying What You Have Learned** Question 2	Sketch of sediment core is useful; measurements of thickness are correct. Estimate of years represented by core is accurate. Paragraph is complete and descriptive, and is based on model and reading.

Chapter 2

Earth System Evolution Climate Change

Activity 2   Paleoclimates

Goals

In this activity you will:

- Understand the significance of growth rings in trees as indicators of environmental change.

- Understand the significance of ice cores from glaciers as indicators of environmental change.

- Investigate and understand the significance of geologic sediments as indicators of environmental change.

- Examine the significance of glacial sediments and landforms as evidence for climate change.

- Investigate and understand the significance of fossil pollen as evidence for climate change.

Think about It

The cross section of a tree trunk shows numerous rings.

- What do you think the light and dark rings represent?
- What might be the significance of the varying thicknesses of the rings?

What do you think? Record your ideas about these questions in your *EarthComm* notebook. Be prepared to discuss your responses with your small group and the class.

Activity Overview

Students examine an image of growth rings from a tree and determine how the rings vary in thickness over time. They correlate changes in the growth rings with temperature data for the past 1000 years to understand how growth rings in trees can be used as indicators of environmental change. Students then use modeling clay to produce a model of sediments deposited in a lake or pond over time. They use a key to determine the types of fossil pollen found in different colors of clay. This allows students to describe how climate has changed near the lake over the time represented by the sediment core. **Digging Deeper** reviews various tools used by scientists to understand paleoclimates, including:

- fossil pollen
- ice cores
- deep-sea sediments
- glacial landforms and sediments
- growth rings in trees

Preparation and Materials Needed

Preparation

Part A

You will need to make copies of the photo of the tree rings. Use **Blackline Master Climate Change 2.1, Tree Rings**. No additional preparation is required.

Part B

Aside from assembling all of the necessary materials, no advanced preparation is needed for this part of the activity.

Materials

- Blue, red, green, and yellow modeling clay
- Rolling pin (optional)
- Waxed paper (optional)
- Beaker or small milk carton
- Small copper pipe or empty barrel from a ballpoint pen
- Thin stick (that fits inside pipe)

Think about It

Student Conceptions

You may wish to provide students with a photo of the cross section of a tree trunk for reference, or refer them to the photo on page E97 in the student text. Most students will likely be aware that the rings are an indicator of the age of the tree, but it is unlikely that they will understand the significance of variations in the size and color of the rings.

Answer for the Teacher Only

Each yearly growth ring is in two parts: a relatively light ring and a relatively dark ring. The light ring represents the spring and early summer time of most rapid growth, and the dark ring represents the late summer and autumn time of maturing of the wood. In a year with weather more favorable for growth, the annual ring is thicker; in a year with weather less favorable for growth, the annual ring is thinner. The thickness of the ring is the outcome of both temperature and precipitation during the growing season.

Assessment Tool

Think about It Evaluation Sheet
Use this evaluation sheet to help students understand and internalize the basic expectations for the warm-up activity.

Blackline Master Climate Change 2.1
Tree Rings

Investigate

Part A: Tree Rings

1. Examine the photo that shows tree growth rings from a Douglas fir. Notice the arrow that marks the growth ring that formed 550 years before the tree was cut.

a) Where are the youngest and the oldest growth rings located?

b) Not all the growth rings look identical. How are the rings on the outer part of the tree different from those closer to the center?

c) Mark the place on a copy of the picture where the change in the tree rings occurs.

2. Using the 550-year arrow as a starting point, count the number of rings to the center of the tree. Now count the number of rings from the arrow to where you marked the change in the way the rings look.

a) Record the numbers.

b) Assuming that each ring represents one year, how old is the tree?

c) Assume that the tree was cut down in the year 2000. What year did the tree rings begin to look different from the rings near the center (the rings older than about 550-years old)?

3. Compare the date you calculated in **Step 2 (c)** to the graph that shows change in temperature for the last 1000 years.

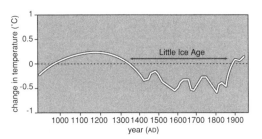

a) What was happening to the climate during that time?

b) From these observations, what would you hypothesize is the correlation between the thickness of tree rings and climate?

4. Examine the diagram that shows temperature change for the last 150,000 years.

a) How does the duration of the Little Ice Age compare to the duration of the ice age (the time between the two interglacial periods) shown in this figure?

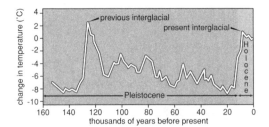

Investigate

Part A: Tree Rings

1. a) The youngest growth rings are located on the outer perimeter of the trunk; the oldest growth rings are located at the center of the trunk.

    b) The rings on the outer part of the tree are much thinner than the rings closer to the center. (Note that there are two possible reasons for this: the climate changed over time to be less favorable for tree growth, or the rate of tree growth slowed merely because the tree was not as vigorous in old age.)

    c) The change in the thickness of the tree rings occurs approximately seven growth rings to the right (towards the outside of the tree) of the arrow that marks the ring that grew 550 years ago. (It is important to note that this arrow does not indicate the ring that was formed when the tree was 550 years old, but instead marks the ring that formed 550 years ago.)

2. a) From the 550-year arrow to the center of the tree, there are 28 rings. From the 550-year arrow to the place where the rings change, there are seven rings.

    b) As mentioned above, the arrow points to the ring that is 550 years old (i.e., the ring that grew 550 years ago), not to the ring that grew when the tree was 550 years old. There are 28 clearly visible rings to the left of the ring with the arrow pointing to it (or inward of that ring). Accordingly, that ring was formed in the tree's 29th year. Because the ring with the arrow pointing to it is 550 years old, the tree is a total of 578 years old. Students' answers may vary a bit because it is difficult to count the rings.

    c) There are an additional seven rings to the right of the arrow before the rings start getting much thinner. (Note that because the ring indicated by the arrow was formed in the tree's 29th year, adding an additional seven years would make the tree 36 years old when the rings started getting thinner.) Subtracting seven years from 550 years gives 543 years. That means that this transition would have taken place 543 years before the tree was cut down. Assuming that the tree was cut down in the year 2000, then the tree rings began to be thinner in the year 1457.

3. a) Temperatures began to drop around the time that the tree rings began to change. (Actually the temperatures began to decrease as much as 100 years before this, but the full magnitude of temperature decreases of the Little Ice Age, as shown in this graph, were not fully realized until the mid-1400s. It is also important to keep in mind that even in places where the Little Ice Age was significant, it did not take effect everywhere simultaneously.)

    b) The information suggests that thinner tree rings correspond to cooler temperatures.

4. a) The Little Ice Age lasted less than 500 years, whereas the last glacial period in the figure lasted several tens of thousands of years. Hence, the Little Ice Age was a minor fraction of the duration of the last glacial period.

Blackline Master Climate Change 2.2
Graphs of Change in Temperature over Time

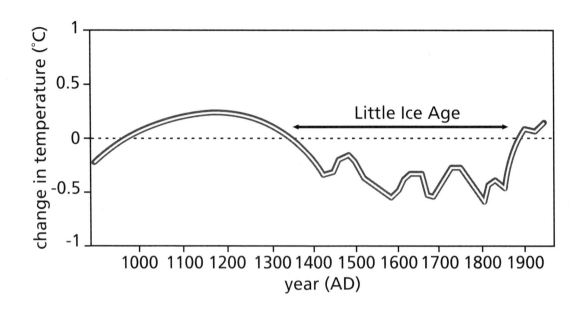

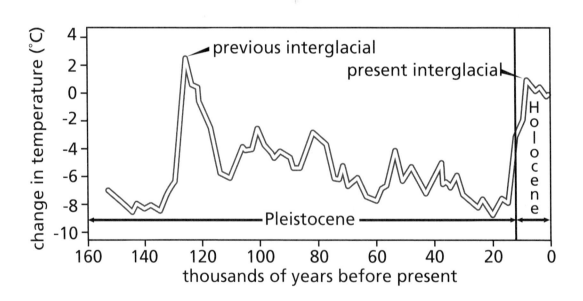

NOTES

Earth System Evolution Climate Change

Part B: Fossil Pollen

1. Using blue, red, green, and yellow modeling clay, put down layers in a small container. You may put them down in any order and thickness. The container represents a lake or pond, and the clay represents sediment that has settled out over a long time.

2. When you have finished laying down your "sediments," use a small pipe to take a core. Push the pipe straight down through all of the layers. Then carefully pull the pipe back up. Use a thin stick to push the core of sediments out of the pipe.

    Place folded paper towels or other padding between your hand and the upper end of the pipe before pushing it into the clay.

   a) Draw a picture of the core in your notebook. Note which end is the top.

   b) Measure and record the thickness of each layer of sediment to the nearest tenth of a centimeter.

3. The different colors of clay represent sediments that have settled out of the lake water at different times. For this exercise, imagine that each centimeter of clay represents the passage of 1000 years.

   a) How many years does your core represent from top to bottom?

b) How many years does each layer represent?

4. Imagine that the different colors of clay represent the following:

   • Blue: sediments containing pollen from cold-climate plants like spruce and alder trees.

   • Red: sediments containing pollen from warm-climate plants, like oak trees and grasses.

   • Green: sediments containing mostly spruce and alder pollen, with a little oak and grass pollen.

   • Yellow: sediments containing mostly oak and grass pollen, with a little spruce and alder pollen.

   a) How do you think the pollen gets into the lake sediments?

   b) Describe what the climate around the lake was like when each layer of sediment was deposited.

   c) Write a paragraph describing the climate changes over the period of time represented by your core. Make sure you say at how many years before the present each climate change occurred. Note whether transitions from one type of climate to another appear to have happened slowly or quickly.

Reflecting on the Activity and the Challenge

You will need to explain some of the ways that geologists know about climates that existed in the geologic past. The activity helped you understand two of the ways that geologists find out about ancient climates. Geologists study tree growth rings. They relate the thickness of the tree rings to the climate. Geologists also collect cores from layers of sediments and study the kinds of pollen contained in the sediments. The pollen shows what kinds of plants lived there in the past, and that shows something about what the climate was like.

Part B: Fossil Pollen

1. Students can make the clay layers by flattening lumps of clay between their hands. It is easier to do that when the clay is warm. A better way to make the layers is to flatten lumps of clay with a kitchen rolling pin. The best way to do that is to put the clay between sheets of waxed paper.

2. **a)** Answers will vary depending upon how students have chosen to build the clay layers. An example is provided below.

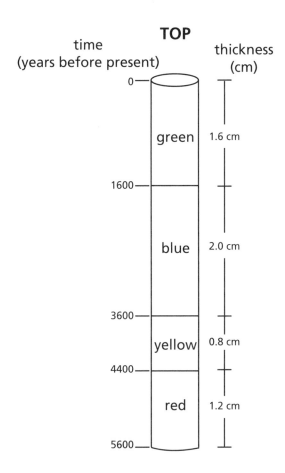

Note: vertical exaggeration is 2x

b) Answers will vary depending on the students' models, but the diagram from **Step 2 (a)** and the measured thicknesses should reflect one another. In the example provided, the first layer (red) is 1.2 cm thick, the next layer (yellow) is 0.8 cm thick, the third layer (blue) is 2.0 cm thick, and the fourth layer (green) is 1.6 cm thick

3. **a)** Answers will vary depending upon how students have chosen to build the clay layers. In the example given, the core is 5.6 cm thick. Given that 1 cm represents 1000 years, the total time represented in the core is 5600 years.

b) Answers will vary depending upon how students have chosen to build the clay layers. In the example given, the red layer represents 1200 years, the yellow layer represents 800 years, the blue layer represents 2000 years, and the green layer represents 1600 years.

4. a) The pollen is carried by the wind and settles out over the lake. It then settles down to the bottom of the lake, where it is incorporated into sediments.

b) - c) Answers will vary depending upon how students have chosen to build the clay layers, and in order to place time constraints on the layers, one must assume a date for the top of the core. The following description assumes that the top of the core represents the present. In the example given, the first layer deposited (red) contains pollen from grasses and oaks. This is indicative of a warm climate. This layer represents a time from about 5600 years before present (yr BP) to about 4400 yr BP. The next layer (yellow) represents the time period from about 4400 yr BP to about 3600 yr BP. In this layer pollen from oak trees and grasses is still predominant, but pollen from alder and spruce trees is also present. This indicates that the climate has cooled somewhat relative to the time when the red layer was deposited. Oak and grass pollen is absent from the next (blue) layer, indicating a further cooling of the climate during the period between 3600 yr BP and 1600 yr BP. The last layer (green) mostly has the pollen from the cold-climate plants (spruce and alder), but it also has a little pollen from oak trees and grasses. This implies that the climate began to warm again after 1600 yr BP, with the warmer climate continuing to the present.

Assessment Tools

EarthComm Notebook Entry-Checklist
Use this checklist as a quick guide for student self-assessment and/or an opportunity to quickly score student work. Add further criteria specific to your classroom needs or to this particular investigation.

Investigate Notebook Entry-Evaluation Sheet
Point out the criteria listed on this evaluation sheet that are relevant to this particular investigation. Encourage students to internalize the criteria by making them part of your "assessment conversations" as you circulate around the classroom.

Reflecting on the Activity and the Challenge

Have students read this brief passage and share their thoughts about the main point of the activity in their own words. Hold a class discussion about how this investigation relates to the **Chapter Challenge**. Discuss with students why understanding the tools that paleoclimatologists use to learn about past climates is important.

Digging Deeper

HOW GEOLOGISTS FIND OUT ABOUT PALEOCLIMATES

Direct Records and Proxies

A **paleoclimate** is a climate that existed sometime in the past: as recently as just a few centuries ago, or as long as billions of years ago. For example, in the previous activity you learned that the world was warmer in the Mesozoic Era (245 – 65 million years ago) and experienced periods of glaciation that affected large areas of the Northern Hemisphere continents during the Pleistocene Epoch (1.6 million–10,000 years ago). At present the Earth is experiencing an interglacial interval—a period of warmer climate following a colder, glacial period. The Earth today has only two continental ice sheets, one covering most of Greenland and one covering most of Antarctica.

The last retreat of continental glaciers occurred between about 20,000 years ago and 8000 years ago. That was before the invention of writing, so there is no direct record of this change. Systematic records of local weather, made with the help of accurate weather instruments, go back only about 200 years. A global network of weather stations has existed for an even shorter time. Historical accounts exist for individual places, most notably in China. For certain places in China records extend back 2000 years. They are useful, but more extensive information is required to understand the full range of climate variability. **Paleoclimatologists** use a variety of methods to infer past climate. Taken together, the evidence gives a picture of the Earth's climatic history.

Unfortunately, nothing gives a direct reading of past temperature. Many kinds of evidence, however, give an indirect record of past temperature. These are called **climate proxies**. Something that represents something else indirectly is called a proxy. In some elections, a voter can choose another person to cast the vote, and that vote is called a proxy. There are many proxies for past climate, although none is perfect.

Fossil Pollen

Pollen consists of tiny particles that are produced in flowers to make seeds. Pollen is often preserved in the sediments of lakes or bogs, where it is blown in by the wind. For example, a layer of sediment may contain a lot of pollen from spruce trees, which grow in cold climates. From that you can infer that the climate around the lake was cold when that layer of sediment was being deposited. Geologists collect sediment from a succession of

Geo Words

paleoclimate: the climatic conditions in the geological past reconstructed from a direct or indirect data source.

paleoclimatologist: a scientist who studies the Earth's past climate.

climate proxy: any feature or set of data that has a predictable relationship to climatic factors and can therefore be used to indirectly measure those factors.

Digging Deeper

Assign the reading for homework, along with the questions in **Check Your Understanding** if desired.

Assessment Opportunity

Use a brief quiz to check comprehension of key ideas and skills. Use a quiz (or a class discussion) to assess your students' understanding of the main ideas in the reading and the activity. A few sample questions are provided below:

Question: What is a proxy, and why do scientists use proxies to learn about past climates?

Answer: A proxy is something that represents something else indirectly. Because it is not possible to directly study a climate that occurred in the past, scientists must rely on indirect evidence that suggests environmental change and make inferences about their findings.

Question: List three ways that scientists can use ice cores to help understand ancient climates.

Answer: Scientists can measure oxygen and hydrogen isotopes in air bubbles trapped in ice cores to determine the relative proportions of the different isotopes. In the case of oxygen, greater amounts of ^{18}O suggest warmer climates. Scientists can also measure CO_2 concentrations in air bubbles trapped in ice cores; higher CO_2 concentrations correspond to warmer climates. They can measure the amount of dust found in ice cores; dust in the ice core indicates stronger winds and colder climates.

Question: How can scientists investigate the prehistoric distribution of plants of a region? How could this information help to understand climate change?

Answer: The prehistoric distribution of plants in a given area can be investigated by looking at the distribution of pollen and spores that have been buried in sediments or preserved in sedimentary rocks. If the ages of the sediments or rocks are known, then changes in preserved plants through time can be interpreted. This data can be used to understand changing climate because floral assemblages often can be related to climatic factors.

Earth System Evolution Climate Change

sediment layers. They count the number of pollen grains from different plants in each layer. Then they make charts that can give an idea of the climate changes that have taken place. (See *Figure 1*.) Pollen is easy to study because there is so much of it. Geologists also study fossil plants and insects to reconstruct past climates.

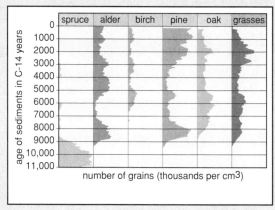

Figure 1 Changes over time in the relative amounts of different types of pollen from various trees and grasses give clues as to how climate has changed in the past.

Ice Cores

Figure 2 Scientists are able to obtain clues into past climatic conditions from air bubbles trapped in ice cores.

In recent years, study of cores drilled deep down into glaciers, like the one shown in *Figure 2*, has become a very powerful technique for studying paleoclimate. Very long cores, about 10 cm (4 in.) in diameter, have been obtained from both the Greenland and Antarctic ice sheets. The longest, from Antarctica, is almost 3400 m (about 1.8 mi.) long. Ice cores have been retrieved from high mountain glaciers in South America and Asia.

Glaciers consist of snow that accumulates each winter and does not melt entirely during the following summer. The snow is gradually compressed into ice as it is buried by later snow. The annual layers can be detected by slight changes in dust content. The long core from Antarctica provides a record of climate that goes back for more than 400,000 years. See *Figure 3*.

Chapter 2

Teaching Tip

Use **Blackline Master Climate Change 2.3, Change over Time in Relative Amounts of Pollen** to make an overhead of the graph shown in *Figure 1*. Ask students to interpret the data. Tell them that spruce trees are found in colder climates than alder, birch, pine, oak, or grasses.

Teaching Tip

Ice cores (like the one shown in *Figure 2*) drilled from an ice sheet provide a sample of all of the layers of snow that have accumulated in a particular area over time. The oldest layers are on the bottom of the core. Selection of a drilling site is based on factors like the underlying bedrock, the thickness of the ice sheet, the location on the ice sheet where the basal ice is likely to be the oldest, and the potential for dangerous crevasses in the ice.

Once scientists have chosen a drilling site, mechanical drills are used to penetrate the ice. It takes about one day to remove 50 to 70 m of ice, which typically covers about 200 years. The ice cores are typically about 10 cm in diameter, and are removed at increments of 1 m.

When the cores are brought to the surface, the first thing scientists do is to count the alternating layers of light and dark layers in the core to determine the age of the ice. The darker ice represents the winter snow season, and the lighter ice represents the summer season. The reason for the banding is in part due to the grain size of the ice crystals, which tend to be finer in the winter. In some glaciers, annual dust banding also plays a role. This situation is a bit more complicated, in that the season where there is the most dust in the atmosphere can vary in different places. In Antarctica, however, there is no strong seasonal difference to the amount of dust in the air. The concentration of dust in the snow is thus greatest in the summer, because the Sun sublimates some of the snow, concentrating the dust. The concentration of dust in the summer ice causes greater scattering of light giving the ice a white appearance, yielding a light-colored layer. The ice in the winter layer is clearer and does not scatter light as greatly. This makes the winter layer appear darker, although it is really just clearer than the summer ice. Scientists can use the alternating layers, much as they can use tree rings, to determine the age of the core. More sophisticated techniques for dating ice cores are used back in the laboratory, where the cores are preserved in deep refrigeration.

Blackline Master Climate Change 2.3
Change over Time in Relative Amounts of Pollen

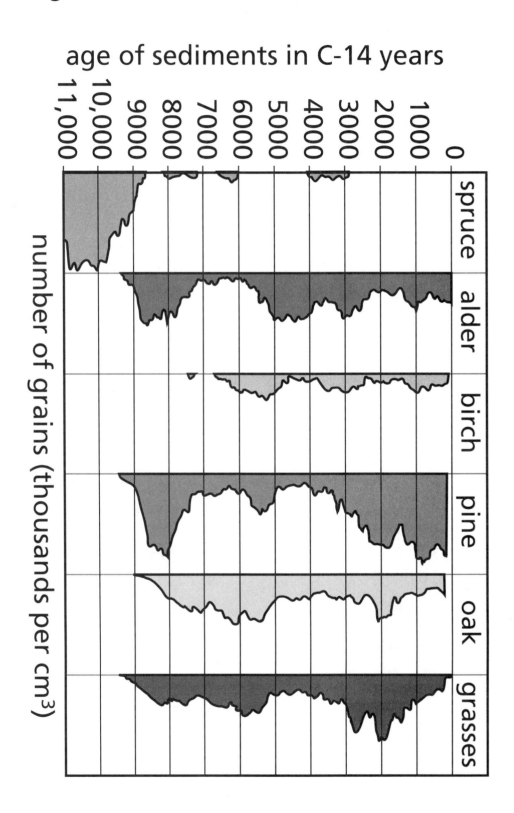

age of sediments in C-14 years

number of grains (thousands per cm³)

spruce   alder   birch   pine   oak   grasses

Blackline Master Climate Change 2.4
Temperature Variations

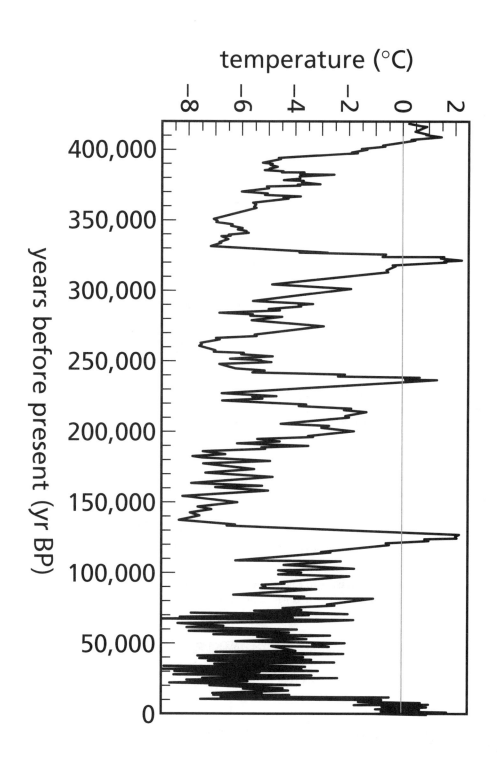

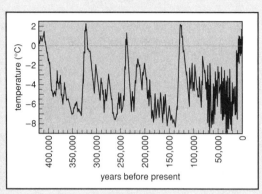

Figure 3 Temperature variation over the past 420,000 years, relative to the modern surface temperature at Vostok (−55.5°C).

Geo Words

isotope: one of two or more kinds of atoms of a given chemical element that differ in mass because of different numbers of neutrons in the nucleus of the atoms.

Bubbles of air trapped in the ice contain samples of the atmosphere from the time when the snow fell. Paleoclimatologists study the oxygen in the water molecules in the ice. Almost all of the oxygen atoms in the atmosphere are in two forms, called **isotopes**. The two isotopes are oxygen-16 (written ^{16}O) and oxygen-18 (written ^{18}O). They are the same chemically, but they have slightly different weights. ^{18}O is slightly heavier than ^{16}O. The proportion of these two isotopes in snow depends on average global temperatures. Snow that falls during periods of warmer global climate contains a greater proportion of ^{18}O, and snow that falls during periods of colder global climate contains a smaller proportion. The ratio of ^{18}O to ^{16}O can be measured very accurately with special instruments. Another important way of using the glacier ice to estimate global temperature is to measure the proportions of the two naturally occurring isotopes of hydrogen: ^{1}H, and ^{2}H (which is called deuterium).

The air bubbles in the ice contain carbon dioxide. The amount of carbon dioxide in the glacier ice air bubbles depends on the amount of carbon dioxide in the air at that time. The amount of carbon dioxide in the atmosphere can be correlated to global temperatures. During times when the paleoclimate is thought to have been warm, the ice core record shows relatively higher levels of atmospheric carbon dioxide compared to times of interpreted colder climate. Measurements of carbon dioxide taken from the cores give a global picture, because carbon dioxide is uniformly distributed in the global atmosphere.

A third component of the ice that yields clues to paleoclimates is dust. During colder climates, winds tend to be stronger. The stronger winds erode more dust, and the dust is deposited in small quantities over large areas of the Earth.

Teaching Tip

Use **Blackline Master Climate Change 2.4, Temperature Variations** to make an overhead of the graph shown in *Figure 3*. Use the overhead to discuss with students how global temperature has changed over time. Ask them if they notice a periodicity in the changes in temperature. Also, ask them to consider why the data from about 75,000 years to present is much more "cluttered." This is a good way to discuss with students the difficulties associated with studying past climate. One of the complications is that records of climate change are better for more recent periods, where the less compacted glacier ice can be sampled with greater temporal resolution.

Assessment Tool

Check Your Understanding Notebook Entry-Evaluation Sheet
Use this sheet to evaluate the extent to which students understand the key concepts explored in **Activity 2** and explained in **Digging Deeper,** and to evaluate the students' clarity of expression.

Chapter 2

Earth System Evolution Climate Change

Deep-Sea Sediments

Sand-size shells of a kind of single-celled animal called **foraminifera** ("forams," for short) accumulate in layers of ocean-bottom sediment. During warm climates, the shells spiral in one direction, but during cold climates, the shells spiral in the opposite direction. Also, the shells consist of calcium carbonate, which contains oxygen. Geologists can measure the proportions of the two oxygen isotopes to find out about paleoclimates. The shells contain more ^{18}O during colder climates than during warmer climates.

Glacial Landforms and Sediments

Glaciers leave recognizable evidence in the geologic record. Glacial landforms are common in northern North America. Glaciers erode the rock beneath, and then carry the sediment and deposit it to form distinctive landforms. Cape Cod, in Massachusetts, and Long Island, in New York, are examples of long ridges of sediment deposited by glaciers. Similar deposits are found as far south as Missouri.

Fine glacial sediment is picked up by the wind and deposited over large areas as a sediment called **loess**. There are thick deposits of loess in central North America. The loess layers reveal several intervals of glaciation during the Pleistocene Epoch.

Glaciers also leave evidence in the ocean. When glaciers break off into the ocean, icebergs float out to sea. As the icebergs melt, glacial sediment in the icebergs rains down to the ocean bottom. The glacial sediment is easily recognized because it is much coarser than other ocean-bottom sediment.

Tree Rings

Paleoclimate is also recorded in the annual growth rings in trees. Trees grow more during warm years than during cold years. A drawback to tree rings is that few tree species live long enough to provide a look very far in the past. Bristlecone pines, which can live as long as 5000 years, and giant sequoias, which are also very long-lived, are most often used.

Figure 4 A glacier carries ground-up pieces of rock and sediment into lakes and oceans.

Check Your Understanding

1. How do preserved tree rings indicate changes in climate?

2. List three ways that sediments in the ocean help scientists understand ancient climates.

3. Imagine an ice core taken from the Antarctic ice sheet. A layer of ice called "A" is 100 m below the surface. A layer of ice called "B" is 50 m below the surface. Explain why layer "A" represents the atmospheric conditions of an older climate than layer "B."

Teaching Tip

The glacier shown in *Figure 4* is losing its ice volume at its terminus by calving large icebergs into the sea.

Check Your Understanding

1. Other conditions being equal, trees grow more during warm years than during cold years, therefore wider growth rings correspond to warm climatic periods. It is important to note, however, that one must also consider other factors controlling growth (like precipitation, soil moisture, and the age of the tree).

2. • Scientists can look at the shells of foraminifera, because during warm climates they spiral in a different direction than during cold climates.

   • Scientists can also measure the proportions of oxygen isotopes in foraminifera shells. All other things being equal, the shells will contain a greater proportion of ^{18}O during colder climates.

   • Scientists can recognize glacial sediment on the ocean floor. An increase in glacial activity in coastal continental areas can cause icebergs to deposit coarse glacial sediment over wide areas of the ocean. At times of decreased glaciation, the same areas of the ocean floor receive nonglacial fine-sediment.

3. The layers were formed by the successive accumulation of ice and snow. Layers near the bottom must have been deposited before layers on top of them. This is the same principle that is applied to sedimentary and volcanic rocks; it is referred to as the principle of superposition.

Assessment Tool

Check Your Understanding Notebook Entry-Evaluation Sheet
Use this sheet to evaluate the extent to which students understand the key concepts explored in **Activity 2** and explained in **Digging Deeper**, and to evaluate the students' clarity of expression.

Understanding and Applying What You Have Learned

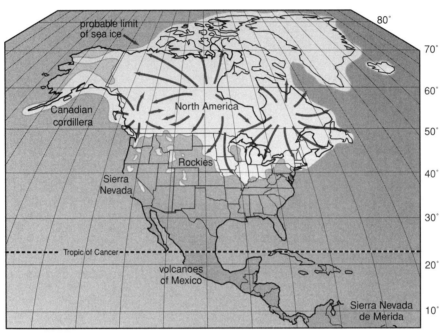

The approximate extent of the ice sheets in North America during the last Pleistocene glaciation.

1. Using evidence from glacial landforms and sediments, geologists have pieced together the maximum advance of glaciers about 18,000 years ago, which is shown in the figure above.

   a) Was your community located under ice during this time period? If not, describe how far your part of the country was from the ice sheet.

   b) What do you think the climate was like in your part of the country during the glacial maximum?

2. Use colored pencils and a ruler to draw a hypothetical series of lake-bottom sediments representing the sequence of climates given on the following page. Use the same colors as you used in the **Investigate** section to represent layers containing different kinds of pollen. Again, assume that it takes 1000 years to deposit 1 cm of sediment.

Understanding and Applying What You Have Learned

1. a) - b) Answers will vary depending upon where your community is located. Students should recognize that the climate in their community was probably colder during the glacial maximum, regardless of whether or not the community was under ice.

2.

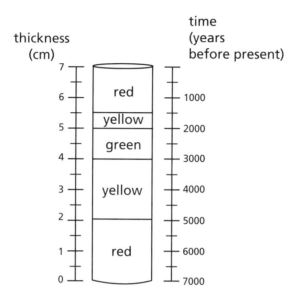

thickness (cm) / time (years before present)

a) Student descriptions of climate should generally correspond to those given in the Student Edition on page E104. Students should also note that the climate around the lake was generally on a cooling trend from about 5000 years ago to about 2000 years ago. The climate around the lake then began a general trend of warming starting at about 2000 years ago that continued to the present.

b) The coldest climate around the lake occurred between 2000 and 3000 years ago.

c) This is hard to answer without really knowing the magnitude of the changes, but assuming that the warm climate of the past 1500 years is roughly equivalent to the warm climate of 5000–7000 years ago, the answer would be no. The cooling trend lasted from about 5000 years ago to about 2000 years ago (3000 years), and the warming trend started about 2000 years ago and has continued to the present (2000 years). The transitional "yellow" zone lasted for about 2000 years during the cooling trend, but it only lasted for about 500 years as the climate warmed.

d) Student hypotheses will vary. One hypothesis might be that the cooling and warming of the climate did, in fact, occur at the same rate, but that the ability of the pollen data to resolve the magnitude of the temperature changes was insufficient to quantify the actual rates of change. Collecting data using a more sensitive indicator of paleotemperature, for example oxygen isotope data from the shells of organisms within the sediments, could test this hypothesis.

Earth System Evolution Climate Change

- 7000–5000 years ago: Warm climate supporting grasses and oaks.
- 5000–3000 years ago: Moderate climate supporting mostly grasses and oaks, with some spruce and alder.
- 3000–2000 years ago: Colder climate supporting mostly spruce and alder, with some grasses and oaks.
- 2000–1500 years ago: Moderate climate supporting mostly grasses and oaks, with some spruce and alder.
- 1500 years ago to the present: Warm climate supporting grasses and oaks.

a) Describe what the climate around the lake was like when each layer of sediment was deposited.

b) From this data, what time period marks the coldest climate recorded in the lake bottom sediments?

c) Did the climate cool at the same rate as it warmed?

d) Think of a hypothesis that might explain your answer to **Part (c)**. What additional observations might help you test this hypothesis?

Preparing for the Chapter Challenge

Using a style appropriate for a newspaper article, write a paragraph or two about each of the following topics, explaining how geologists use them to find out about paleoclimates:

- deep-sea sediments;
- glacial landforms and sediments;
- ice cores from Antarctica;
- pollen studies, and
- tree rings.

Inquiring Further

1. **GISP2 (Greenland Ice Sheet Project)**

   Research GISP2 (Greenland Ice Sheet Project), a project that is collecting and analyzing ice cores. What are the most recent discoveries? How many ice cores have been collected? How many have been analyzed? How far into the past does the data currently reach? Visit the *EarthComm* web site to help you start your research.

2. **Dating deep-sea sediments**

   Investigate some of the techniques geologists use to date deep-sea sediments: carbon-14 dating, isotope dating of uranium, fission-track dating of ash layers, and geomagnetic-stratigraphy dating.

3. **Paleoclimate research in your community**

   Investigate whether any of the paleoclimate techniques discussed in this activity have been used near your community or in your state to research paleoclimates.

Preparing for the Chapter Challenge

In **Activity 1**, students became familiar with the style of writing used in newspapers. They can now build upon this knowledge to write a second newspaper article. Remind students that they should write a paragraph or two about each of the topics listed. They will likely want to do some additional research to supplement their knowledge of these tools.

For assessment of this section (see **Assessment Rubric for Chapter Challenge on Climate Change** on pages 296 to 298), students should include discussion about how geologists find out about past climates.

Assessment Opportunity

You may want to use a rubric like the one shown below to assess student articles. Review these criteria with your students so that they can be certain to include the appropriate information in their article.

Item	Missing	Incomplete/Inaccurate	Complete/Accurate
Deep-sea sediments			
Glacial landforms and sediments			
Ice cores from Antarctica			
Pollen studies			
Tree rings			

Inquiring Further

1. GISP2 (Greenland Ice Sheet Project Two)
On July 1, 1993, the Greenland Ice Sheet Project Two (GISP2) recovered an ice core 3053.44 m deep. The GISP2 core has yielded a high-resolution, 200,000-year history of global change. Eighty percent of the sampling and some analyses were performed directly on site after recovery of the core. Analyses include stable isotopes (including carbon and oxygen), methane concentrations, carbon dioxide concentrations, electrical conductivity, and much more. Between 20% and 67% of the core is maintained as an archive for future investigations. Students can visit the *EarthComm* web site to learn more about the coring. They can also find information about scientific studies that are making use of the core.

2. Dating deep-sea sediments
Carbon-14 dating is a way of determining the age of an organic material. Carbon-14 (conventionally written ^{14}C) is a radioactive isotope of carbon, with a half-life of

about 5700 years. ^{14}C is produced when cosmic rays entering the Earth's atmosphere strike an atom of nitrogen-14 (7 protons, 7 neutrons) and convert it into an atom of ^{14}C (6 protons, 8 neutrons) and an atom of hydrogen (1 proton, 0 neutrons). The ^{14}C combines with oxygen to form CO_2, which is incorporated into living organisms through respiration (plants) and ingestion (animals eating plants). Eventually, the ^{14}C decays back to ^{14}N.

The ratio of ^{14}C to the most abundant stable isotope of carbon (^{12}C) is almost exactly the same in every living organism at any given time, because the carbon in the atmosphere and the oceans is maintained in a thoroughly mixed state. When an organism dies, it stops taking in new carbon. The ratio of ^{14}C to ^{12}C at the time of death is the same as that of every living thing. Over time, however, the ratio of ^{14}C to ^{12}C decreases as the ^{14}C decays. Scientists can determine the ratio in a sample by means of sophisticated machines called mass spectrometers, and compare this ratio to its modern value. Scientists can then use a decay equation to determine the age of the sample with good precision.

Because the half-life of ^{14}C is only about 5700 years, however, this dating technique is good only for samples up to about 60,000 years old. The essential requirement for a valid date is that there must be no exchange of carbon between the organic matter and its environment after the organism dies.

The deep ocean is a repository for sediment derived from the continents and from the upper layers of the ocean. Most areas of the deep ocean are places of continuous, slow sedimentation. Marine geologists have developed various techniques to date deep-ocean sediments. A few of these are mentioned in **Question 2** in the student text:
- isotopic dating of uranium
- fission-track dating of ash layers
- geomagnetic-stratigraphy dating

Another standard way, not mentioned, is study of the planktonic organisms, like foraminifera, in the sediments. The assemblage of different species preserved in the sediments can be used to constrain the age of those sediments. Forams are especially valuable for study because they yield not only a date but also the ratio of oxygen isotopes in the calcite shells, which varies as a function of temperature.

3. Paleoclimate research in your community
Answers will vary depending upon your community. Students might try investigating the web sites of the Earth science departments (sometimes also called geology departments or geological science departments) of local universities in their research. Another useful resource might be the web site for the geological survey in your state. Links to your state geological survey's web site, as well as other sites related to paleoclimatic research can be found on the *EarthComm* web site.

Blackline Master Climate Change 2.5
Extent of North American Glaciation

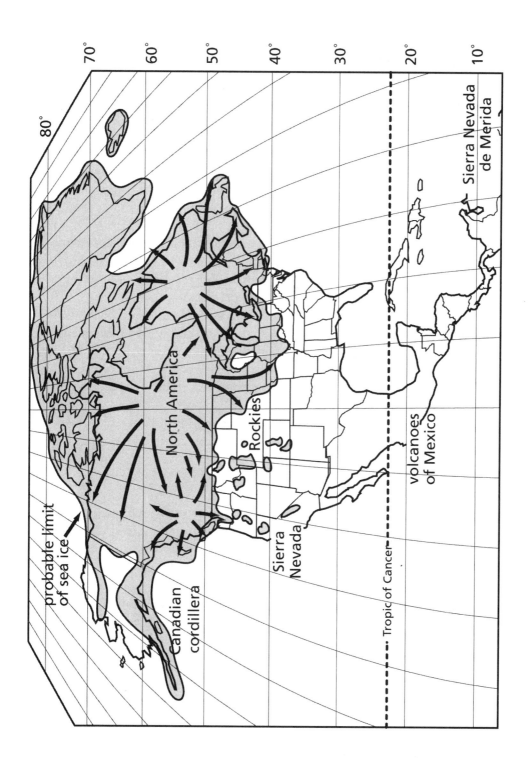

ACTIVITY 3 — HOW DO EARTH'S ORBITAL VARIATIONS AFFECT CLIMATE?

Background Information

Causes of Climate Change

It is now almost universally accepted among climatologists that long-term climate change leading to glaciation and deglaciation is linked to cyclic changes in insolation (solar radiation received by the Earth) related to variations in the Earth's orbital parameters. These cycles of solar radiation received by the Earth are called Milankovitch cycles, after the Serbian scientist who first systematized the effect of the Earth's orbital parameters on insolation cycles.

Three aspects of the Earth's movement in space account for variations in insolation over the course of a given year: eccentricity, obliquity, and precession. These phenomena are caused by the slight but significant effects of the gravitational attractions of the other planets in the solar system, as they change their positions relative to the Earth.

Eccentricity

The Earth's orbit around the Sun is an ellipse, with the Sun at one of the foci of the ellipse. This ellipse is very close to being a circle. The eccentricity of the orbit (i.e., the degree to which the ellipse is elongated, and therefore differs from a circle) varies significantly on a time scale of about 100,000 years. When the orbit is more eccentric, insolation varies more over the course of a year than when the orbit is less eccentric.

Obliquity

The Earth's axis of rotation is tilted relative to the plane of its orbit around the Sun. At present, the angle is about 23.5°. The angle changes significantly, however, with a period of about 40,000 years. At times of smaller obliquity (a small angle of inclination of the axis), insolation varies less from season to season; at times of greater obliquity, insolation varies more from season to season.

Precession

The Earth's axis of rotation wobbles in the same way that a spinning top wobbles. With both the Earth and a spinning top, the wobble is far slower than the rotation of the body. This phenomenon is known as axial precession. There is another important kind of precession related to the Earth as well: the precession of the Earth's orbit. As the Earth moves around the Sun in its elliptical orbit, the major axis of the Earth's orbital ellipse is rotating about the Sun. In other words, the orbit itself rotates around the Sun!

The combined effect of the two types of precession causes the timing of the seasons to vary relative to the times when the Earth is closest to the Sun and farthest away from the Sun. This combined effect is called the precession of the equinoxes, and it goes through one complete cycle approximately every 22,000 years. At present, the Earth is closest to the Sun during the Northern Hemisphere winter (on 5 January!). Don't let your students think that winter happens because the Earth is farthest from the Sun: it happens because the North Pole is tilted away from the Sun rather than toward it. In another 11,000 years, when the Earth is farthest from the Sun during the Northern Hemisphere winter, our winters will be more severe (other factors being equal) just because we will be receiving less solar radiation in winter than now.

Season-to-season changes in insolation are governed by the three effects noted above, which are ever changing in their combination because of the three different periodicities. These changes in insolation lead to changes in season-to-season climate, in complex ways that are not yet entirely clear. What is clear is that glaciation and deglaciation are somehow triggered by these changes. The time scales of glaciation and deglaciation match the periodicities of variation in insolation extremely well.

The Causes of Ice Ages

Many theories for the causes of the ice ages have been proposed. Clearly, glaciers form because the Earth's climate changes in such a way as to make possible the development of continent-scale ice sheets. But that just pushes the question back further: what causes the climate to change? Keep in mind that it's not just a matter of the climate becoming colder. To be conducive to glaciation, the climate must be such that the conditions for winter snowfall are enhanced and/or the conditions for summer melting are lessened. Temperature is a major factor, but precipitation is also important. A warmer climate means more summer melting, but it can also be conducive to greater winter snowfall.

Two separate aspects of glaciation need to be considered when trying to develop a theory for glaciation. On the one hand, the evidence is incontrovertible that the ice sheets, when present, expanded and contracted regularly on time scales of the order of several tens of thousands of years. On the other hand, there have been only a few such periods of fluctuating ice sheets during Earth history. At other times, there were no major volumes of glacier ice on Earth.

Because glaciers can form only on land, we know that for ice sheets to exist there must be large landmasses at high latitudes. By the operation of plate tectonics, the continents have shifted their positions greatly through geologic time. The geologic record shows clearly that at times of major glaciation, large continents were located at high latitudes. The Antarctic ice sheet did not begin to develop until the continent of Antarctica moved into high latitudes, long after the original breakup of the supercontinent of Pangea, of which Antarctica was a part. Plate movements can account for the major periods of glaciation in Earth history, but they cannot explain the repetitive growth and shrinkage of ice sheets during those periods.

More Information – on the Web

Visit the *EarthComm* web site www.agiweb.org/earthcomm/ to access a variety of links to web sites that will help you deepen your understanding of content and prepare you to teach this activity. Many of the sites also contain images that you can download.

Chapter 2

Goals and Assessment

Clarify that the goals indicate what students should understand and be able to do as a result of the activity. Make sure students understand that Chapter Assessments are based upon these goals.

Goal	Location in Activity	Assessment Opportunity
Understand that Earth has an axial tilt of about 23.5°.	Investigate Parts A and B Digging Deeper	Drawing is completed correctly. Responses reflect finding of investigation.
Use a globe to model the seasons on Earth.	Investigate Parts B and C	Temperature recordings are complete and accurate. Responses to questions are correctly based on observations.
Investigate and understand the cause of the seasons in relation to the axial tilt of Earth.	Investigate Part A and C Digging Deeper; Check Your Understanding Question 1 Understanding and Applying What You Have Learned Questions 1 and 3	Temperature recordings are complete and accurate. Responses to questions are correctly based on observations.
Understand that the shape of the Earth's orbit around the Sun is an ellipse, and that the shape of the ellipse influences climate.	Investigate Part D Digging Deeper; Check Your Understanding Questions 2 – 3	Ellipses are drawn correctly; data are complete and accurately recorded. Responses to questions are reasonable.
Understand that insolation received by the Earth varies as the inverse square of the distance to the Sun.	Investigate Part E Digging Deeper; Check Your Understanding Question 3 Understanding and Applying What You Have Learned Question 2	Measurements and calculations are completed correctly. Responses accurately reflect findings.

NOTES

Chapter 2

Activity 3 How Do Earth's Orbital Variations Affect Climate?

Goals

In this activity you will:

- Understand that Earth has an axial tilt of about 23 1/2°.

- Use a globe to model the seasons on Earth.

- Investigate and understand the cause of the seasons in relation to the axial tilt of the Earth.

- Understand that the shape of the Earth's orbit around the Sun is an ellipse and that this shape influences climate.

- Understand that insolation to the Earth varies as the inverse square of the distance to the Sun.

Think about It

When it is winter in New York, it is summer in Australia.

- Why are the seasons reversed in the Northern and Southern Hemispheres?

What do you think? Write your thoughts in your *EarthComm* notebook. Be prepared to discuss your responses with your small group and the class.

E 105

Activity Overview

Students begin the investigation with a series of exercises on paper that lead them to understand how the tilt of the Earth on its axis causes the seasons. They complete an experiment using a globe to model how the axial tilt affects the temperature of their community. Students then repeat the experiment using an axial tilt of 10° and a tilt of 35°. They draw a series of ellipses to help them understand that the shape of the Earth's orbit is not circular. Finally, students complete an experiment to understand how energy from the Sun varies with distance. **Digging Deeper** explains how the Earth's orbital variations affect climate and introduces the concept of Milankovitch cycles.

Preparation and Materials Needed

Preparation

Part A

You will need to assemble the materials for this part of the investigation. Otherwise, no advance preparation is required.

Part B

For this part of the activity, students will be using globes to measure the difference in temperature caused by the change in axial tilt. If you have the resources available, you can supply one globe for each group of students. Your school may have these available on mobile carts. You may also be able to borrow them from other teachers for the day.

If you do not have access to several globes, you may choose to set up this part of the activity as a series of stations or complete the exercise as a class.

You may wish to attach the thermometer to the globe, as described in the student text. Alternatively, you can have the students attach the thermometer.

Part C

As with **Part A,** you will need to assemble the materials for this part of the investigation. Otherwise, no advance preparation is required.

Part D

You will need a piece of rope, about 4 m long, for each student group. Students should tie loops into the ends of the rope to make the rope 3 m long from the outer corner of one loop to the outer corner of the other loop (see illustration on page E108 of the student text). You may want to do this for them before class. It may take some trial and error to tie the loops accurately. Given the space requirements for this activity, you may wish to conduct this experiment in the school gymnasium or outside on the school grounds. Alternatively, this investigation can be conducted as a class or by a few larger groups of students.

Chapter 2

Part E

This part of the activity requires a darkened room, so you will most likely want to do this as a class activity. Preparation does not extend beyond assembling the necessary materials.

Materials

Parts A and C

- Compass (to draw a circle)
- Ruler
- Protractor
- Blue, red, and black colored pencils

Part B

- Globe
- Small thermometer
- Duct tape
- Black marker
- Lamp with a reflector or halogen bulb
- Meter stick

Part D

- Piece of rope, 4 m long
- Two wooden dowels
- Ruler
- Chalk, at least four different colors

Part E

- Scissors
- Ruler
- Poster board
- Lamp without a shade
- Meter stick
- Chalk

Think about It

Student Conceptions

Students are likely to say that the orbit of the Earth around the Sun causes the succession of the seasons.

Answers for the Teacher Only

It is widely believed among the general public that the seasons are caused by differences in the distance from the Sun to the Earth. The common belief is that the Earth is farther from the Sun in winter and closer in summer. (A recent survey of graduating seniors from a prestigious Ivy League university revealed that a large majority of supposedly well-educated college students were under that impression!) The logical problem with such an explanation for the seasons is that the seasons are reversed between the Northern Hemisphere and the Southern Hemisphere.

In reality, as your students will now learn, the seasons are caused by the tilt of the Earth's axis relative to the plane of the Earth's orbit around the Sun. At times when the Northern Hemisphere is tilted away from the Sun, it receives less solar radiation, resulting in winter; at the same time, the Southern Hemisphere is tilted toward the Sun, and experiences summer. Students will probably be surprised to learn that the Earth is closest to the Sun on January 5, in the middle of the Northern Hemisphere winter, and is farthest from the Sun on July 5, in the middle of the summer!

Assessment Tool

Think about It Evaluation Sheet
Use this evaluation sheet to help students understand and internalize the basic expectations for the warm-up activity.

Chapter 2

Blackline Master Climate Change 3.1
An Experiment on Paper

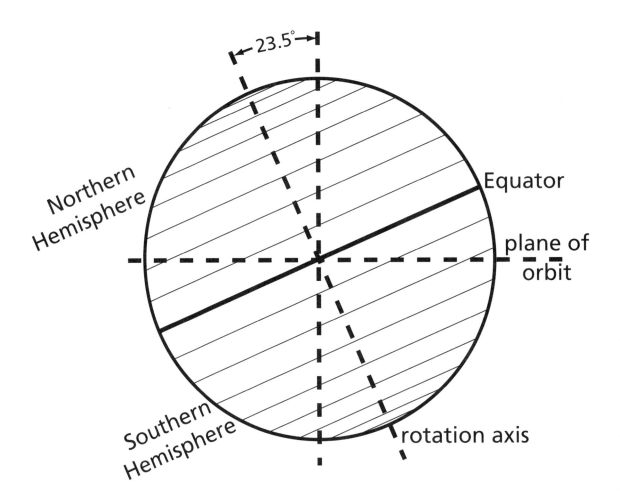

NOTES

Chapter 2

Earth System Evolution Climate Change

Investigate

Part A: What Causes the Seasons? An Experiment on Paper

1. In your notebook, draw a circle about 10 cm in diameter in the center of your page. This circle represents the Earth.

   Add the Earth's axis of rotation, the Equator, and lines of latitude, as shown in the diagram and described below. Label the Northern and Southern Hemispheres.

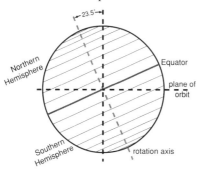

   Put a dot in the center of the circle. Draw a dashed line that goes directly up and down from the center dot to the edge of the circle. Use a protractor to measure 23 1/2° from this vertical dashed line. Use a blue pen or pencil to draw a line through the center of the "Earth" at 23 1/2° to your dashed line. This blue line represents the Earth's axis of rotation. Use your protractor to draw a red line that is perpendicular to the axis and passes through the center dot. This red line represents the Equator of the Earth. Label the Northern and Southern Hemispheres. Next you need to add lines of latitude. To do this, line your protractor up with the dot in the center of your circle so that it is parallel with the Equator.

Now, mark off 10° increments starting from the Equator and going to the poles. You should have eight marks between the Equator and pole for each quadrant of the Earth. Use a straight edge to draw black lines that connect the marks opposite one another on the circle, making lines that are parallel to the Equator. This will give you lines of latitude in 10° increments so you can locate your latitude fairly accurately. Note that the lines won't be evenly spaced from one another because latitude is measured as an angle from the center of the Earth, not a linear distance.

2. Imagine that the Sun is directly on the left in your drawing. Draw horizontal arrows to represent incoming Sun rays from the left side of the paper.

3. Assume that it is noon in your community. Draw a dot where your community's latitude line intersects the perimeter of the circle on the left. This dot represents your community.

   a) Explain why this represents noon.

   b) At any given latitude, both north and south, are the Sun's rays striking the Northern Hemisphere or the Southern Hemisphere at a larger angle relative to the local vertical to the Earth's surface?

   c) Which do you think would be warmer in this drawing—the Northern Hemisphere or the Southern Hemisphere? Write down your hypothesis. Be sure to give a reason for your prediction.

   d) What season do you think this is in the Northern Hemisphere?

Investigate

Part A: What Causes the Seasons? An Experiment on Paper

1. Student drawings should resemble the illustration in the text. If your students have not used a compass before to draw a circle, give a demonstration beforehand.

2.

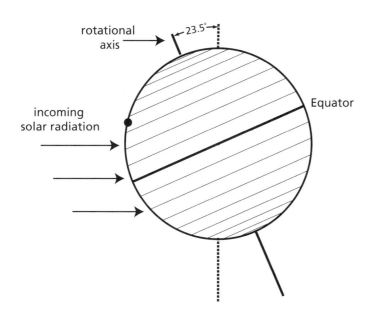

3. a) This represents noon in your community because your community (the dot in your diagram) is directly facing the Sun. Noon is when the Sun reaches the highest point in its path across the sky. Before the advent of satellite technology to find position at sea, seafarers could tell their latitude by measuring the position of the noontime Sun with an instrument called a sextant. What is dealt with here is local noon. It differs from "official" noon, which the students might see on the classroom clock. "Official" noon is at the same time throughout an entire time zone, whereas local noon varies continuously with longitude. At most points on Earth, the Sun is not quite at its highest point in the sky at "official" noon. It was only in the early 19th century, with the advent of long-distance railroad travel, that clocks were universally synchronized between distant points within a time zone.

    b) The smaller the angle, the higher in the sky the Sun appears. In the drawing, at any given latitude the angle is smaller in the Northern Hemisphere than in the Southern Hemisphere. This is because in this configuration the Sun is directly overhead at 23.5° north latitude where the plane of orbit (which for Earth is the same as the plane of the ecliptic) intersects the surface of the Earth that faces towards the Sun. To find the local vertical at any point on the surface of the Earth, draw a line tangent to the Earth's surface at that point. The local vertical will be perpendicular to that line. The angle of the Sun's incoming radiation to that local vertical can then be calculated by measuring the angle that the local vertical forms with a horizontal line.

c) Students should be sure to justify the reasoning for their hypothesis. The Northern Hemisphere would be warmer. With the Sun higher in the sky, the Northern Hemisphere is receiving more solar radiation per unit area of the Earth's surface. There are generally two reasons for this. Firstly, as the angle of incoming solar radiation relative to the local vertical to the Earth's surface increases, the same amount of incoming solar energy is spread over a greater area. If your students need to be convinced of this more directly, have them shine a flashlight on the floor of a darkened room. When the flashlight is directed straight down toward the floor, the illumination is much brighter per unit area of the floor than when the flashlight is directed at a very small angle to the floor. Secondly, all other things being equal, less incoming radiation is reflected back to space as the angle between the incoming radiation and local vertical approaches zero.

d) This would be summer in the Northern Hemisphere.

NOTES

Chapter 2

4. Now consider what happens six months later. The Earth is on the opposite side of its orbit, and the sunlight is now coming from the right side of the paper. Draw horizontal arrows to represent incoming Sun's rays from the right.

5. Again, assume that it is noon. Draw a dot where your community's latitude line intersects the perimeter of the circle on the right.

   a) Explain why the dot represents your community at noon.

   b) Are the Sun's rays striking the Northern Hemisphere or the Southern Hemisphere more directly?

   c) Which do you think would be warmer in this drawing—the Northern Hemisphere or the Southern Hemisphere? Why?

   d) What season do you think this is in the Northern Hemisphere?

Part B: What Causes the Seasons? An Experiment with a Globe

1. Test the hypothesis you made in **Part A** about which hemisphere would be warmer in which configuration. Find your city on a globe. Using duct tape, tape a small thermometer on it. The duct tape should cover the thermometer bulb, and the thermometer should be over the city. With a permanent black marking pen, color the duct tape black all over its surface.

2. Set up a light and a globe as shown.

   ⚠ Use only alcohol thermometers. Place a soft cloth on the table under the thermometer in case it falls off. Be careful not to touch the hot lamp.

3. Position the globe so that its axis is tilted 23 1/2° from the vertical, and the North Pole is pointed in the direction of the light source.

4. Turn on the light source.

   a) Record the initial temperature. Then record the temperature on the thermometer every minute until the temperature stops changing.

5. Now position the globe so that the axis is again tilted 23 1/2° from the vertical but the North Pole is pointing away from the light source. Make sure the light source is the same distance from the globe as it was in **Step 4**. Turn on the light source again.

   a) Record the temperature every minute until the temperature stops changing.

6. Use your observations to answer the following questions in your notebook:

   a) What is the difference in the average temperature when the North Pole was pointing toward your "community" and when it was pointing away?

   b) What caused the difference in temperature?

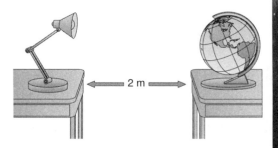

4.

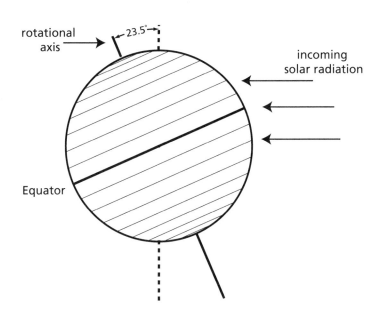

5. a) See the commentary on **Step 3(a)**, on page 379.

    b) The Sun's rays are striking the Southern Hemisphere more directly. See commentary on **Step 3(b)** above.

    c) The Southern Hemisphere would be warmer, for the same reason described in **Step 3(c)**.

    d) This would be winter in the Northern Hemisphere.

Part B: What Causes the Seasons? An Experiment with a Globe

1. - 3. The point of this exercise is to show that less solar radiation is received, per unit area of the Earth's surface, when the Sun is low in the sky than when it is high in the sky. This would be best done with a "flat" thermometer that could be pasted flat onto the surface of the globe. The cylindrical shape of real alcohol thermometers might interfere with the experiment. It would be best to have the students tape the thermometer snugly onto the surface of the globe with the long axis of the thermometer oriented north–south. That will minimize the effect of the bulge of the thermometer.

4. a) Temperatures will vary. The temperature should stop changing within several minutes, but that will depend upon the size of the thermometer bulb used, the thickness and composition of the tape, and the blackness of the tape.

5. a) Temperatures will vary, but they should be at least slightly less than in **Step 4(a)**.

6. a) Although actual temperatures will vary, students should observe that temperatures are at least slightly higher when the North Pole is pointed towards the light source than when the North Pole is pointed away from the light source.

   b) When the North Pole is pointed towards the light source, the surface of the globe is receiving more radiation per unit surface area. Equilibrium (the condition for which as much heat is lost from the thermometer bulb and covering tape as is gained by them, so the temperature remains the same) is struck at a lower temperature.

NOTES

Earth System Evolution Climate Change

Part C: What Would Happen if the Earth's Axial Tilt Changed? An Experiment on Paper

1. Repeat the experiment you did in **Part A**, except this time use an axial tilt of 10°.

   a) Compared to an axial tilt of 23 1/2°, would your hemisphere experience a warmer or colder winter?

   b) Compared to an axial tilt of 23 1/2°, would your hemisphere experience a hotter or cooler summer?

2. Repeat the experiment you did in **Part A**, except this time use an axial tilt of 35°.

   a) Compared to an axial tilt of 23 1/2°, would your hemisphere experience a warmer or colder winter?

   b) Compared to an axial tilt of 23 1/2°, would your hemisphere experience a hotter or cooler summer?

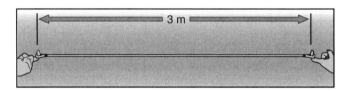

Part D: The Earth's Elliptical Orbit around the Sun

1. Tie small loops in each end of a rope, as shown in the diagram above.

2. Pick a point in about the middle of the floor, and put the two loops together over the point. Put a dowel vertically through the loops, and press the dowel tightly to the floor.

3. Stretch out the rope from the dowel until it is tight, and hold a piece of chalk at the bend in the rope, as shown in the diagram below. While holding the chalk tight against the rope, move the chalk around the dowel.

   a) What type of figure have you constructed?

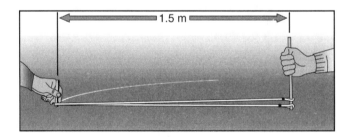

Part C: What Would Happen if the Earth's Axial Tilt Changed? An Experiment on Paper

1. a) With a smaller axial tilt, the Northern Hemisphere would have a warmer winter.

    b) With a smaller axial tilt, the Northern Hemisphere would have a cooler summer.

2. a) With a greater axial tilt, the Northern Hemisphere would have a colder winter.

    b) With a greater axial tilt, the Northern Hemisphere would have a warmer summer.

Part D: The Earth's Elliptical Orbit around the Sun

1. Refer to the comment about making the loops in the **Preparation** section.

2. You might have the students practice making their circle, and then erasing, until they are skilled at making an accurate circle.

3. a) Students should have constructed a circle.

Assessment Tools

EarthComm Notebook Entry-Checklist
Use this checklist as a quick guide for student self-assessment and/or an opportunity to quickly score student work. Add further criteria specific to your classroom needs or to this particular investigation.

Investigate Notebook Entry-Evaluation Sheet
Point out the criteria listed on this evaluation sheet that are relevant to this particular investigation. Encourage students to internalize the criteria by making them part of your "assessment conversations" as you circulate around the classroom.

4. Draw a straight line from one edge of the circle that you just made to the opposite edge, through the center of the circle. This line represents the diameter of the circle. Mark two points along the diameter, each a distance of 20 cm from the center of the circle. Put the loops of the rope over the two points, hold them in place with two dowels, and use the chalk to draw a curve on one side of the straight line. Move the chalk to the other side of the line and draw another curve.

a) What type of figure have you constructed?

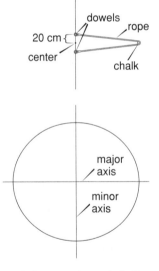

5. Using different colors of chalk, make a few more curved figures in the same way. Choose sets of dowel points that are farther and farther away from the center.

a) Describe the shapes of the figures you constructed. Make sketches in your *EarthComm* notebook.

b) What would be the shape of the curve when the dowel points are spaced a distance apart that is just equal to the length of the rope between the two loops?

Part E: How Energy from the Sun Varies with Distance from the Sun

1. Using scissors and a ruler, cut out a square 10 cm on a side in the middle of a poster board.

2. Hold the poster board vertically, parallel to the wall and exactly two meters from it.

3. Position a light bulb along the imaginary horizontal line that passes from the wall through the center of the hole in the poster board. See the diagram.

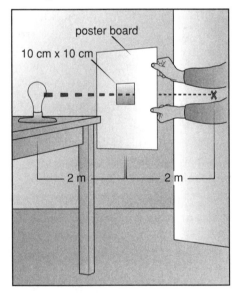

 Avoid contact with the hot light bulb. Do not look directly into the light.

4. You might have the students practice making their ellipse, and then erasing, until they are skilled at making an accurate ellipse.

    a) Students should have constructed an ellipse.

5. a) The farther away from the center the sets of dowel points are, the more "flattened" (i.e., the more eccentric) the resulting ellipse is.

    b) When the dowel points are spaced at a distance apart that is just equal to the length of the rope between the loops, the result will be a straight line.

Part E: How Energy from the Sun Varies with Distance from the Sun

1. – 3. See diagram on page E109 of the student text.

Earth System Evolution Climate Change

4. Turn on the light bulb, and turn off the lights in the room. If the room is not dark enough to see the image on the wall, close any curtains or shades, or cover the windows with dark sheets or blankets.

5. With the chalk, trace the edge of the image the hole makes on the wall.

   a) Measure and record the length of the sides of the image you marked with the chalk.

   b) Divide the length of the image on the wall by the length of the sides of the square in the poster board. Now divide the distance of the light bulb from the wall by the distance of the poster board from the wall. What is the relationship between the two numbers you obtain?

   c) Compute the area of the image on the wall, and compute the area of the square hole in the poster board. Divide the area of the image by the area of the hole. Again, divide the distance of the light bulb from the wall by the distance of the poster board from the wall. What is the relationship between the two numbers you obtain?

6. Repeat **Part E** for other distances.

   a) What do you notice about the relationship between the area of the image and the area of the hole?

Reflecting on the Activity and the Challenge

In this activity you modeled the tilt of the Earth's axis to investigate the effect of the angle of the Sun's rays. You discovered that the axial tilt of the Earth explains why there are seasons of the year. You also discovered that if the tilt were to vary, it would affect the seasons.

You also modeled the Earth's elliptical orbit around the Sun. This will help you to understand one of the main theories for explaining why the Earth's climate varies over time. You will need to explain this in your newspaper articles.

4. The room should be as dark as possible.

5. **a)** The square image on the wall should measure about 20 cm by 20 cm. Because the bulb is not a pinpoint of light, however, the edges of the image will not be sharp, so the measurements made by the students will probably vary somewhat.

b) The ratio of the length of the image on the wall to the length of the sides of the square hole in the poster board should be about 2:1. As per the figure, the ratio of the distances of the light bulb to the wall and the light bulb to the poster board is 2:1.

c) The ratio of the area of the image on the wall to the area of the hole in the poster board should be about 4:1. This ratio is equal to [(20 cm)(20 cm)]/ [(10 cm)(10 cm)], or (400 cm²)/(100 cm²). Although this ratio is twice as large as that in **Step 5(b)**, the important point is the ratio of the area of the image on the wall to the area of the hole in the poster board is equal to the square of the ratio of the distances mentioned above (four is equal to the square of two, the ratio of the distances mentioned above). This is an illustration of the inverse-square law.

6. If, for example, the distance from the poster board to the wall had been 4 m instead of 2 m (and the distance from the light to the poster board remains at 2 m), the ratio of the distance from the light to the wall and the distance from the light to the poster board would be 3 (6 m / 2 m). As such, the ratio of the areas would have been 9; 9 is the square of 3, in accordance with the inverse-square law. It might help students' thinking to have them represent the situation in the form of a vertical cross section through the line from the light bulb to the center of the image on the wall and label the relevant distances and dimensions. The significance of this exercise is to show that the amount of solar radiation per unit area received by a solid surface facing the Sun decreases as the square of the distance away from the Sun.

Reflecting on the Activity and the Challenge

This is a long investigation, with many parts. Take this opportunity to pull together all of the information students have gained. Have students revisit their responses to the **Think about It** question. They should now be able to explain the cause of the seasons. Discuss with students how the information from this activity can be applied to the **Chapter Challenge**.

Digging Deeper

THE EARTH'S ORBIT AND THE CLIMATE

The Earth's Axial Tilt and the Seasons

The Earth's axis of rotation is tilted at about 23 1/2° away from a line that is perpendicular to the plane of the Earth's orbit around the Sun (*Figure 1*). This tilt explains the seasons on Earth. During the Northern Hemisphere summer, the North Pole is tilted toward the Sun, so the Sun shines at a high angle overhead. That is when the days are warmest, and days are longer than nights. On the summer solstice (on or about June 22) the Northern Hemisphere experiences its longest day and shortest night of the year. During the Northern Hemisphere winter the Earth is on the other side of its orbit. Then the North Pole is tilted away from the Sun, and the Sun shines at a lower angle. Temperatures are lower, and the days are shorter than the nights. On the winter solstice (on or about December 22) the Northern Hemisphere experiences its shortest day and longest night of the year.

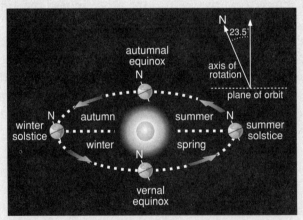

Figure 1 The tilt of the Earth's axis explains the seasons. Note that the Earth and Sun are not shown to scale.

How Do Earth's Orbital Variations Affect Climate?

In **Activity 2** you saw that paleoclimatologists have developed a good picture of the Earth's climatic history. The advances and retreats of the continental ice sheets are well documented. Nevertheless, questions remain.

Digging Deeper

Assign the reading for homework, along with the questions in **Check Your Understanding** if desired.

Assessment Opportunity

Reword or restructure the questions in **Check Your Understanding** for a brief quiz. Use the quiz (or a class discussion of the questions in the textbook) to assess your students' understanding of the main ideas in the reading and the activity.

Teaching Tip

Use **Blackline Master Blackline Master Climate Change 3.2, What Causes Seasons?** to make an overhead of the illustration shown in *Figure 1*. Use this overhead in a class discussion about how the tilt of the Earth's axis explains the seasons.

Chapter 2

Earth System Evolution Climate Change

Geo Words

eccentricity: the ratio of the distance between the foci and the length of the major axis of an ellipse.

obliquity: the tilt of the Earth's rotation axis as measured from the perpendicular to the plane of the Earth's orbit around the Sun. The angle of this tilt varies from 22.5° to 24.5° over a 41,000-year period. Current obliquity is 23.5°.

precession: slow motion of the axis of the Earth around a cone, one cycle in about 26,000 years, due to gravitational tugs by the Sun, Moon, and major planets.

orbital parameters: any one of a number of factors that describe the orientation and/or movement of an orbiting body or the shape of its orbital path.

insolation: the direct or diffused shortwave solar radiation that is received in the Earth's atmosphere or at its surface.

inverse-square law: a scientific law that states that the amount of radiation passing through a specific area is inversely proportional to the square of the distance of that area from the energy source.

Why does Earth's climate sometimes become cold enough for ice sheets to advance? Why does the climate later warm up and cause ice sheets to retreat? The answers to these questions are not yet entirely clear. Most climatologists believe that variations in the geometry of the Earth's orbit around the Sun are the major cause of the large variations in climate. These variations have caused the advance and retreat of ice sheets in the past couple of million years.

If the Earth and the Sun were the only bodies in the solar system, the geometry of the Earth's orbit around the Sun and the tilt of the Earth's axis would stay exactly the same through time. But there are eight other known planets in the solar system. Each of those planets exerts forces on the Earth and the Sun. Those forces cause the Earth's orbit to vary with time. The Moon also plays a role. The changes are slight but very important. There are three kinds of changes: **eccentricity**, **obliquity**, and **precession**. These three things are called the Earth's **orbital parameters**.

Eccentricity

The Earth's orbit around the Sun is an ellipse. The deviation of an ellipse from being circular is called its eccentricity. A circle is an ellipse with zero eccentricity. As the ellipse becomes more and more elongated (with a larger major diameter and a smaller minor diameter), the eccentricity increases. The Earth's orbit has only a slight eccentricity.

Even though the eccentricity of the Earth's orbit is very small, the distance from the Earth to the Sun varies by about 3.3% through the year. The difference in **insolation** is even greater. The word insolation (nothing to do with insUlation!) is used for the rate at which the Sun's energy reaches the Earth, per unit area facing directly at the Sun. The seasonal variation in insolation is because of what is called the **inverse-square law**. What you found in **Part E** of the investigation demonstrates this. The area of the image on the wall was four times the area of the hole, even though the distance of the wall from the bulb was only twice the distance of the hole from the bulb. Because of the inverse-square law, the insolation received by the Earth varies by almost 7° between positions on its orbit farthest from the Sun and positions closest to the Sun.

Because of the pull of other planets on the Earth–Sun system, the eccentricity of the Earth's orbit changes with time. The largest part of the change in eccentricity has a period of about 100,000 years. That means that one full cycle of increase and then decrease in eccentricity takes 100,000 years. During that time, the difference in insolation between the date of

Blackline Master Climate Change 3.2
What Causes Seasons?

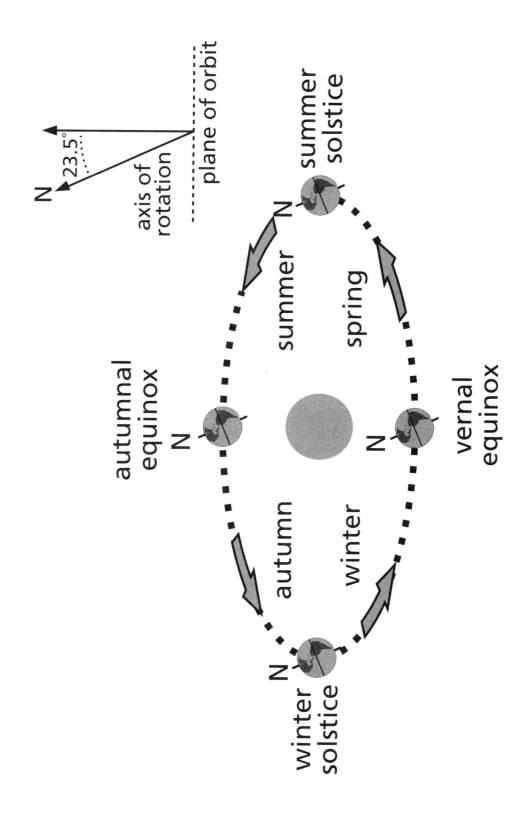

the shortest distance to the Sun and the date of the farthest distance to the Sun ranges from about 2° (less than now) to almost 20° (much greater than now!).

The two points you used to make ellipses in **Part D** of the investigation are called the foci of the ellipse. (Pronounced "FOH-sigh". The singular is focus.) The Sun is located at one of the foci of the Earth's elliptical orbit. The Earth is closest to the Sun when it is on the side of that focus, and it is farthest from the Sun when it is on the opposite side of the orbit. (See *Figure 2*.) Does it surprise you to learn that nowadays the Earth is closest to the Sun on January 5 (called **perihelion**) and farthest from the Sun on July 5 (called **aphelion)?** That tends to make winters less cold and summers less hot, in the Northern Hemisphere.

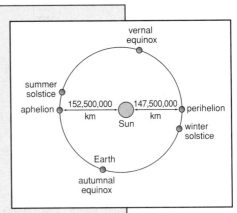

Figure 2 Schematic diagram showing occurrence of the aphelion and perihelion.

Obliquity

The tilt of the Earth's axis relative to the plane of the Earth's orbit is called the obliquity. The axis is oblique to the plane rather than perpendicular to it. In the investigation you discovered that a change in the obliquity would cause a change in the nature of the seasons. For example, a smaller obliquity would mean warmer winters and cooler summers in the Northern Hemisphere. This might result in more moisture in the winter air, which would mean more snow. In cooler summers, less of the snow would melt. You can see how this might lead to the buildup of glaciers.

Geo Words

perihelion: the point in the Earth's orbit that is closest to the Sun. Currently, the Earth reaches perihelion in early January.

aphelion: the point in the Earth's orbit that is farthest from the Sun. Currently, the Earth reaches aphelion in early July.

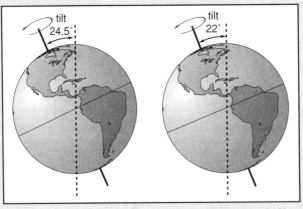

Figure 3 The angle of tilt of the Earth's axis varies between about 24.5° and 22°, causing climatic variations over time.

Teaching Tip

Use **Blackline Master Climate Change 3.3, Occurrence of the Aphelion and Perihelion** to make an overhead of the diagram shown in *Figure 2*. Note that, at this time, perihelion and aphelion do not exactly correspond with either of the equinoxes or solstices. The solstices and equinoxes relate to the orientation of the Earth's rotation axis relative to the Sun, not the Earth's distance from the Sun. Use this overhead in a class discussion about the shape of the Earth's orbit, pointing out that the eccentricity of the Earth's orbit is greatly exaggerated in *Figure 1*. Discuss with students the findings of **Part D** of the investigation. How elliptical is the Earth's orbit relative to the ellipses that they drew? What variations would we see in climate if the Earth's orbit were a perfect circle? What if the orbit were more elliptical?

Teaching Tip

Figure 3 illustrates the range of variation in axial tilt that occurs throughout the 41,000-year obliquity cycle. Currently, the Earth is in the part of the cycle where the angle of the axial tilt is decreasing. Knowing this, students can use the extremes of the cycle that are shown in this diagram and the current tilt of the Earth's axis to approximate where the Earth currently is in this 41,000 year cycle. Students can then calculate approximately how long will it be before the axial tilt reaches its minimum angle of 22°.

Teaching Tip

Figure 4 on page E114 contains more information than is presented in the **Digging Deeper** reading section. The idea is to combine the effect of the variations in each of the orbital parameters to obtain a curve that shows the variation in total solar radiation received by the Earth as a function of time. The units in the fourth column are percentage deviations from the long-term average. The fifth column, derived from measurements of the oxygen-isotope ratio ($^{18}O/^{16}O$) from marine planktonic foraminifera in seafloor sediments (which is a proxy for global temperature), is supplied in order to show the good correlation between solar radiation and global temperature.

Earth System Evolution Climate Change

Again, because of the varying pull of the other planets on the Earth–Sun system, the Earth's obliquity changes over a period of about 40,000 years. The maximum angle of tilt is about 24 1/2°, and the minimum angle is about 22°. At times of maximum tilt, seasonal differences in temperature are slightly greater. At times of minimum tilt angle, seasonal differences are slightly less.

Precession

Have you ever noticed how the axis of a spinning top sometimes wobbles slowly as it is spinning? It happens when the axis of the top is not straight up and down, so that gravity exerts a sideways force on the top. The same thing happens with the Earth. The gravitational pull of the Sun, Moon, and other planets causes a slow wobbling of the Earth's axis. This is called the Earth's **axial precession**, and it has a period of about 26,000 years. That's the time it takes the Earth's axis to make one complete revolution of its wobble.

There is also another important kind of precession related to the Earth. It is the precession of the Earth's orbit, called **orbital precession**. As the Earth moves around the Sun in its elliptical orbit, the major axis of the Earth's orbital ellipse is rotating about the Sun. In other words, the orbit itself rotates around the Sun! The importance of the two precession cycles for the Earth's climate lies in how they interact with the eccentricity of the Earth's orbit. This interaction controls how far the Earth is from the Sun during the different seasons. Nowadays, the Northern Hemisphere winter solstice is at almost the same time as perihelion. In about 11,000 years, however, the winter solstice will be at about the same time as aphelion. That will make Northern Hemisphere winters even colder, and summers even hotter, than today.

Milankovitch Cycles

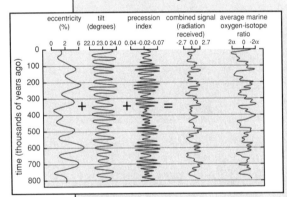

Figure 4 Interpretation of Milankovitch cycles over the last 800,000 years.

Early in the 20th century a Serbian scientist named Milutin Milankovitch hypothesized that variations in the Earth's climate are caused by how insolation varies with time and with latitude. He used what is known about the Earth's orbital parameters (eccentricity, obliquity, and precession) to compute the variations in insolation. Later scientists have refined the computations. These insolation cycles are now called **Milankovitch cycles**. (See *Figure 4*.)

Blackline Master Climate Change 3.3
Occurrence of the Aphelion and Perihelion

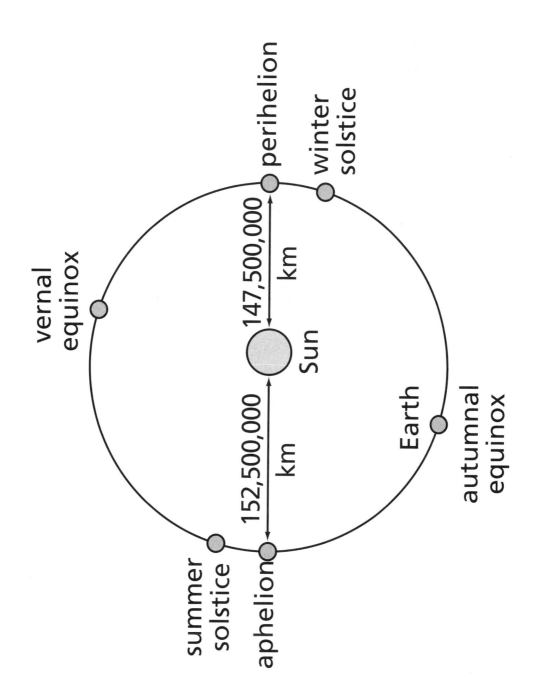

Chapter 2

Climatologists now generally agree that Milankovitch cycles are closely related to the glacial–interglacial cycles the Earth has experienced in recent geologic time. *Figure 3* in **Activity 2**, based on a long Antarctic ice core, shows how temperature has varied over the past 420,000 years. There is a clear 100,000-year periodicity, which almost exactly matches the eccentricity cycle. Temperature variations seem to have been controlled by the 100,000-year eccentricity cycle during some time intervals but by the 41,000-year obliquity cycle during other time intervals.

Climatologists are trying to figure out how Milankovitch cycles of insolation trigger major changes in climate. The Milankovitch cycles (the "driver" of climate) are just the beginning of the climate story. Many important climate mechanisms must be taken into account. They involve evaporation, precipitation, snowfall, snowmelt, cloud cover, greenhouse gases, vegetation, and sea level. What makes paleoclimatology difficult (and interesting!) is that these factors interact with one another in many complicated ways to produce climate.

Check Your Understanding

1. Explain why the days are longer than the nights during the summer months. Include a diagram to help you explain.

2. What are the three factors in Milankovitch cycles?

3. Explain how Milankovitch cycles might cause changes in global climate.

Understanding and Applying What You Have Learned

1. You have made a drawing of winter and summer in the Northern Hemisphere showing the tilt of Earth.

   a) Make a drawing showing Earth on or about March 21 (the vernal equinox). Indicate from which direction the Sun's rays are hitting the Earth.

   b) Explain why the daytime and nighttime last the same length of time everywhere on the Earth on the vernal equinox and on the autumnal equinox.

2. In **Part E** of the **Investigate** section you explored the relationship between energy from the Sun and distance to the Sun. How would you expect the area of light shining on the wall to change if the light source was moved farther away from the wall (but the cardboard was left in the same place)?

3. The tilt of the Earth varies from about 22° to about 24.5° over a period of 41,000 years. Think about how the solar radiation would change if the tilt was 24.5°.

   a) What effect would this have on people living at the Equator?

   b) What effect would this have on people living at 30° latitude?

   c) What effect would this have on people living at 45° latitude?

   d) What modifications in lifestyle would people have to make at each latitude?

Check Your Understanding

1. When it is summer in one of the Earth's hemispheres, the days are longer than the nights because that hemisphere is tilted toward the Sun. Then, the Sun illuminates a given point in that hemisphere for a longer period of time during the 24-hour day. A good diagram to show this effect is not easy to draw. See the example below.

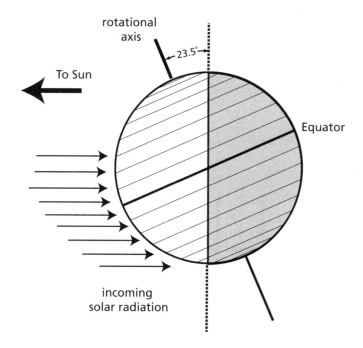

A clear three-dimensional sketch to show that the days are longer than the nights in the summer.

2. The three factors in Milankovitch cycles are eccentricity of the Earth's orbit, the obliquity of the Earth's axial tilt, and the precession of the equinoxes.

3. Milankovitch cycles cause variations in insolation with time and with latitude. Climatologists are trying to determine how Milankovitch cycles of insolation trigger major changes in climate.

Assessment Tool

Check Your Understanding Notebook Entry-Evaluation Sheet
Use this sheet to evaluate the extent to which students understand the key concepts explored in **Activity 3** and explained in **Digging Deeper**, and to evaluate the students' clarity of expression.

Understanding and Applying What You Have Learned

1. a) The perspective of the example drawing below can lead to some confusion in its interpretation. In this drawing, the Sun is above the plane of the paper, oriented directly above the small dot in the middle of the Earth's Equator. Thus, the Northern and Southern Hemispheres of the Earth receive equal amounts of sunlight because the Earth's rotation axis is neither tilted toward nor away from the Sun.

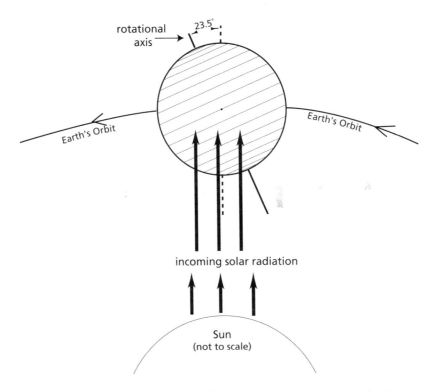

b) During the vernal equinox and the autumnal equinox, the Sun rises exactly in the east, travels through the sky for 12 hours, and sets exactly in the west. Every point on the Earth has 12 hours of light and 12 hours of darkness (although at or near the poles, this is manifested in a different way: the Sun circles the horizon, making a complete circuit in 24 hours). This is because at each of these two times the Sun is directly overhead at the Equator because the tilt of the Earth's axis is neither toward nor away from the Sun, but is instead tilted in the direction of the Earth's orbit. This is illustrated in *Figure 1* on page E111 of the Student Edition.

2. The area of the image would become more nearly the same as the hole. This is because as the light source is moved farther away, the ratio of the distance between the light and the wall and the distance between the light and the cardboard decreases (because the cardboard and the wall are not moving). In other words, the distance between the light and the cardboard becomes a greater

percentage of the overall distance between the light and the wall. This can be shown with a simple example. If the light source were moved to a position 40 m from the wall, then the distance to the cardboard would be 38 m and the ratio of distances would be 40/38 or ~1.05. The square of 1.05 is ~1.1. So in this case, the image on the wall would only have an area 10% larger than the hole in the cardboard.

3. a) This would not have a large effect on the people living at the Equator, although the Sun would be less directly overhead during times near the summer and winter solstices than it is now. Climate change at higher latitudes could also have a slight effect on climate near the Equator.

   b) People living at 30° latitude would receive more solar radiation during the summer and less solar radiation during the winter, meaning warmer summers and colder winters.

   c) People living at 45° latitude would receive more solar radiation during the summer and less solar radiation during the winter, meaning warmer summers and colder winters. The difference would be greater than that for people living at 30° latitude.

   d) Answers will vary. One example might be that they would have to budget more money for the heating and cooling of their home.

Chapter 2

Earth System Evolution Climate Change

Preparing for the Chapter Challenge

Use a style of writing appropriate for a newspaper to discuss the following topics:

• How does the tilt of the Earth's axis produce seasons?

• How does a variation in this tilt affect the nature of the seasons?

• How might a variation in the tilt affect global climate?

• How does the shape of the Earth's orbit influence climate?

Inquiring Further

1. **Sunspots and global climate**

   Do sunspots affect global climate? There is disagreement over this in the scientific community. Do some research to find out what sunspots are and how the activity of sunspots has correlated with global climate over time. Why do some scientists think sunspot activity affects global climate? Why do some scientists think that sunspot activity does NOT affect global climate?

2. **Milutin Milankovitch**

   Write a paper on Milutin Milankovitch, the Serbian scientist who suggested that variations in the Earth's orbit cause glacial periods to begin and end. How were his ideas received when he first published them?

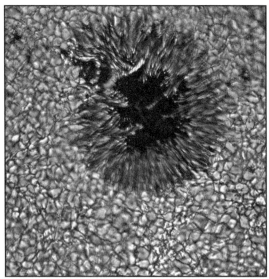

False color telescope image of a sunspot.

Preparing for the Chapter Challenge

Students should now be able to explain how the tilt of the Earth's axis and the shape of the Earth's orbit are related to climate. Encourage students to use their findings during the investigation to support their writing.

Criteria for assessment (see **Assessment Rubric for Chapter Challenge on Climate Change** on page 296 to 298) should include a discussion of how Milankovitch cycles can cause changes in climate.

Inquiring Further

1. Sunspots and global climate

If students have already completed the Astronomy chapter, they should be familiar with sunspots and some of the problems associated with them. Sunspots may also have a long-term connection with changes in the Earth's climate. Scientists are currently debating whether ice ages on Earth are related to the Sun having fewer sunspots than usual. Variability in the amount of energy from the Sun has caused climate changes in the past.

2. Milutin Milankovitch

Milutin Milankovitch was born in 1879 in Dalj, Serbia. Milankovitch received his doctorate in technical sciences in 1904. In 1909, he accepted a position at the University of Belgrade where he spent the rest of his career. Milankovitch dedicated his career to developing a mathematical theory of climate based on the seasonal and latitudinal variations in solar radiation received on Earth. For nearly 50 years, Milankovitch's theory was largely ignored. In 1976, a study published in the journal *Science* (Hays et al.) investigated deep-sea sediment cores and found that changes in climate could be linked to Milankovitch's theory. Since that study, the model of the Milankovitch cycles has been widely accepted by the scientific community.

ACTIVITY 4 — HOW DO PLATE TECTONICS AND OCEAN CURRENTS AFFECT GLOBAL CLIMATE?

Background Information

Plate Tectonics

The theory of plate tectonics developed in earnest beginning in the early 1960s. Up until that time, most geoscientists had rejected the idea that continents "drifted" through the ocean basins and around the planet, a hypothesis put forth in 1912 by the German meteorologist Alfred Wegener. As evidence amassed in the 1950s and 1960s, it became clear that Wegener had initiated a unifying theory of geology—the theory of plate tectonics.

At the heart of the concept of plate tectonics is the idea that large segments of Earth's lithosphere (consisting of the crust and the rigid, uppermost part of the mantle) move relative to one another. New lithosphere is continuously created at the mid-ocean spreading ridges. This lithosphere consists of a thin layer of ocean crust above and a layer of rigid mantle below. As the lithosphere moves away from its place of origin at the spreading ridge, it becomes thicker. Thickening occurs by progressive downward rigidification of underlying mantle asthenosphere as the material cools.

Because the Earth is not becoming larger, the addition of surface material in one place must be accompanied by an equal loss of surface material somewhere else on the planet. This is manifested as subduction zones. Also, in certain places, two lithospheric plates slide parallel to each other.

There are thus three basic kinds of plate boundaries:
- divergent boundaries, at spreading ridges
- convergent boundaries, at subduction zones and also at sites of continent–continent collision
- transform boundaries (faults), where two plates slide parallel to each other

Divergent Boundaries

Divergent boundaries exist where two plates are moving apart. This occurs where convection currents in the mantle rise toward the surface and spread apart. Divergent boundaries are characterized by basaltic volcanism, intrusion of gabbro dikes, and shallow-focus earthquakes. The magma is formed by partial melting of mantle rock as the rocks rise slowly upward. Because the melting temperature of rock decreases with decreasing pressure, at a certain level some of the rock melts to form magma, which rises buoyantly upward toward the crest of the mid-ocean ridge.

As the plates move away from each other, a geographic feature known as a rift valley is often produced. Rift valleys are usually found atop mid-ocean ridges like the Mid-Atlantic Ridge. The mid-ocean ridges are prominent features of the world's deep oceans, but don't picture them as rugged mountain ranges. Although in places they can have fairly dramatic features, they are generally very broad, with fairly gentle side slopes descending gradually to the deep ocean basins that flank them.

Convergent Boundaries

Convergent boundaries exist where two plates move toward each other. The processes associated with convergent boundaries differ depending upon the nature of the lithosphere involved. Crustal material can be continental (with largely granitic composition) or oceanic (with largely basaltic composition). Granitic rock is less dense than basaltic rock and therefore cannot subside into the higher-density mantle. As a result, convergent boundaries can be divided into three subtypes:

• ocean–ocean
• ocean–continent
• continent–continent

Ocean–Ocean Subduction Zones
If both of the plates are capped by oceanic crust, one of the plates moves downward under the other plate at some angle, which ranges from rather shallow (in the case of fast convergence) to almost vertical (in the case of slow convergence). This pattern of convergent motion is called subduction.

A deep trench is formed on the sea floor at the point where the one plate bends downward under the other. Earthquakes occur along the subducting plate, ranging from shallow-focus earthquakes near the trench to deep-focus earthquakes at depths as great as several hundred kilometers. The earthquakes are the result of stresses within the plates themselves, in response to pushing and pulling of the plate rather than to slippage between the subducting plate and the enclosing mantle. Also, an arcuate chain of volcanic islands, called a volcanic island arc, develops above the subduction zone by melting of some of the mantle rock overlying the subducting plate, as the subducting plate releases some of its water content. The Marianas Trench and island arc is an excellent example of a modern subduction zone of this kind, which could be called an ocean–ocean subduction zone.

Ocean–Continent Subduction Zones
If one plate is capped by oceanic crust and the other is capped by much thicker and less dense continental crust, the subducting plate is always the one carrying oceanic crust, because it has a higher density. A trench develops just offshore of the continental plate. The pattern of earthquakes is similar to that associated with subduction of an oceanic plate beneath another oceanic plate, but the earthquakes occur beneath the continent and the volcanism forms a chain of volcanic mountains inboard of the margin of the continent. The subduction zone along the west coast of South America is an excellent example of a modern subduction zone of this kind, which could be called an ocean–continent subduction zone.

Continent–Continent Collisions
The final possibility is the convergence of two plates, both of which carry continental crust. Because continental crust is much less dense than the mantle, neither of these plates is subducted. Instead, one is pushed horizontally beneath the other for some distance, until the friction forces along the boundary between the two plates builds up to be greater than the compressive force that is pushing the plates together. The thickening of the lithosphere in this way results in very high land elevations over very large areas. Folding of rock layers and compressive fault movements of large masses or sheets of continental crust result in the formation of high mountains at the site of the continent–continent collision. Although volcanic activity is not common, earthquakes are common and can occur far from the plate boundary beneath either plate. The outstanding modern example of continent–continent collision is the collision of the Indian Peninsula with the main mass

of southern Asia, which has resulted in the Himalayas mountain chain and the Tibetan Plateau.

Transform Faults

Transform faults exist where two plates slide past each other in parallel, but opposite, motion. A look at any map of mid-ocean ridges shows that the ridges are offset by numerous transform faults. Movement along transform faults generates frequent earthquakes but usually no volcanic activity. The San Andreas Fault in California is an unusually long transform fault.

Ocean Currents

What Drives Water Movements in the Ocean?

With the exception of the tides, virtually all of the large-scale water motions in the oceans are caused by global temperature differences, either directly or indirectly. All materials radiate energy from their surfaces; the hotter the surface, the stronger the radiation and the shorter the wavelength of the radiation. The relatively cool materials of the Earth's surface—rock, soil, vegetation, water— radiate at a much longer wavelength than that coming from the Sun. At any point on the Earth's surface, there is both heating from the Sun and loss of heat by means of long-wave radiation back out to space. In low-latitude areas, there is a long-term excess of incoming energy over outgoing energy, meaning a tendency for temperature to increase. In high-latitude areas, the situation is just the opposite.

To maintain the long-term average temperature distribution on the Earth's surface, heat must be transported from low-latitude areas to high-latitude areas. This is done in part by the wind systems of the Earth and in part by ocean currents; these two agents of heat transport are about equal in their effects. The circulation in the oceans, which is responsible for the oceanic part of the necessary latitudinal heat transport, consists of two largely separate parts:
- the deep circulation, which is driven directly by latitudinal temperature differences
- the surface circulation, which is driven by the Earth's wind systems; these winds are in turn driven by the latitudinal temperature differences

The Coriolis Effect

The Earth's rotation introduces great complexity to the circulation of the oceans, as well as to the Earth's wind systems. In accordance with Newton's first law of motion, air in the atmosphere or water in the oceans, once set in motion, tend to move at constant speed and in a straight line, to the extent that they are not acted upon by frictional forces. (They are indeed acted upon by friction, but the magnitude of the friction forces is sufficiently small that the air or water can move for long distances without being slowed appreciably.)

The key point, however, is that this straight line has to be seen from the viewpoint of an observer in outer space. Meanwhile, the Earth is rotating relative to that outer-space observer; moreover, the Earth is rotating relative to the moving water. To observers fixed upon the rotating Earth—us, that is— the moving air and water seem to be moving in a curved path rather than a straight line— even though in reality it is us, not the air or water, that is moving in a curved path. In the science of mechanics, it takes an external, sideways acting force to change the direction of motion of a body. We therefore invoke a fictitious force, called the Coriolis force, to account for the seemingly curved path of the mass of air or water.

The Coriolis force is not really a force: it is a consequence of viewing a moving body from a rotating frame of reference. Even though it is not a real force, however, it has the effect of a real force, in terms of explaining the patterns of winds and ocean currents that we, as rotating observers, see.

Deep Currents
Currents exist in the deep ocean worldwide, although these currents are generally much slower (a few centimeters per second, or even less) than typical surface currents. Deep currents originate by production of very dense ocean surface waters at high latitudes in both the Northern Hemisphere and (especially) the Southern Hemisphere. The density of the surface waters is increased in two ways: it is cooled, and its salinity is increased by freezing out of fresh-water ice. The water then sinks and flows slowly at or near the ocean bottom by circuitous routes to low latitudes. It is then warmed and rises up slowly to the surface, to be carried by surface currents back to high-latitude areas.

It is easy to imagine that this deep circulation of the oceans is a gigantic convection cell. It was proved long ago, however, and is universally accepted by oceanographers, that convection is almost nonexistent in a system, like the oceans, that is both heated and cooled at the surface rather than from beneath. The key element that drives the deep circulation is vertical mixing through the water column at low latitudes. Vertical mixing has the effect of carrying warm surface water downward and cold deep water upward. In that way, and in that way only, the cold deep water at low latitudes is warmed sufficiently to rise slowly upward toward the surface, and then to flow toward higher latitudes as part of surface currents, completing the circulation cell.

Oceanographers are still debating the relative importance of various possible mechanisms for this mixing. It is becoming widely accepted that the tides are mainly responsible for the mixing, and therefore the existence of the deep circulation of the oceans! Because of the common misconception that the deep circulation is a convection phenomenon, it is somewhat misleadingly called the thermohaline circulation. A better term for the deep circulation, which is coming into more general use, is the meridional overturning circulation. (The term meridional implies movements parallel to longitude lines: i.e., north–south or south–north movement.)

Surface Currents
The Ekman Drift.—
Surface currents are produced by the frictional forces exerted by the wind. Except near the Equator, where the effect of the Earth's rotation is unimportant, the water is deflected by the Coriolis force (to the right in the Northern Hemisphere, and to the left in the Southern Hemisphere). This leads to the highly counterintuitive result that the net transport of water is at right angles to the wind direction! This effect is called the Ekman drift.

A qualitative, nonmathematical explanation of the dynamics of the Ekman drift is not simple. The theory of the Ekman drift predicts that at the surface, the water moves at an angle of 45° to the wind. This represents a balance among three forces:
• the downwind force of the wind
• the Coriolis force, at right angles to the motion
• friction by the underlying water, acting opposite to the direction of movement

To understand this balance clearly, you would need to draw the three forces as vectors, as in basic physics. At each successively lower level of water below the surface, the current direction keeps turning. It moves clockwise in the Northern Hemisphere, because it turns out that this is the only way to accommodate the balance of forces. This balance involves two friction forces, with slightly different magnitudes and directions, exerted on the water layer by the overlying and underlying layers.

Gyres.—
One of the most prominent features of the pattern of surface ocean currents is the existence of large rotatory currents, called gyres, in the ocean basins. The gyre in the North Atlantic is a good example. These gyres are caused by combined action of the trade winds, which blow from east to west at low latitudes, and the westerly winds at mid-latitudes. Because of the Coriolis effect, however, the mechanism is less direct than it seems. The Ekman drift caused by these two wind systems piles up water (literally) in the center of the basin, and then, as the water tends to flow back out radially from the center, it is turned by the Coriolis force to flow in a circular pattern. This concept of gyres involves two separate effects of the Coriolis force: piling up of water between the two opposing wind systems, and the deflection of the water as it attempts to "relax" back to a level surface.

Ocean Currents and Climate
There is a very close connection between ocean currents and climate (global and local). The efficiency of latitudinal heat transport by ocean currents has a direct effect on the average temperature difference between low-latitude regions and high-latitude regions. At times when ocean currents are less efficient in heat transport, there tends to be a greater temperature difference between low latitudes and high latitudes. At such times, the Earth's wind systems tend to become more vigorous, transporting the necessary heat to maintain the balance.

The pattern of ocean surface currents is also a major factor in the climate of land areas of subcontinental size. Land areas adjacent to, and downwind of, warm ocean currents tend to have warmer and usually wetter climates, whereas areas downwind of cold ocean currents tend to have colder and usually drier climates.

The specific heat capacity of a material is a measure of how much heat must be added, per unit mass, to raise the temperature of the material by one unit of temperature (one degree Celsius, say). Water has one of the highest specific heat capacities of any substance, much higher than that of common Earth-surface materials like rock and soil. For that reason, proximity to the oceans has a moderating effect on the climate of land areas: maximum temperatures tend to be lower and minimum temperatures tend to be higher.

Warm ocean waters are the main suppliers of water vapor to the atmosphere. Land areas affected by winds that blow from areas of warm ocean waters have much wetter climates than areas far removed from—or upwind of—areas of warm ocean waters. For example, most of the precipitation along the eastern seaboard of the United States, in both winter and summer, originates from the warm waters of the Gulf of Mexico and the western Atlantic. The much drier areas of the west central United States experience an influx of such moisture-laden air much less often.

While ocean circulation has a major effect on climate, climate can also have a major effect on ocean circulation. At times in Earth history when global climate was unusually warm (much warmer than now!), temperatures in high-latitude oceans were higher than they are now. There was much less tendency for production of the dense water masses that now sink, flow along the deep ocean bottom, and are warmed as they rise in low-latitude areas to complete the circulation cell. At such times, most of the deep ocean below the thermocline became stagnant. Without a continuing supply of dissolved oxygen from high-latitude surface waters, the deep ocean became anoxic. Oxygen-breathing flora disappeared, to be replaced by accumulation of dead organic matter raining down from the near-surface ocean.

More Information – on the Web

Visit the *EarthComm* web site www.agiweb.org/earthcomm/ to access a variety of links to web sites that will help you deepen your understanding of content and prepare you to teach this activity. Many of the sites also contain images that you can download.

Chapter 2

Goals and Assessment

Clarify that the goals indicate what students should understand and be able to do as a result of the activity. Make sure students understand that Chapter Assessments are based upon these goals.

Goal	Location in Activity	Assessment Opportunity
Model present and ancient landmasses and oceans to determine the flow of ocean currents.	**Investigate;** **Check Your Understanding** Question 1 **Understanding and Applying What You Have Learned** Question 1	Landmasses are correctly constructed; paragraph describing observations is accurate.
Explain how ocean currents affect regional and global climate.	**Investigate;** **Digging Deeper;** **Understanding and Applying What You Have Learned** Question 3	Paragraph and presentation of results of model is accurate. Responses to questions are accurate.
Understand how ocean currents are affected by Earth's moving plates.	**Investigate;** **Digging Deeper**	Responses to questions are reasonable and closely match those given in Teacher's Edition.
Understand the relationship between climate and such Earth processes as plate movements, mountain building, and weathering.	**Investigate;** **Digging Deeper;** **Check Your Understanding** Question 3 **Understanding and Applying What You Have Learned** Question 2	Responses to questions are reasonable and closely match those given in Teacher's Edition.

NOTES

Activity 4

How Do Plate Tectonics and Ocean Currents Affect Global Climate?

Goals

In this activity you will:

- Model present and ancient land masses and oceans to determine current flow.

- Explain how ocean currents affect regional and global climate.

- Understand how ocean currents are affected by Earth's moving plates.

- Understand the relationship between climate and Earth processes like moving plates, mountain building, and weathering.

Think about It

Ocean currents help to regulate global climate by transferring heat and moisture around the globe.

- How would a change in the position of a land mass influence global climate?

What do you think? Record your ideas about this question in your *EarthComm* notebook. Be prepared to discuss your responses with your small group and the class.

Activity Overview

Students build models of the continents at the South Pole in the Eocene Epoch, the Miocene Epoch, and the present. They model how the patterns of ocean circulation around Antarctica might have differed during each of these times. Students then consider how a change in ocean circulation might cause a change in the climate of their community. **Digging Deeper** explains how ocean currents affect regional and global climate, and also how plate tectonics affects global climate.

Preparation and Materials Needed

Preparation

You will need to make blue ice cubes. You will also need to make copies of the continental configurations around Antarctica during the Eocene, the Miocene, and the present using **Blackline Master Climate Change 4.1 A - C, Continental Configurations**. You may need to enlarge or reduce the copies to best fit the container being used.

Materials

- Copies of the Southern Hemisphere continental configurations: Eocene, Miocene, and present (**Blackline Master Climate Change 4.1 A - C, Continental Configurations**)
- Cake pan, 8 inches round
- Modeling clay
- Water
- Blue-colored ice cubes

Think about It

Student Conceptions

From previous activities, students are likely to recognize the importance of latitude in determining climate. Therefore, they may say that if a landmass changed its latitude, it would become warmer or colder, depending upon whether the landmass moved toward higher or lower latitude. Ask students:

- how they think ocean circulation affects climate
- how a change in the position of landmasses would affect patterns of circulation
- how this change would affect climate

Answer for the Teacher Only

This question has a number of ramifications. Perhaps the most important of them is the effect on continental glaciation. Continental ice sheets can develop only if large landmasses are located at high latitudes. In recent geologic times, the Antarctic ice sheet has persisted through major changes in climate because the large Antarctic continent is located at the South Pole. There is no corresponding ice sheet at the

North Pole because the area is occupied by the Arctic Ocean rather than a large continent. Periods of major glaciation earlier in Earth history correspond to times when large continental areas have been located at high latitudes.

Aside from the effect of glaciation, a shift in the position of a continent is likely to change the pattern of major ocean currents, thereby affecting climate over large areas of the Earth.

Assessment Tool

Think about It Evaluation Sheet
Use this evaluation sheet to help students understand and internalize the basic expectations for the warm-up activity.

Blackline Master Climate Change 4.1A
Continental Configuration: Present

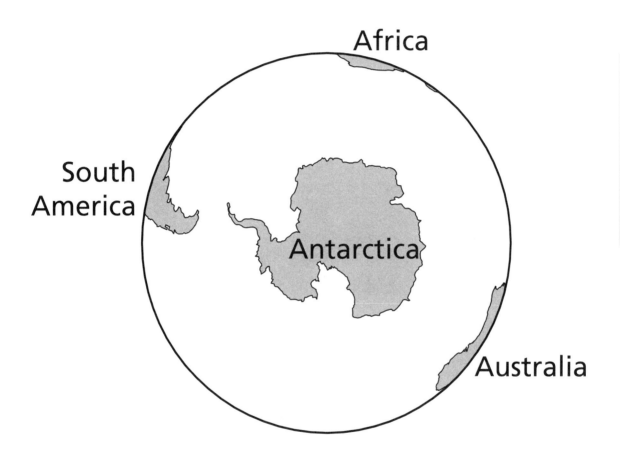

Present

Chapter 2

Blackline Master Climate Change 4.1B
Continental Configuration: Miocene Epoch

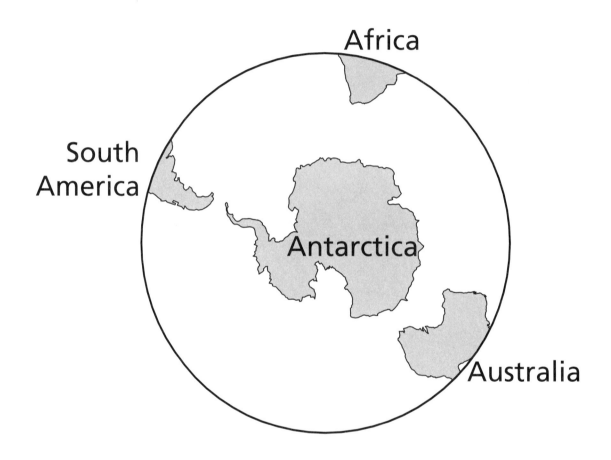

Africa

South
America

Antarctica

Australia

26 Million Years ago
Miocene Epoch

Blackline Master Climate Change 4.1C
Continental Configuration: Eocene Epoch

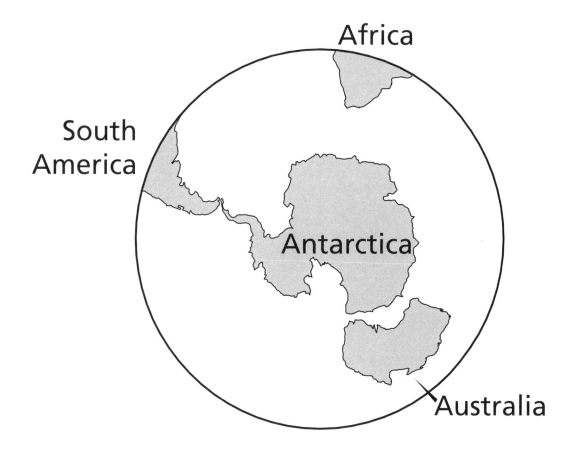

54 Million Years ago
Eocene Epoch

Blackline Master Climate Change 4.2
Map of Ocean Currents

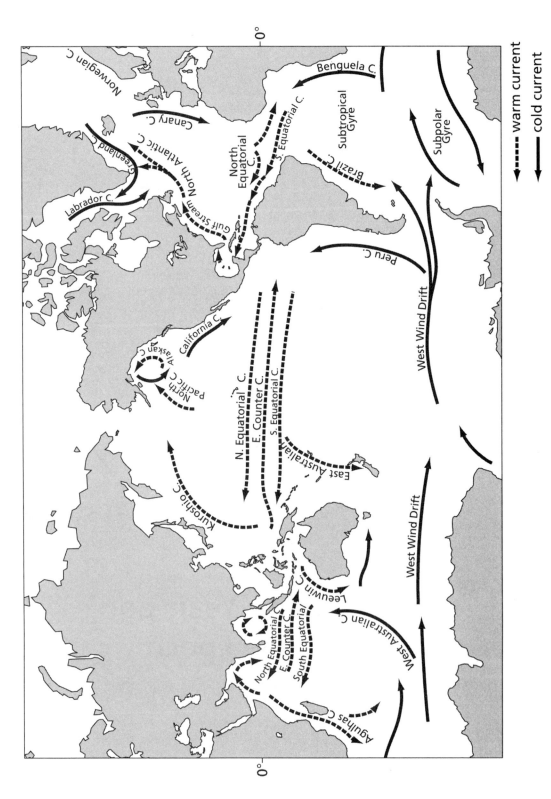

NOTES

Chapter 2

Earth System Evolution Climate Change

Investigate

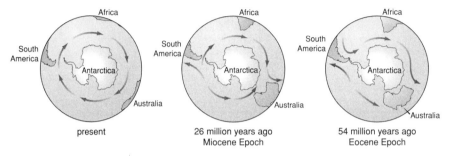

present | 26 million years ago
Miocene Epoch | 54 million years ago
Eocene Epoch

1. Divide your class into three groups. Each group will investigate the flow of water at one of these three periods of time:

   Group 1: The present.

   Group 2: During the Miocene Epoch, 26 million years ago.

   Group 3: During the Eocene Epoch, 54 million years ago.

2. Obtain a copy of the map for your assigned time period. Put this map under a clear plastic container.

3. Using clay, construct the correct land masses inside the container using the map as a template. Make the land masses at least 3 cm high.

4. Remove the map and add water to the container up to the level of the land masses.

5. Obtain a blue-colored ice cube. Place this as close to the South Pole as possible.

   a) As the ice cube melts and the cold water flows into the clear water, draw arrows on the map to record the direction of flow.

    Use only food coloring to dye the ice. Clean up spills immediately. Dispose of the water promptly.

   b) Write a paragraph in which you describe your observations.

6. Present your group's data to the class.

   a) Compare the direction of current flow on your map to those presented by the other groups. How does the current change as the land masses change from the Eocene Epoch to the Miocene Epoch to the present?

7. Use the results of your investigation to answer the following:

   a) Based on the map that shows the position of Australia in the Eocene Epoch and your investigation, what tectonic factors do you think are most important to consider when contemplating Australia's climate in the Eocene Epoch? How might Australia's Eocene Epoch climate have been different from its climate today?

   b) What tectonic factor(s) (for example, the position of the continents, occurrence of major ocean currents, mountain ranges, etc.) is/are most important in affecting the climate in your community today?

E 118

Investigate

1. – 4. Circulate around the room to see that students are constructing the landmasses correctly.

5. a) Direction of flow of the water should resemble the arrows drawn in the illustrations in student text.

b) Students should observe that during the Eocene Epoch, the South American and Australian continents blocked the flow of water around Antarctica. In the Miocene Epoch, South America and Australia diverted some water, while some of the water flowed entirely around Antarctica. At present, water flows completely around Antarctica and is not diverted by any landmasses. The present-day model may not perfectly illustrate this, but students should observe that circumpolar circulation was greatly improved relative to the Eocene Epoch and Miocene Epoch models.

6. a) Students should observe that with the movement of South America and Australia away from Antarctica, more complete circumpolar flow was possible.

7. a) During the Eocene Epoch, Australia was located closer to the South Pole and the climate was colder. The colder Eocene Australian climate was further reinforced by the northward diversion of the cold Antarctic currents. As Australia moved north, away from Antarctica and the South Pole, its climate warmed. The warming of the Australian continent was further bolstered by the development of the circumpolar current, which reduced the transport of cold Antarctic water northward around Australia.

b) Answers will vary.

Chapter 2

8. Examine the map of ocean surface
currents carefully.

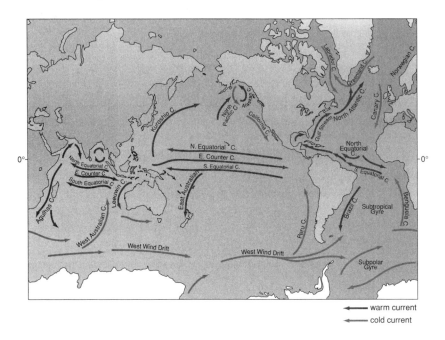

← warm current
← cold current

a) How do you think surface ocean
currents modify the climatic
patterns in the United States?

b) What changes in the surface-
current patterns might arise if
North America and South
America split apart from one
another, leaving an open
passageway from the Atlantic
Ocean to the Pacific Ocean?

c) How might the climate of the East
Coast of the United States change
if tectonic forces changed the
positions of the continents so that
the Gulf Stream no longer flowed
north?

d) How might some other parts of the
world (including your community)
be affected if the Gulf Stream
stopped flowing north?

Reflecting on the Activity and the Challenge

Ocean currents play a large role in
global climate. This activity helped
you see how ocean currents change in
response to movements of the Earth's
lithospheric plates.

8. a) The oceans transport heat around the globe. An example of how current patterns affect the climate of the United States is that cold currents along its western coast cause temperatures to be cooler than on its eastern coast, along which warm currents flow northward.

   b) Answers will vary. Most likely, students will suggest that an exchange of water between the Pacific and Atlantic oceans would occur between North and South America. Given the investigation they just undertook, they might also hypothesize that the Gulf Stream would be affected by such a tectonic rearrangement.

   c) The Gulf Stream is a warm current. If its flow were diverted the eastern coast of the United States would most likely be cooler.

Assessment Tools

EarthComm **Notebook Entry-Checklist**
Use this checklist as a quick guide for student self-assessment and/or an opportunity to quickly score student work. Add further criteria specific to your classroom needs or to this particular investigation.

Investigate Notebook Entry-Evaluation Sheet
Point out the criteria listed on this evaluation sheet that are relevant to this particular investigation. Encourage students to internalize the criteria by making them part of your "assessment conversations" as you circulate around the classroom.

Reflecting on the Activity and the Challenge

Have a student read this section aloud to the class, and discuss the major points raised. Students should recognize that the movement of landmasses could alter the patterns of ocean circulation. **Digging Deeper** will help them understand that ocean circulation affects climate; therefore, a change in ocean circulation can lead to a change in climate. Discuss with students how this activity will help them as they prepare their **Chapter Challenge.**

Earth System Evolution Climate Change

Geo Words

thermohaline circulation:
the vertical movement of
seawater, generated by density
differences that are caused by
variations in temperature and
salinity.

Digging Deeper

CHANGING CONTINENTS, OCEAN CURRENTS, AND CLIMATE

How Ocean Currents Affect Regional Climates

A community near an ocean has a
more moderate climate than one
at the same latitude inland
because water has a much higher
heat capacity than rocks and soil.
Oceans warm up more slowly and
cool down more slowly than the
land. Currents are also an
important factor in coastal
climate. A coastal community near
a cold ocean current has cooler
weather than a coastal community
near a warm ocean current. For
example, Los Angeles is located on
the Pacific coast near the cold
California Current. The city has an
average daily high temperature in
July of 75°F. Charleston, South
Carolina, is located at a similar

Figure I This thermal infrared image of the
northwest Atlantic Ocean was taken from an
NOAA satellite. The warm temperatures (25°C)
are represented by red tones, and the cold
temperatures (2°C) by blue and purple tones.

latitude, but on the Atlantic coast near the warm Gulf Stream. (See *Figure I*.)
Charleston's average daily high temperature in July is 90°F.

How Ocean Currents Affect Global Climate

Patterns of ocean circulation have a strong effect on global climate, too. The
Equator receives more solar radiation than the poles. However, the Equator
is not getting warmer, and the poles are not getting colder. That is because
oceans and winds transfer heat from low latitudes to high latitudes. One of
the main ways that the ocean transfers this heat is by the flow of North
Atlantic Deep Water (abbreviated NADW). It works like this: In the
northern North Atlantic, the ocean water is cold and salty, and it sinks
because of its greater density. It flows southward at a deep level in the
ocean. Then at low latitudes it rises up toward the surface as it is forced
above the even denser Antarctic Bottom Water. Water from low latitudes
flows north, at the ocean surface, to replace the sinking water. As it moves
north, it loses heat. This slow circulation is like a "conveyor belt" for
transferring heat. This kind of circulation is usually called **thermohaline
circulation**. (*thermo* stands for temperature, and *haline* stands for saltiness.)

Digging Deeper

Assign the reading for homework, along with the questions in **Check Your Understanding** if desired.

Assessment Opportunity

Use a quiz to assess student understanding of the concepts presented in this activity. Some sample questions are listed below:

Question: Explain how the climate of a coastal community near a cold ocean current differs from the climate of a coastal community near a warm ocean current.

Answer: Other things being equal, the climate of the community near the cold ocean current is cooler and most likely drier than that of the community near the warm ocean current.

Question: What are two ways that plate tectonics can affect the climate of a coastal community?

Answer: Plate tectonics can change the latitudinal position of the community and also can result in a change in ocean circulation, which can affect ocean currents near the community.

Question: How can volcanic activity be linked to climate change?

Answer: Volcanoes release carbon dioxide into the atmosphere; an increase of carbon dioxide is thought to lead to global warming. Volcanoes also add dust to the atmosphere, which can result in global cooling.

Teaching Tip

Ask students to share their ideas about what the satellite image of the northwest Atlantic Ocean (*Figure 1*) has to do with the subheading of the **Digging Deeper** reading section: **How Ocean Currents Affect Regional Climates**. The image shows clearly how cold water from the north moves south, and how warm water from the south moves north (note how the red coloring thins to the north and how the blue strand of color thins to the south).

Teaching Tip

Make an overhead of the circulation cell in *Figure 2* on page E121 using **Blackline Master Changing Climate 4.3, A Circulation Cell in the Ocean** and incorporate this overhead into a class discussion or lecture about the circulation of water masses in the ocean. Deep-water masses in the ocean form a circulation cell, which cycles slowly through the ocean. One aspect of the mechanism for movement of deep currents is the difference in density between water masses; therefore, deep circulation is often referred to as the thermohaline circulation, in reference to the two factors that determine water density: temperature (*thermo-*) and salinity (*-haline*).

Chapter 2

Activity 4 How Do Plate Tectonics and Ocean Currents Affect Global Climate?

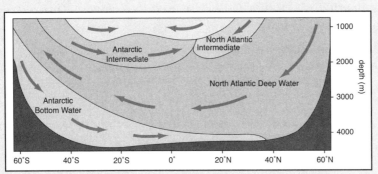

Figure 2 Circulation cell formed by the movement of deep-water masses in the ocean.

Geo Words

lithospheric plate: a rigid, thin segment of the outermost layer of the Earth, consisting of the Earth's crust and part of the upper mantle. The plate can be assumed to move horizontally and adjoins other plates.

plate tectonics: the study of the movement and interaction of the Earth's lithospheric plates.

When this conveyor belt is disturbed, the entire global climate is affected. For example, about 12,000 years ago, glaciers were melting rapidly, because the Earth was coming out of a glacial age. The melting glaciers discharged a lot of fresh water into the North Atlantic in a short time. The fresh water decreased the salinity of the ocean water thus reducing its density. This decrease was so much that the production of NADW was decreased. This seems to have plunged the world back into a short cold period, which lasted about 1000 years.

How Plate Tectonics Affects Global Climate

The positions of the continents on the Earth change as the Earth's **lithospheric plates** move. (**Plate tectonics** is the study of the movement and interaction of the Earth's lithospheric plates.) The arrangement of the continents has a strong effect on the Earth's climate. Think about the requirements for the development of large continental ice sheets. Glaciers form only on land, not on the ocean. For an ice sheet to develop there has to be large land areas at high latitudes, where snow can accumulate to form thick masses of ice. Where oceans occupy polar areas, accumulation of snow is limited by melting in the salty ocean waters. Polar oceans, like the Arctic Sea, around the North Pole, are mostly covered by pack ice. This ice is no more than several meters thick.

Today, the two continental landmasses with permanent ice sheets are Antarctica, in the Southern Hemisphere, and Greenland, in the Northern Hemisphere. The continent of Antarctica has not always been centered on the South Pole. About two hundred million years ago, all of the Earth's continents were welded together. They formed a single continent,

Blackline Master Changing Climate 4.3
A Circulation Cell in the Ocean

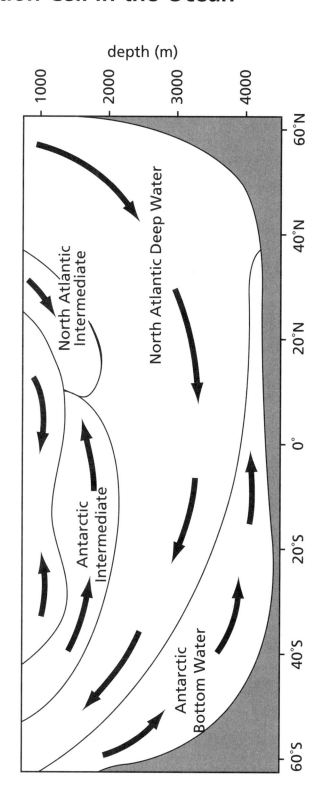

Chapter 2

Earth System Evolution Climate Change

Geo Words

Pangea: Earth's most recent supercontinent, which was rifted apart about 200 million years ago.

called the supercontinent of **Pangea**. Pangea was eventually rifted apart into several large pieces. One of the pieces, the present Antarctica, moved slowly southward. Eventually it moved close enough to the South Pole for ice sheets to form. In recent geologic time, Antarctica has been

Figure 3 An ice-core station in Antarctica.

directly over the South Pole, so the Antarctic ice sheet has remained in existence even during interglacial periods. We know that because otherwise global sea level would have been much higher during the interglacial periods of the past million years.

At present, most of the Earth's continental land area is in the Northern Hemisphere. Much of North America and Eurasia is at a high latitude. Ice sheets can form during the parts of Milankovitch cycles that are favorable for decreased global temperatures. During times of increased global temperatures, the North American and Eurasian ice sheets have melted away completely. The picture is very different in the Southern Hemisphere. Except for Antarctica, there is not enough continental land area at high latitudes for large continental glaciers to form.

Figure 4 Mt. St. Helens is an example of a volcano associated with a plate boundary.

Plate tectonics affects climate in other ways besides changing the positions of the continents. Volcanoes like the one shown in *Figure 4* from along active plate margins. Increased activity at these margins causes increased volcanic activity. Volcanoes release carbon dioxide, which is a gas that traps heat in the atmosphere. (You'll learn more about carbon dioxide in the following activity.) In this way, plate tectonics might cause global climate to warm. However, volcanic eruptions also add dust to the atmosphere, which blocks out some solar radiation. This tends to decrease global temperatures.

EarthComm

Teaching Tip

The ice cores drilled and recovered at stations like the one pictured in *Figure 3* have led to great advances in scientists' understanding of climate change. This picture can be used as an opportunity to have students review what they learned about ice cores and climate proxies. Ask students if ice cores might be useful in understanding tectonically driven changes in climate, like those investigated in this activity, that can take many millions of years to occur. This is an excellent opportunity to discuss the limitations of using ice-core data, which currently extends back only to several hundred thousand years before present, as a climate proxy.

Teaching Tip

Mt. St. Helens Volcano, in southern Washington State, last erupted in 1980. This eruption, and the Cascades mountain range in general, was the result of the subduction of the Juan de Fuca lithospheric plate beneath the North American lithospheric plate. Volcanoes related to subduction zones constitute the majority of volcanoes that occur on land, and their eruptions can launch enormous amounts of gas and ash high up into the atmosphere. Check the *EarthComm* website to see if there has recently been a volcanic eruption that can be used to focus a discussion on plate tectonics and climate.

Chapter 2

Continent–continent collisions create huge mountain ranges. The Himalayas and the Alps are modern examples. Many scientists believe that the weathering of such mountain ranges uses up carbon dioxide from the Earth's atmosphere because some of the chemical reactions that break down the rock use carbon dioxide from the atmosphere. This causes global climate to cool. For example, the collisions between continents that produced the supercontinent Pangea resulted in high mountain ranges like the one at the present site of the Appalachian Mountains. The Appalachians were much taller and more rugged when they first formed—perhaps as tall as the Himalayas shown in *Figure 5.* Three hundred million years of erosion have given them their lower and well-rounded appearance. On a global scale, all that weathering (which uses up carbon dioxide) may have contributed to the period of glaciation that began about 300 million years ago and ended about 280 million years ago.

Figure 5 The geologically young Himalayan mountains formed when India collided with Asia.

Check Your Understanding

1. Explain how North Atlantic Deep Water circulates.

2. Why do glaciers form only on continents and not in oceans?

3. Explain how plate tectonics can affect global climate.

Understanding and Applying What You Have Learned

1. In the **Investigate** section, you made models designed to demonstrate how ocean currents were different during the Eocene, Miocene, and today.

   a) Assuming that the maps were accurate, in what ways was your model helpful in exploring possible differences in oceanic currents?

   b) What are some of the drawbacks or problems with your model (how is a model different than the "real world"?).

   c) What improvements could you make to the model so that it would behave in a more accurate way?

Teaching Tip

The Tibetan Plateau and Himalayan Mountains (pictured in *Figure 5*) have been formed by the relatively recent (in geological terms) collision of two continental landmasses. The collision between the India and the Eurasian continent began about 40–50 million years ago, and convergence continues today. The Himalayan Mountains host several of the world's highest mountains, including Mount Everest which exceeds an elevation of 29,000 feet above sea level. The Himalayas continue to rise even today, currently rising at a rate of about 1 cm/year.

Check Your Understanding

1. In the North Atlantic, the ocean water is relatively cold and salty, and it sinks because of its greater density. It flows southward at a deep level in the ocean. As it reaches the south Atlantic, it is forced above even denser Antarctic deep waters. As it is forced upwards and ages, it is warmed and rises up to the surface. Water from low latitudes flows north at the surface of the ocean, losing heat as it travels north. Eventually, that water will become sufficiently dense and sink to the bottom again, to complete the cycle.

2. For glaciers to form there must be large land areas at high latitudes, where snow can accumulate to form thick masses of ice. Where oceans occupy polar areas, accumulation of snow is limited by melting in the salty ocean waters.

3. The arrangement of the continents has a strong effect on the Earth's climate. The latitude of a landmass can influence its climate: landmasses at higher latitudes have cooler climates. Also, the position of the continents can influence patterns of ocean circulation that play a large role in the distribution of heat around the globe.

Assessment Tool

Check Your Understanding Notebook Entry-Evaluation Sheet
Use this sheet to evaluate the extent to which students understand the key concepts explored in **Activity 4** and explained in **Digging Deeper,** and to evaluate the students' clarity of expression.

Chapter 2

Understanding and Applying What You Have Learned

1. a) Although the models were not exactly accurate, they provided a good approximation for understanding how the positions of South America and Australia relative to Antarctica influence ocean circulation.

   b) The landmasses could be constructed to reflect the bathymetry and topography of the continents more accurately. Also, it is not realistic to think that the water from the ice cube represents water in the ocean accurately; in reality, it is a much more complicated system.

   c) Answers will vary but should relate to the inaccuracies identified in **Question 1(b)**.

NOTES

Chapter 2

Earth System Evolution Climate Change

2. Increased weathering of rocks uses up carbon dioxide. Decreasing carbon dioxide in the atmosphere contributes to global cooling. Reconstructions of the collision between India and Asia suggest that India first collided with Asia during the Late Eocene but that most of the mountain building took place during the Miocene and later.

   a) When would you expect to observe the greatest changes in weathering rate?

   b) Why?

3. Melting glaciers discharged a lot of fresh water into the North Atlantic in a short period of time about 12,000 years ago. Adding fresh water "turned off" the North Atlantic Deep Water current for about 1000 years.

   a) What other changes could disturb the NADW?

   b) In the event that this happened, what effect would it have had on global climate?

   c) How would this change affect your community? Even if you don't live near the Atlantic Ocean, your physical environment might still be greatly affected.

Preparing for the Chapter Challenge

Using a style of writing appropriate for a newspaper, write several paragraphs containing the following material:

- Explain how the locations of the continents on the Earth affect global climate.

- Explain how ocean currents affect global climate.
- Explain how moving continents change ocean currents.

Inquiring Further

1. **Modeling North Atlantic Deep Water flow**

   Make a physical model of the flow of North Atlantic Deep Water. Experiment with several ideas. Think of how you might do it using actual water, and how you

   might do it using other materials. If your model idea is large and expensive, draw a diagram to show how it would work. If your model idea is small and simple, construct the model and see if it works.

2. **a)** The greatest changes in weathering rates were during the Miocene, when mountains were rapidly rising.

   b) The rate of weathering is controlled by both climate and the rate of uplift. The Miocene (and later) mountain-building events (and the associated uplift of the land) greatly increased the rates of weathering in this region.

3. **a)** The formation of NADW is driven by density. Salinity and temperature control density. An increase in temperature in the North Atlantic could alter NADW circulation. Also, the position of landmasses that could alter the flow of water in the North Atlantic could disrupt the formation of NADW.

   b) Movement of NADW is one of the primary loops in the global ocean circulation. Disruption of NADW would alter the transfer of heat on the globe and therefore cause changes in climate.

   c) Student responses to this question will likely vary greatly, but should contain some reasoning that supports the hypothesized changes.

Teaching Tip

The situation described in **Step 3** of **Understanding and Applying What You Have Learned** is known as the Younger Dryas Cooling. Ice core data from Greenland (and other data as well) indicate that sometime around 12,000 years ago a rapid climatic cooling event took place. This cooling reversed the trend of climatic warming and glacial retreat that had been ongoing for a few thousand years. The North Atlantic polar front readvanced southward, and several lines of evidence suggest that the climatic cooling affected regions well beyond the vicinity of the North Atlantic. The Younger Dryas Cooling lasted for about 1000 years, and it has been hypothesized that its cause can be related to a shut down of North Atlantic Deep Water (NADW) production triggered by glacial melting and retreat. The reasoning for this hypothesis goes something like this. During the early stages of glacier retreat, much of the meltwater from the Laurentide (North American) Ice Sheet emptied primarily into the Gulf of Mexico. Once the ice margin had retreated sufficiently to open up drainage into the St. Lawrence Seaway, the outflow of meltwater into the North Atlantic could have created a low-salinity lens in the region of NADW production. Such low-salinity water would not be dense enough to sink into the deep ocean, reducing or even temporarily shutting down production of NADW. Because the production of NADW results in an export of cold deep water from the North Atlantic basin and an import of warm South Atlantic surface waters, a reduction in NADW production rates might result in decreased flow of what is today known as the Gulf Stream. This would cool waters (and climate) north of the Equator.

More information on the Younger Dryas Cooling can be found on the *EarthComm* web site.

Chapter 2

Preparing for the Chapter Challenge

Students should now be able to explain how plate tectonics and ocean circulation affect global climate. Being able to understand the causes of climate change will help them to better understand changes in climate that have occurred in the past. Students should now be comfortable with writing in a style that is appropriate for a newspaper.

Relevant criteria for assessing this section (see **Assessment Rubric for Chapter Challenge on Climate Change** on pages 296 to 298) include discussions on causes of climate change, including plate tectonics and ocean currents.

Assessment Opportunity

As in the previous activities, a rubric like the one shown below may be helpful in assessing student articles. Review these criteria with your students so that they can be certain to include the appropriate information in their article.

Item	Missing	Incomplete/Inaccurate	Complete/Accurate
Explanation of how the locations of the continents on the Earth affect global climate.			
Explanation of how ocean currents affect global climate.			
Explanation of how moving continents change ocean currents.			

Inquiring Further

1. Modeling North Atlantic Deep Water flow

There are many possibilities for this model, but common ideas will probably have to do with using colored waters (or other liquids) of differing densities to model the different water masses. Fans or other cooling devices (like a container with ice) could be used to cool water at one end to make it sink. Students should be encouraged to develop models that could realistically be constructed, but they should not necessarily limit themselves to the space, time, and monetary constraints of the classroom environment.

NOTES

ACTIVITY 5 — HOW DO CARBON DIOXIDE CONCENTRATIONS IN THE ATMOSPHERE AFFECT GLOBAL CLIMATE?

Background Information

Glaciation

The most striking climatic effect in the past million years of Earth history has been the repeated growth and shrinkage of enormous ice sheets on the Northern Hemisphere continents of North America and Eurasia, as well as on Greenland and Antarctica. Periods during which these ice sheets have existed are called glacial periods, or simply glacials, the intervening, briefer periods during which only the Greenland and Antarctic ice sheets have remained are called interglacial periods, or simply interglacials. The glacial periods have lasted several tens of thousands of years, whereas the interglacials have lasted for only of the order of 20,000 years. The Earth is due for the onset of another glacial period in only a few thousand years—unless humankind has so altered the Earth's climate that renewed glaciation does not come to pass!

Glacier Movement: Internal Deformation
A glacier is a large body of ice and snow, resting on land (or, if floating in the ocean, then anchored to land at a number of points) and flowing by internal deformation under its own weight. That carefully worded definition emphasizes that a glacier is different from a large block of ice sliding down a sloping tabletop. Glacier ice must accumulate on the land surface to a minimum thickness of a few tens of meters before the pressure is great enough to allow the ice to behave plastically: that is, to flow by internal deformation under its own weight. This effect of plasticity is somewhat like that of Silly Putty®, which acts as a solid when it is struck with a hammer; when left on the tabletop as a lump, however, it sags downward by flowing under its own weight. You can think of glacier ice, then, as an extremely viscous liquid in terms of how it flows.

In an approximate way, the flow of ice in a valley glacier can be compared to the flow in a river. The vertical distribution of ice velocity is qualitatively similar to the velocity distribution in a river. The maximum velocity is at the surface of the glacier, and the minimum velocity is at the base. The reason is that the downglacier driving force—the weight of the glacier—acts throughout the ice mass, whereas the counterbalancing force— the friction at the base—acts only at the base. In a continental-scale ice sheet, the glacier flows even though it does not rest on a sloping land surface. It flows simply because the elevation of the glacier surface is greater in the center of the ice sheet than around its edges.

Glacier Movement: Basal Slip
In addition to flow by internal deformation, there is another component of glacial velocity: solid-body motion by slip at the glacier base. Glaciers can be classified by the temperature of the ice at the base: there are cold-based glaciers, for which the ice at the base is below the melting temperature, and there are warm-based glaciers, for which the ice at the base is at the melting temperature.

Only glaciers in which the ice at the base is at its melting point can undergo basal slip. In such glaciers, a thin film of lubricating water, typically a few millimeters thick, develops in two ways: heat flow from the interior of the Earth, and frictional heat generated by the friction of sliding. If the ice at and near the base of the glacier is at the melting temperature, then all of the heat added to the base of the glacier goes toward melting of ice. This heat can't be conducted upward, because there is no vertical gradient of temperature, as is the case for cold-based glaciers. (Keep in mind that heat conduction can happen only when the temperature of the material is different from place to place in the material.)

Cold-based glaciers are frozen fast to their rock beds, and do little or no geological work. Warm-based glaciers, on the other hand, are very effective in eroding, transporting, and depositing rock and mineral particles, large and small.

Glacier Movement: Regimen and Economy
There are two related, but different, aspects of glacier movement. On the one hand, new glacier ice is formed in the zone of accumulation, moves downglacier, and is melted (or calved into the ocean) in the zone of ablation. On the other hand, the terminus of the glacier may advance, retreat, or stay in about the same position, as glacier ice moves to the terminus and melts (or is calved).

The activity of a glacier is described by its regimen. A glacier with an active regimen has large values of both accumulation and ablation; such glaciers are usually fast moving. A glacier with an inactive regimen has small values of both accumulation and ablation; such glaciers are usually slow moving. High-latitude glaciers, like the

northern part of the Greenland ice sheet or the central part of the Antarctic ice sheet, have a very inactive regimen, because both winter snowfall and summer melting are slight. The mid-latitude Pleistocene ice sheets must have had a very active regimen.

The balance between accumulation and ablation is described by a glacier's economy. In a glacier with a positive economy, accumulation is greater than ablation, and the glacier grows in volume (and the terminus usually advances). In a glacier with a negative economy, ablation is greater than accumulation, and the glacier shrinks in volume (and the terminus usually recedes). There are thus four combinations of regimen and economy that can characterize a glacier, because regimen and economy are independent of one another.

Ice Ages
We think of the existence of major ice sheets in high-latitude land areas as normal, because in the recent past there have always been high-latitude ice sheets. If we look at all of Earth history, however, we find that for long periods—for much of the greater part of Earth history, in fact—there were no ice sheets!

There have been four distinct periods of glaciation in North America in the last 1.6 million years, with briefer interglacial periods. The most recent glacial period ended about 10,000 years ago. The time between 1.6 million years ago and 10,000 years ago is called the Pleistocene Epoch (one of the standard time divisions of Earth history). Ice covered approximately 22 million km² of the Earth's surface during the height of the Pleistocene glaciations. In contrast, the areas of the present ice sheets in Greenland and Antarctica are 2,175,600 km² and 14,200,000 km² respectively. The Pleistocene

Chapter 2

Epoch is particularly interesting because the features of many present landscapes reflect the work of Pleistocene glaciers. Before the Pleistocene, there seem to have been at least three other periods of glacial activity, at about 2 billion, 600 million, and 250 million years ago.

The Greenhouse Effect

One hears a lot about the greenhouse effect nowadays, in connection with the contentious issue of global warming. It's likely that most of you have had the experience of stepping into a large commercial greenhouse, or your own home greenhouse, on a cool but sunny spring day. The warmth inside the greenhouse on such a day is impressive. It's easy to understand the cause of this warmth— assuming, of course, that the greenhouse is not being heated artificially. The glass panes of the greenhouse roof transmit nearly all of the incoming solar radiation in the visible range of the spectrum; although some is reflected back to space, the proportion reflected is small if the Sun's rays strike the glass panes at a large angle. The sunlight warms the exposed surfaces within the greenhouse, and those surfaces in turn heat the overlying air by conduction and convection. Heat is lost mainly by conduction through the walls and roof of the greenhouse, and the balance between heat gained and heat lost is struck at a temperature well above the ambient outside temperature.

The so-called greenhouse effect in the Earth's atmosphere is somewhat similar to that of an actual greenhouse, but the processes involved are in part fundamentally different. The clear atmosphere is almost transparent to visible sunlight, and the Earth's surface is warmed as it absorbs the energy delivered by the sunlight, just as in a greenhouse. Likewise, the air near the ground is warmed by conduction, and the heat gained is distributed upward in the atmosphere by vertical motions of the air. The difference in the processes lies in the way the Earth loses the heat gained from the Sun to maintain the long-term balance between incoming and outgoing energy.

All materials at temperatures above absolute zero radiate electromagnetic energy. The wavelength of the radiation decreases with increasing temperature of the radiating body. Cool materials, like those of the Earth's surface, radiate at (mostly infrared) wavelengths much longer than those of visible light. In contrast, the radiation from the far hotter surface of the Sun is at its peak in the visible range. The gases of the atmosphere are nearly transparent to incoming solar radiation. However, each of the constituent gases absorbs some of the long-wavelength radiation emitted by the Earth's surface as it passes upward through the atmosphere. This absorption has to do with how the electromagnetic waves interact, or do not interact, with the thermal vibrations of the atoms (or molecules) of the various atmospheric gases.

Each gas has a characteristic picture of energy absorption, whereby there is strong absorption in one or more narrow wavelength bands. The combined effect of such absorption acts to absorb a large proportion of the outgoing long-wave radiation. Absorbed energy is re-radiated both back to the surface and out to space.

The budget of radiation is balanced when for some period of time the magnitude of the outgoing long-wave radiation that escapes to space is just balanced with the incoming solar radiation. The key point in the greenhouse effect is that because of the absorption and re-radiation of the long-wave radiation from the Earth's surface, the balance is struck at a much higher temperature of the Earth's surface and atmosphere than would be the case if there were no absorption of the outgoing radiation. The effect is thus like that of a greenhouse, although the higher temperature is maintained by a different physical process.

Greenhouse Gases

The gases of the atmosphere that play a significant role in interception of outgoing long-wave radiation from the Earth's surface are called greenhouse gases. The atmosphere includes several important greenhouse gases. It is not universally appreciated by the general public that the most important of these is water vapor, which is ubiquitous in the atmosphere in variable concentrations that range from a small fraction of 1%, in the coldest and driest air, to as much as 4% to 5% where the atmosphere is warmest and most humid. Carbon dioxide, which is so much in the news these days, is also a very important greenhouse gas. As its concentration gradually increases, it absorbs a greater and greater proportion of the outgoing long-wave radiation.

The budget of carbon dioxide in the Earth's near-surface zone (the atmosphere and the solid and liquid materials at or just below the Earth's surface) is intricate and not yet entirely well understood. As with other materials, the budget of carbon dioxide has to be studied from the standpoint of reservoirs (places where carbon dioxide is temporarily stored) and pathways (routes along which carbon dioxide is moved from place to place, and the processes involved in this transport). Climatologists attempt to identify the important reservoirs and pathways. They also try to estimate the mass stored in the various reservoirs and the fluxes (rates transport of mass per unit time) associated with the pathways.

The Earth's Carbon Budget

Let's review some of the important elements of the Earth's carbon budget.

Enormous quantities of carbon are stored in the Earth's crust in two forms: as organic matter in coal and petroleum hydrocarbons, and as carbonate minerals in sediments and sedimentary rocks. Some of the carbon in coal and petroleum hydrocarbons is released slowly by natural weathering, but far more is released by burning of fossil fuels. Weathering of carbonate rocks releases carbon dioxide into the Earth's surface waters and atmosphere, and this is in part balanced by precipitation of carbonate minerals to form new sediments in the warm and shallow waters of the oceans.

Plant growth fixes carbon dioxide in plant tissues temporarily, and decay and burning of vegetation releases carbon dioxide into the atmosphere. This increased release of carbon dioxide is one of the reasons that massive clearing of tropical forests is a serious concern.

The oceans constitute another large storehouse for carbon dioxide, because the carbon dioxide of the atmosphere dissolves in water. The increase in atmospheric carbon dioxide concentrations due to burning of fossil fuels is lessened by additional solution in the oceans; it is estimated that about one-third of the carbon dioxide added to the atmosphere is ultimately absorbed by the oceans.

Chapter 2

There have been proposals to sequester atmospheric carbon dioxide on a large scale by dissolving it in the oceans; the downside is that in the more distant future, much of that carbon dioxide will again be released into the atmosphere. Another method for decreasing the rate of addition of carbon dioxide into the atmosphere is to put it into deep storage, in liquefied form, in the rocks of the Earth's crust. Pilot operations of this kind are already underway.

More Information – on the Web
Visit the *EarthComm* web site www.agiweb.org/earthcomm/ to access a variety of links to web sites that will help you deepen your understanding of content and prepare you to teach this activity. Many of the sites also contain images that you can download.

Goals and Assessment

Clarify that the goals indicate what students should understand and be able to do as a result of the activity. Make sure students understand that Chapter Assessments are based upon these goals.

Goal	Location in Activity	Assessment Opportunity
Compare data to understand the relationship between carbon dioxide and global temperature.	**Investigate** Parts A and B **Digging Deeper**	Data is correctly graphed; axes are properly labeled. Glacial and interglacial periods are correctly identified. Questions are answered correctly.
Evaluate given data to draw a conclusion.	**Investigate** Parts A, B, and C **Understanding and Applying What You Have Learned** Question 4	Data is correctly interpreted.
Recognize a pattern of information plotted on a graph in order to predict future temperature.	**Understanding and Applying What You Have Learned** Question 4	Extrapolations are reasonable and based on the trends of existing data in the graph.
Understand some of the causes of global warming.	**Investigate** Part C **Digging Deeper; Understanding and Applying What You Have Learned** Questions 1, 5 – 6	Experimental designs are appropriate; data is relevant and useful.

Activity 5

How Do Carbon Dioxide Concentrations in the Atmosphere Affect Global Climate?

Goals

In this activity you will:

- Compare data to understand the relationship of carbon dioxide to global temperature.

- Evaluate given data to draw a conclusion.

- Recognize a pattern of information graphed in order to predict future temperature.

- Understand some of the causes of global warming.

Think about It

"What really has happened to winter?" You may have heard this type of comment.

- What causes "global warming?"

What do you think? Write down your ideas to this question in your *EarthComm* notebook. Be prepared to discuss your responses with your small group and the class.

E 125

EarthComm

Activity Overview

Students start the investigation by graphing data of atmospheric CO_2 concentration and global average temperature. They look for relationships between the data. Students then examine data on methane, CO_2, and temperature from an ice core in Antarctica and identify glacial and interglacial periods on the graph. Finally, they design an experiment to model the greenhouse effect. **Digging Deeper** reviews what greenhouse gases are, how the carbon cycle functions, and how greenhouse gases are linked to changes in climate. The reading also examines how changes in CO_2 concentrations are linked to temperature changes.

Preparation and Materials

Preparation

Part A
No special preparation is required for this part of the investigation.

Part B
You will need to make copies of the graph of methane, CO_2, and temperature using **Blackline Master Climate Change 5.1, Graph of Data from Ice Core in Antarctica**. Otherwise, no special preparation is required for this investigation.

Part C
This part of the investigation can be approached in different ways. If the list of materials from which the students must design their experiments is provided, then this part of the investigation requires little advance preparation beyond the gathering of those materials. Alternatively, if students are given the freedom to design their own experiments from a wider range of materials, then this part of the investigation is probably best done in two sittings, with the design and approval phase being done at least the day before the implementation of the experiments. This will allow students the time to gather the materials needed for their experiments.

Materials

Part A
  • Graph paper (**Blackline Master Astronomy 2.2** can be used to photocopy graph paper)

Part B
  • Copy of graph showing ice core data (use **Blackline Master Climate Change 5.1, Graph of Data from Ice Core in Antarctica.**)

Part C
Suggested materials for greenhouse-effect experiment:
  • two identical 2-L plastic bottles or beakers or small aquarium tanks or dish tubs with glass "lids"
  • water

- clear plastic bags
- thermometers
- ice cubes
- lamps
- graph paper
- See the *EarthComm* web site for further suggestions on modeling the greenhouse effect.

Think about It

Student Conceptions

Students are likely to be familiar with the concept of global warming from the news media. They are likely to know that global warming has been linked, by some, to emissions from factories, cars, trucks, etc.

Answers for the Teacher Only

From study of various proxies for global temperature, it is known that the Earth's average temperature has varied by many degrees Celsius in the past, on time scales of centuries, millennia, and longer. Climatologists are not entirely certain about the causes of such temperature changes.

Warming and cooling on time scales of tens of thousands of years has occurred in connection with the advance and retreat of continental ice sheets. It is generally agreed that these changes are ultimately driven by differences in solar radiation received by the Earth, and that these differences are caused by cyclic variations in the Earth's orbital parameters (eccentricity of the orbit; obliquity of the rotation axis; precession of the rotation axis and the Earth's orbit).

It is also generally agreed that the average global temperature has been increasing in the past few decades. There are differences of opinion about how much of that increase is due to natural effects and how much is due to humankind's activities. There is good reason to believe that the increase in CO_2 concentrations in the atmosphere contribute to global warming, although the magnitude of that effect is not yet certain. Because we are adding so much CO_2 to the atmosphere by burning fossil fuels, humankind is engaging in an experiment unprecedented in Earth history! It is also important to note, however, that a strong correlation between global temperature and atmospheric CO_2 concentration extends back far beyond humankind's ability to influence atmospheric CO_2, and the cause-and-effect relationship of this correlation has not been well established This concept is addressed to some extent in the **Digging Deeper** reading section of this activity.

Blackline Master Climate Change 5.1
Graph of Data from Ice Core in Antarctica

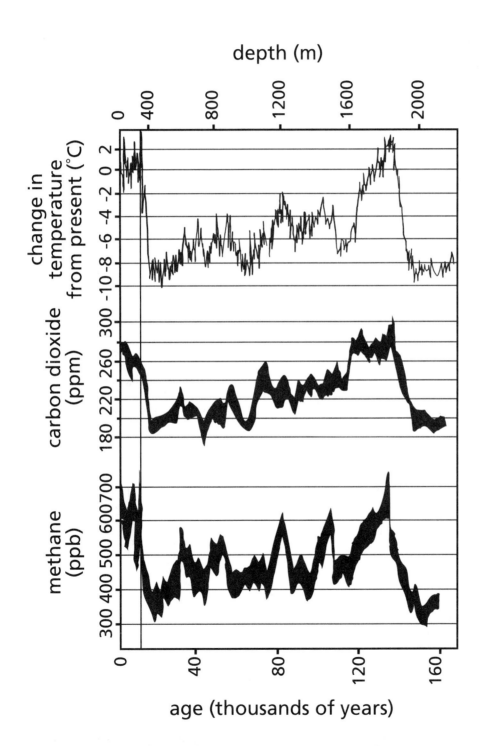

Earth System Evolution Climate Change

Investigate

Part A: Atmospheric Carbon Dioxide Concentrations over the Last Century

Data on 10-year Average Global Temperature and Atmospheric Carbon Dioxide Concentration		
time interval	average global temperature (°F)	atomospheric carbon dioxide (ppm)
1901–1910	56.69	297.9
1911–1920	56.81	301.6
1921–1930	57.03	305.19
1931–1940	57.25	309.42
1941–1950	57.24	310.08
1951–1960	57.20	313.5
1961–1970	57.14	320.51
1971–1980	57.26	331.22
1981–1990	57.71	345.87
1991–2000*	57.87	358.85

*carbon dioxide data only through 1998.

1. Graph the concentration of carbon dioxide in the atmosphere from 1900 to 2000. Put the year on the x axis and the CO_2 levels (in parts per million) on the y axis.

2. On the same graph, plot the global average temperature for the same period. Put another y axis on the right-hand side of the graph and use it for global average temperature.

   a) Is there a relationship between carbon dioxide concentration and global average temperature? If so, describe it.

   b) What do you think is the reason for the relationship you see?

Part B: Atmospheric Carbon Dioxide Concentrations over the Last 160,000 Years

1. Look at the figure showing data from an ice core in Antarctica. The graph shows changes in concentrations of carbon dioxide and methane contained in trapped bubbles of atmosphere within the ice, and also temperature change over the same

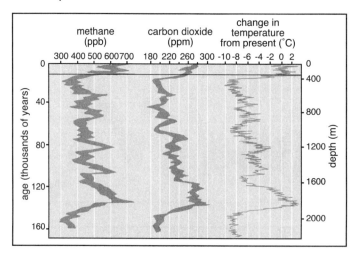

Investigate

Part A: Atmospheric Carbon Dioxide Concentrations over the Last Century

1.

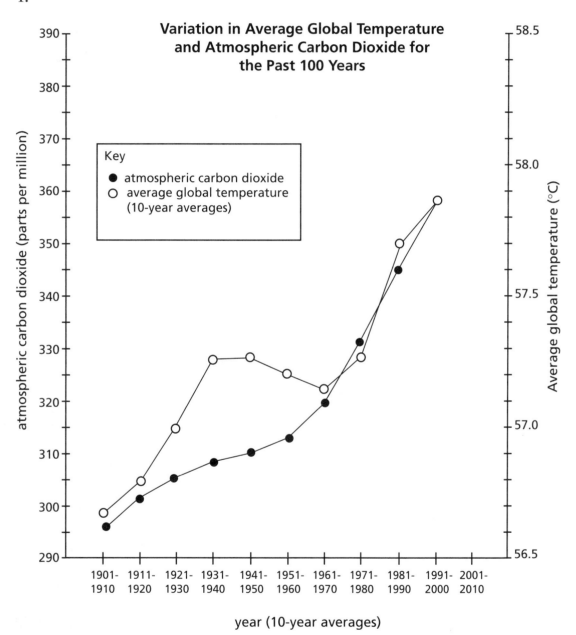

Variation in Average Global Temperature and Atmospheric Carbon Dioxide for the Past 100 Years

Key

● atmospheric carbon dioxide
○ average global temperature (10-year averages)

atmospheric carbon dioxide (parts per million)

Average global temperature (°C)

year (10-year averages)

2. **a)** As CO_2 concentrations have increased over time, so have global temperatures.

   b) CO_2 is a greenhouse gas. On shorter time scales, increased concentrations of CO_2 in the atmosphere may intensify the greenhouse effect and cause increases in global temperature.

Assessment Tools

EarthComm Notebook Entry-Checklist

Use this checklist as a quick guide for student self-assessment and/or an opportunity to quickly score student work. Add further criteria specific to your classroom needs or to this particular investigation.

Investigate Notebook Entry-Evaluation Sheet

Point out the criteria listed on this evaluation sheet that are relevant to this particular investigation. Encourage students to internalize the criteria by making them part of your "assessment conversations" as you circulate around the classroom.

NOTES

Chapter 2

time interval. Data was obtained from the study of an ice core from the Antarctic Ice Sheet. The core was approximately 2200 m long. It was analyzed for methane concentrations (in parts per billion—left graph), carbon dioxide concentrations (in parts per million—middle graph), and inferred change in temperature from the present (in °C—right graph), over the last 160,000 years.

2. Obtain a copy of the graph. Use a straightedge to draw horizontal lines across the three maximum temperatures and the three minimum temperatures.

a) Describe the correlation between these temperature events and changes in levels of carbon dioxide.

b) Label likely glacial intervals (low temperatures) and interglacial intervals (higher temperatures).

c) When did the most recent glacial interval end, according to these graphs?

Part C: The Greenhouse Effect

The phrase "greenhouse effect" is used to describe a situation in which the temperature of an environment (it could be any environment like a room, a car, a jar or the Earth) increases because incoming solar energy gets trapped because heat energy cannot easily escape. The incoming energy easily enters into the environment, but then, once it has been absorbed and is being re-radiated, it is harder for the energy to escape back out of the environment.

1. Work as a group to design an experiment to demonstrate the greenhouse warming in the atmosphere. The experiment should be simple in design, include a control element, and be performed in a short period of time (for example, a class period). The experiment will be presented to the community as a way to show the greenhouse effect.

a) Record your design in your *EarthComm* notebook. Remember to include a hypothesis. Be sure to also include any safety concerns.

2. Decide on the materials you will use. The materials should be inexpensive and easy to get. The following is a possible list:

• two identical 2-L plastic bottles with labels removed and tops cut off or two identical beakers

• water

• a clear plastic bag

• a thermometer

• ice cubes

• a sunny windowsill or two similar lamps

a) Record your list in your *EarthComm* notebook.

3. Decide on the measurements that you will make.

a) Prepare a data table to record your observations.

4. With the approval of your teacher, conduct your experiment.

 Have the design of your experiment checked carefully by your teacher for any safety concerns.

Part B: Atmospheric Carbon Dioxide Concentrations over the Last 160,000 Years

1. Assure that the students understand what the graph is showing before they attempt to proceed with **Part B.**

2.

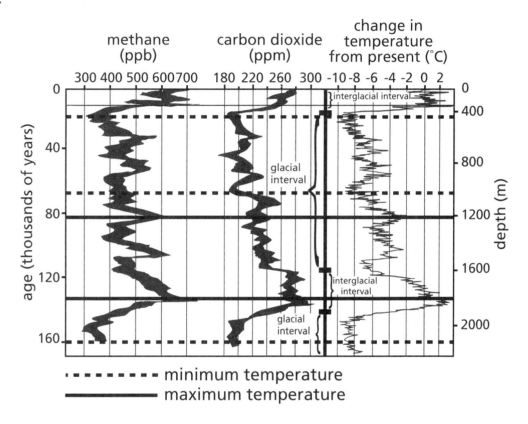

a) Although they are not exactly identical, the CO_2 concentration, methane concentration, and global temperature curves are all very closely correlated to one another (methane perhaps to a lesser degree than CO_2 and temperature). For example, in places where the temperature increases rapidly, peaks in the carbon dioxide and methane concentrations are also observed.

b) See graph above.

c) The most recent glacial period ended about 10 ka (10,000 years ago).

Part C: The Greenhouse Effect

1. a) Student designs will vary. One simple and effective design would be to create an actual small-scale greenhouse, using a small aquarium tank or even just a dish tub with a tight-fitting pane of glass as a lid and a thermometer inside. The control could be an identical container but with the glass supported above

the rim of the container, to allow air to pass. The effect is similar to, but not actually the same as, the atmospheric greenhouse effect: the pane of glass is a solid obstacle to the escape of heated air, rather than an absorber of outgoing long-wave radiation, as in the atmosphere. This same concept could be explored using two 2-liter bottles with their tops cut off. One would be sealed in a plastic bag (the greenhouse model), and the control would be placed in an unsealed plastic bag. The difficulty of developing a model that incorporates the actual atmospheric greenhouse effect is obvious. Other examples of possible experimental designs can be found on the *EarthComm* web site.

2. a) Materials needed by students will vary depending upon experimental design.

3. a) Data tables will vary depending upon experimental design.

4. Make sure that you have approved student experiments before they begin.

NOTES

Earth System Evolution Climate Change

5. Use the results of your experiment to answer the following questions:

a) How did this experiment demonstrate (or fail to demonstrate) the greenhouse effect?

b) How can this experiment serve as an analogy for atmospheric greenhouse effects?

c) Was there any difference observed between the greenhouse experiment and the control?

d) If there was a difference (or differences) describe it (them) in both qualitative and quantitative terms.

e) How did the data in each case change through time during the experiment?

f) Did the experiment reach a point of equilibrium where continuing changes were no longer observed? (Note: To answer this question, it may take longer than the class period, or, alternatively, you could hypothesize an answer to this question based on the trends of the data that you were able to gather.)

Reflecting on the Activity and the Challenge

In this activity you designed an experiment to demonstrate the greenhouse effect. You also examined the concentration of atmospheric carbon dioxide to see if it is correlated with changes in global average temperature. You discovered that an increase in carbon dioxide seems to be correlated with an increase in global average temperature. You will need this information to begin writing your article on "What is Global Warming?"

Geo Words

correlation: a mutual relationship or connection.

Digging Deeper

CARBON DIOXIDE AND GLOBAL CLIMATE

Correlation Studies

The relationship between carbon dioxide and global climate was mentioned in previous activities. When there is more carbon dioxide in the atmosphere, global temperatures are higher. When there is less carbon dioxide in the atmosphere, temperatures are lower. A scientist would say that there is a **correlation** between carbon dioxide concentration and global temperature. You might think, "Oh, that's because carbon dioxide concentration affects global temperature." And you might be right—but you might be wrong.

E 128

5. a) – f) Answers to these questions will vary depending upon the kind
              of experiment students have designed.

Reflecting on the Activity and the Challenge

Through the **Investigate** section of **Activity 5**, students should have gained a good
understanding of how carbon dioxide concentrations in the atmosphere are believed
to be linked to changes in global climate. Review with students how this information
can be used to help them to complete the **Chapter Challenge**.

Digging Deeper

Assign the reading for homework, along with the questions in **Check Your
Understanding** if desired.

Chapter 2

Assessment Opportunity

Use a brief quiz (or a class discussion) to assess your students' understanding of
the main ideas in the reading and the activity. A few sample questions are provided
below:

Question: Does the observed correlation between average global temperature and
atmospheric CO_2 levels prove that carbon dioxide levels in the atmosphere affect
global temperature? Explain why or why not.

Answer: The correlation itself does not prove a causal relationship between CO_2
and global temperature. It only suggests that there is some connection between the
two. It is possible that atmospheric CO_2 levels control global temperature, but it is
also possible that temperature controls CO_2 levels, or that both temperature and
CO_2 are related to a third variable.

Question: How can CO_2 be removed from the "active pool" of the carbon cycle
and be stored for long periods of time?

Answer: Many organisms (marine plankton, for example) use CO_2 that is in the
atmosphere and dissolved in the Earth's hydrosphere to produce organic matter.
This organic matter acts as a storehouse for CO_2. If it gets buried rapidly in
sediments without being oxidized first, this stored CO_2 can be sequestered for
geologically long times. This effectively removes it from the "active" CO_2 pool
for long periods of time.

Question: Why does incoming solar radiation pass more easily through the
atmosphere than radiation from the Earth, which gets absorbed (in part) by
greenhouse gases?

Answer: The wavelength (and associated energy) of incoming radiation (mostly
visible light and other shorter wavelength radiation) is very different from the heat
energy radiated by the Earth (mostly infrared radiation). Compared to the incoming
solar radiation, the longer-wavelength (lower-energy) infrared radiation radiated by
the Earth is absorbed to a greater extent by clouds and greenhouse gases.

Activity 5 How Do Carbon Dioxide Concentrations in the Atmosphere Affect Global Climate?

It is important to keep in mind always that a correlation does not, by itself, prove cause and effect. There are three possibilities: (1) carbon dioxide affects temperature; (2) temperature affects carbon dioxide; and (3) both are affected by a third factor, and are independent of one another! Any one of these three possibilities is consistent with the observations. It is the scientists' job to try to figure out which is the right answer. There are good reasons to think that the first possibility is the right one. That is because carbon dioxide is a "greenhouse gas."

What Are Greenhouse Gases?

The reason that the Earth is warm enough to support life is that the atmosphere contains gases that let sunlight pass through. Some of these gases absorb some of the energy that is radiated back to space from the Earth's surface. These gases are called **greenhouse gases**, because the effect is in some ways like that of a greenhouse. Without greenhouse gases, the Earth would be a frozen wasteland. Global temperatures would be much lower. Water vapor is the most important contributor to the greenhouse effect. Other greenhouse gases include carbon dioxide, methane, and nitrogen oxides.

How do greenhouse gases work? Most solar radiation passes through the clear atmosphere without being absorbed and is absorbed by the Earth's surface (unless it's reflected back to space by clouds first). There is a law in physics that states that all objects radiate electromagnetic radiation. The wavelength of the radiation depends on the objects' surface temperature. The hotter the temperature, the shorter the wavelength. The extremely hot surface of the Sun radiates much of its energy as visible light and other shorter-wavelength radiation. The much cooler surface of the Earth radiates energy too, but at much longer wavelengths. Heat energy is in the infrared range (*infra-* means "below," and the color red is associated with the longest wavelength in the color spectrum.) See *Figure 1*.

> **Geo Words**
>
> **greenhouse gases:** gases responsible for the greenhouse effect. These gases include: water vapor (H_2O), carbon dioxide (CO_2), methane (CH_4), nitrous oxide (N_2O), chlorofluorocarbons (CF_xCl_x), and tropospheric ozone (O_3).

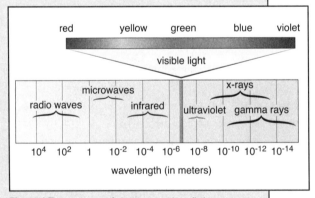

Figure 1 The spectrum of electromagnetic radiation.

Blackline Master Climate Change 5.2
The Spectrum of Electromagnetic Radiation

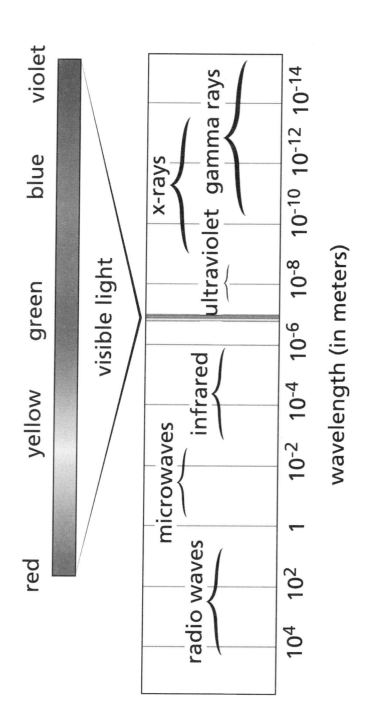

Chapter 2

Earth System Evolution Climate Change

Greenhouse gases are those that absorb some of the outgoing infrared radiation. None of them absorb all of it, but in combination they absorb much of it. They then re-radiate some of the absorbed energy back to the Earth, as shown in *Figure 2*. That is what keeps the Earth warmer than if there were no greenhouse gases.

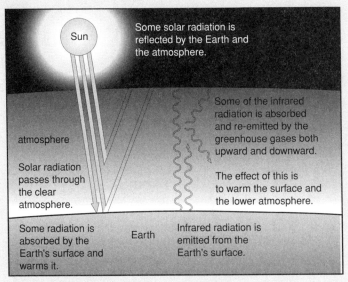

Figure 2 Schematic diagram illustrating how the greenhouse effect works.

The Carbon Cycle

Carbon dioxide is put into the atmosphere in two main ways: during volcanic eruptions, and by oxidation of organic matter. Oxidation of organic matter happens naturally in the biosphere. It occurs when plant and animal tissue decays. The organic matter is converted back to carbon dioxide and water. It also happens when animals breathe (and when plants respire too!). When you breathe, you take in oxygen, which you use to oxidize organic matter—your food. Then you breathe out carbon dioxide. Organic matter is also oxidized (more rapidly!) when it is burned. Carbon dioxide is released into the atmosphere whenever people burn wood or fossil fuels like gasoline, natural gas, or coal.

Plants consume carbon dioxide during photosynthesis. It is also consumed during the weathering of some rocks. Both land plants and algae in the ocean

Teaching Tip

Use the **Blackline Master Climate Change 5.2, The Spectrum of Electromagnetic Radiation** to make an overhead of the illustration shown in *Figure 1* on page E129. If you have already completed the Astronomy chapter, students should be familiar with the electromagnetic spectrum. Take this opportunity to review the spectrum.

Teaching Tip

Use **Blackline Master Climate Change 5.3, The Greenhouse Effect** to make an overhead of the drawing shown in *Figure 2*. Use this overhead to initiate a class discussion about the greenhouse effect. Ask students to consider how the experiments that they designed in **Part C** of the investigation relate to what is shown in the illustration. Would students change any aspects of their experiments to make them better model the real thing? If so, how?

Chapter 2

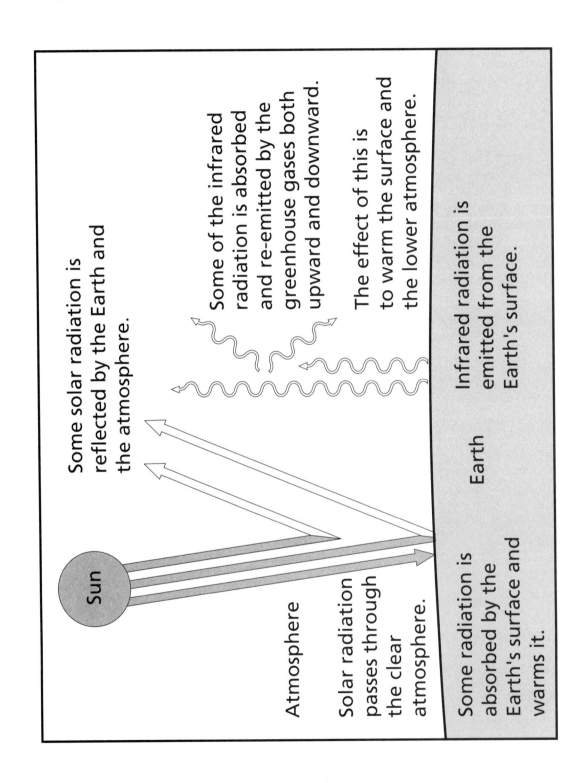

Some solar radiation is reflected by the Earth and the atmosphere.

Some of the infrared radiation is absorbed and re-emitted by the greenhouse gases both upward and downward.

The effect of this is to warm the surface and the lower atmosphere.

Infrared radiation is emitted from the Earth's surface.

Sun

Earth

Atmosphere

Solar radiation passes through the clear atmosphere.

Some radiation is absorbed by the Earth's surface and warms it.

Blackline Master Climate Change 5.4
The Global Carbon Cycle

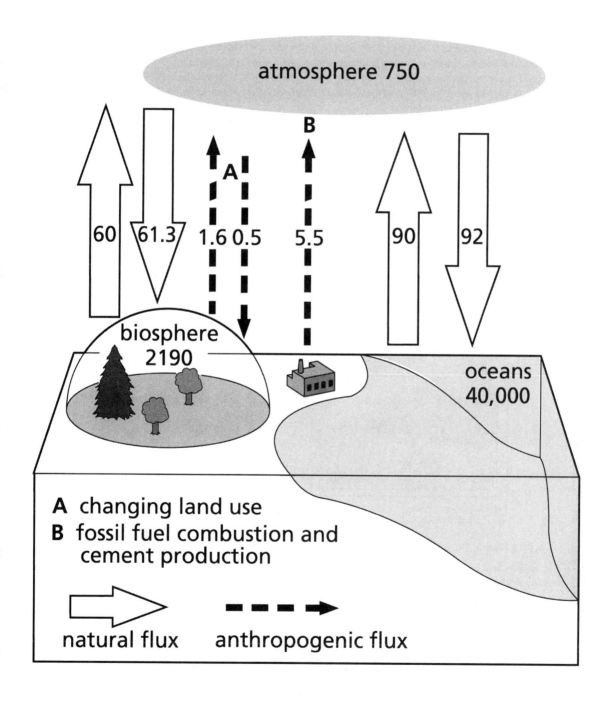

atmosphere 750

B

A

60 61.3 1.6 0.5 5.5 90 92

biosphere
2190

oceans
40,000

A changing land use
B fossil fuel combustion and
   cement production

natural flux   anthropogenic flux

use the carbon dioxide to make organic matter, which acts as a storehouse for carbon dioxide. Carbon dioxide is constantly on the move from place to place. It is constantly being transformed from one form to another. The only way that it is removed from the "active pool" of carbon dioxide at or near the Earth's surface is to be buried deeply with sediments. Even then, it's likely to reenter the Earth–surface system later in geologic time. This may be a result of the uplift of continents and weathering of certain carbon-rich rocks. This transfer of carbon from one reservoir to another is illustrated in the carbon cycle shown in *Figure 3*.

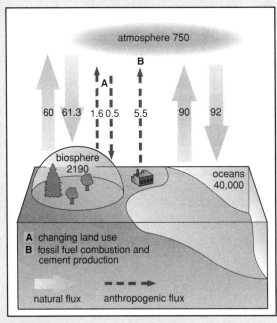

Figure 3 Global carbon cycle. Fluxes are given in billion metric tons per year and reservoirs in billion metric tons.

Carbon Dioxide and Climate

It appears that the more carbon dioxide there is in the environment, the warmer global temperatures are. Scientists have determined this from geologic data like the kind you worked with in the investigation. To what extent is this because carbon dioxide in the atmosphere acts as a greenhouse gas?

Teaching Tip

Figure 3 is a simplified version of the global carbon cycle. It could be made much more complicated by adding other reservoirs and pathways, but it gives a good accounting of the overall carbon budget. Use the **Blackline Master Climate Change 5.4, The Global Carbon Cycle** to make an overhead of the illustration. This can be used in a discussion about the carbon cycle and the ways that it has been simplified in this diagram.

Teaching Tip

Figure 4 on page E132 shows the actual data on CO_2 concentration and temperature from the longest and best ice core yet recovered. The temperature shown is the local average annual temperature from the area where the ice core is located. It is assumed, in a general way, that local temperature is representative of global average temperature, although such an assumption carries many uncertainties.

The correlation between CO_2 concentration and temperature is extremely good. Remind your students, however, that a correlation never proves cause and effect: did the changes in CO_2 concentration drive the temperature, or did the changes in temperature drive the changes in CO_2 concentration, or were they both the consequence of changes in a third factor? The data in *Figure 4* don't tell us that! Climatologists still don't know the answer.

Chapter 2

Earth System Evolution Climate Change

It is valuable to look at this question on two different time scales. On a scale of hundreds of thousands of years, carbon dioxide and global temperature track each other very closely. This correlation occurs through several glacial–interglacial cycles (*Figure 4*). It is not easy to develop a model in which carbon dioxide is the cause and global temperature is the effect. It's much more likely that variations are due to Milankovitch cycles. They may well explain the variation in both global temperature and carbon dioxide. On a scale of centuries, however, the picture is different. It seems very likely that the increase in carbon dioxide has been the cause of at least part of the recent global warming.

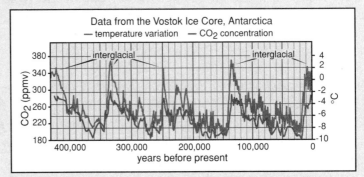

Figure 4 Variations in temperature and carbon dioxide (CO_2) concentration in parts per million by volume (ppmv) over the past 420,000 years interpreted from Antarctic ice cores. Temperature change is relative to the modern surface temperature at Vostok (−55.5°C).

Human emissions of greenhouse gases contribute significantly to the total amount of greenhouse gases in the atmosphere. For a long time humans have been adding a lot of carbon dioxide to the atmosphere by the burning of fossil fuels. This has especially increased in the past couple of centuries. Before the Industrial Revolution, carbon dioxide concentrations in the atmosphere were approximately 300 ppm (parts per million). As of 1995, carbon dioxide concentrations were almost 360 ppm. Scientists are concerned that the temperature of the Earth may be increasing because of this increasing concentration of carbon dioxide in the atmosphere.

Many nations have a commitment to reduce the total amount of greenhouse gases produced. It is their effort to reduce the risk of rapid global temperature increase. The trouble is that the size of the effect is still uncertain. Some people take the position that the increase in carbon dioxide should be reversed. They believe this is necessary even though the size of the contribution to global warming is not certain. It is their belief that the

E 132

Blackline Master Climate Change 5.5

Variations in Temperature and Carbon Dioxide Concentrations

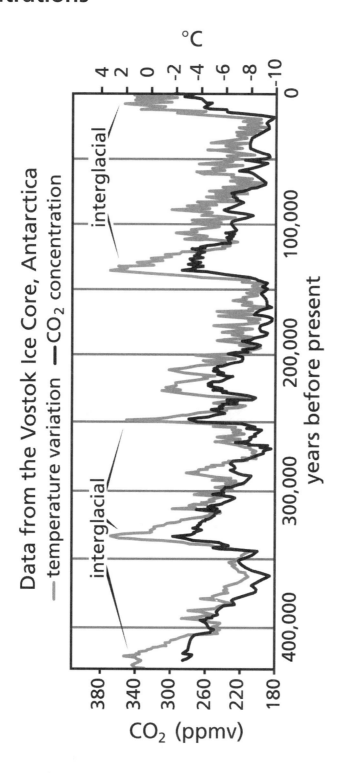

Activity 5 How Do Carbon Dioxide Concentrations in the Atmosphere Affect Global Climate?

consequences would be very difficult to handle. Other people take a different position. They consider that it would be unwise to disrupt the world's present economy. They consider the future danger to be questionable. The big problem is that no one is certain that rapid global warming will take place. If it does, it may be too late to do anything about it!

Not all of the carbon dioxide released by burning of fossil fuels stays in the atmosphere. Carbon dioxide is also dissolved in ocean water. As carbon is put into the atmosphere, some of it is absorbed by the oceans. That lessens the impact of burning of fossil fuels on climate. Some people have even suggested that enormous quantities of carbon dioxide should be pumped into the oceans. That would tend, however, to just postpone the problem until later generations. Carbon dioxide is also stored by **reforestation**. Reforestation is the growth of forests on previously cleared farmland. Did you know that there is a lot more forested land in the eastern United States now than at the time of the Civil War? The Civil War took place almost 150 years ago. By some estimates, the United States is a sink, rather than a source, for carbon dioxide. Extensive reforestation is occurring east of the Mississippi, despite the continuing expansion of suburbs and shopping malls!

Geo Words

reforestation: the replanting of trees on land where existing forest was previously cut for other uses, such as agriculture or pasture.

Figure 5 Clear-cut forest area in Olympic National Forest, Washington.

Check Your Understanding

1. List four greenhouse gases. Which gas contributes most to the greenhouse effect?

2. Explain how greenhouse gases make it possible for humans to live on Earth.

3. What are two ways in which carbon dioxide is put into the Earth's atmosphere?

E 133

EarthComm

Teaching Tip

Most of the carbon stored in the woody tissues of plants is returned slowly to the atmosphere by decay of the plant material, but some is buried and removed from the Earth-surface system. Deforestation, as pictured in *Figure 5*, removes carbon from "storage," although a large percentage of that carbon remains in wood products. The carbon in these products thus continues to be sequestered by being locked into the wood structures of buildings for long periods of time.

Check Your Understanding

1. Greenhouse gases include water vapor, carbon dioxide, methane, and nitrogen oxides. Of these, water vapor is the most important.

2. Greenhouse gases absorb long-wave radiation that is radiated back to space from the Earth's surface. Without the greenhouse gases, global temperatures would be too cold to support life as we know it.

3. Carbon dioxide is put into the atmosphere during volcanic eruptions and by the oxidation of organic matter (including the burning of fossil fuels).

Assessment Tool

Check Your Understanding Notebook Entry-Evaluation Sheet
Use this sheet to evaluate the extent to which students understand the key concepts explored in **Activity 5** and explained in **Digging Deeper**, and to evaluate the students' clarity of expression.

Chapter 2

Earth System Evolution Climate Change

Understanding and Applying What You Have Learned

1. Which of the following activities produce carbon dioxide? Which consume carbon dioxide? Explain how each can influence global climate.

   a) cutting down tropical rainforests
   b) driving a car
   c) growing shrubs and trees
   d) breathing
   e) weathering of rocks
   f) volcanic eruptions
   g) burning coal to generate electricity
   h) heating a house using an oil-burning furnace

2. Describe the carbon cycle in your community. List the ways that carbon dioxide is produced and used up and the organisms responsible for cycling.

3. What are some difficulties involved with predicting concentrations of atmospheric carbon dioxide into the future?

4. Examine the graph your group prepared. You have gathered data through the year 2000. You have seen that this data has changed over time. Using additional graph paper, try to continue this pattern for the next 10 years.

5. The United States has a population of about 280 million people (according to the 2000 census) and uses about 70 billion gigajoules of energy a year. India has a population of about 835 million people (1990) and uses about 7 billion gigajoules of energy a year.

   a) Divide the United States' total yearly energy use by its population to find out the yearly energy use per person.
   b) Calculate the yearly energy use per person for India.
   c) Give as many reasons as you can to explain the difference.
   d) Do you think you use more or less energy than the typical American? Explain.
   e) If you wanted to use less energy, what would you do?
   f) Why is how much energy you use important when considering how much carbon dioxide is in the air?

6. Determine one source of greenhouse gas emission in your community.

   a) What gas is being produced?
   b) How is it produced?
   c) Can you think of a way to determine the level of the gas that is being produced by your community?
   d) Propose a means for limiting emissions of this gas.

Understanding and Applying What You Have Learned

1. Activities that cause increased carbon dioxide production could result in increases in global temperature, whereas activities that cause increased CO_2 consumption could result in decreases in global temperature. The scale of the activities would determine how much of an effect they would have on global climate.

   a) Cutting down tropical rainforests would result in fewer plants using CO_2, and thus would cause an increase in CO_2 in the atmosphere.

   b) Driving a car involves the burning of fossil fuels and production of CO_2.

   c) Growing shrubs and trees would consume CO_2.

   d) Breathing produces CO_2.

   e) Weathering of rocks (silicate rocks) consumes CO_2, but weathering of certain carbon-rich rocks (limestone, for example) can produce CO_2.

   f) Volcanic eruptions produce CO_2.

   g) Burning coal to generate electricity produces CO_2.

   h) Heating a house using an oil-burning furnace produces CO_2.

2. Answers will vary. Encourage students to be specific with regard to local reservoirs and sinks of CO_2 in the community.

3. A large part of the difficulty is not so much scientific as sociopolitical: How much fossil fuel will human society consume, and on what time scale? How far will deforestation proceed? Addition of CO_2 to the atmosphere by volcanic activity is variable and difficult to predict. The rate of consumption of atmospheric CO_2 in chemical weathering of surface rocks is not well constrained. The time scale of absorption of CO_2 by the waters of the oceans is fairly well known but is still subject to uncertainties.

4. Students should use the graph they prepared in **Part A** of the investigation. They should find that the data do not all fall on a line but that portions of each data set do approximate a linear trend. Students will likely either fit a single line through all of the data to extrapolate the trend, or extrapolate the most recent portion of the data that do form a good linear trend. There is no correct answer to this exercise, but students should in some way justify the reasoning for their extrapolation. An example is given on the following page.

Chapter 2

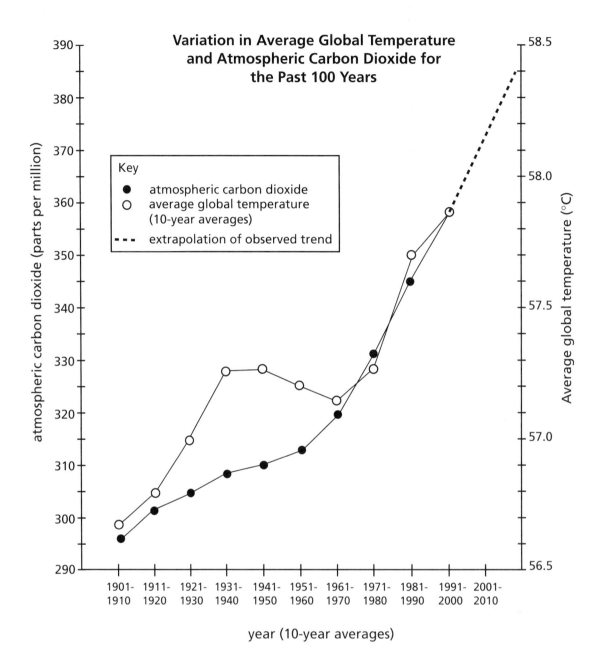

Variation in Average Global Temperature and Atmospheric Carbon Dioxide for the Past 100 Years

Key
● atmospheric carbon dioxide
○ average global temperature (10-year averages)
- - - extrapolation of observed trend

x-axis: year (10-year averages)

left y-axis: atmospheric carbon dioxide (parts per million)

right y-axis: Average global temperature (°C)

5. a) The yearly energy use per person in the United States is 250 gigajoules/person (70 billion/280 million).

   b) The yearly energy use per person in India is about 8.4 gigajoules/person (7 billion/835 million).

   c) Answers will vary. Students will most likely be aware that the United States is more highly industrialized than India.

   d) Answers will vary.

e) Answers will vary. There are several ways to practice energy conservation, including driving fuel-efficient vehicles, using fuel-efficient appliances, turning off electrical and electronic devices when they are not in use, etc.

f) Most of the energy used in the United States comes from the burning of fossil fuels, which adds CO_2 to the atmosphere.

6. Answers will vary by community. Sources of greenhouse gas emissions include power plants, use of automobiles, etc.

Chapter 2

Preparing for the Chapter Challenge

1. Using a newspaper style of writing, write several paragraphs in which you:

   • explain how humans have increased the concentration of carbon dioxide in the atmosphere;
   • explain why scientists think that increased carbon dioxide levels might lead to global climate change.

2. Clip and read several newspaper articles containing quotations.

3. Interview a member of your community about global warming. Is this person concerned about global warming? What does he or she think people should do about it? Look over your notes from your interview. Pick out several quotations from the community member that might work well in a newspaper article.

Inquiring Further

1. **Intergovernmental Panel on Climate Change (IPCC)**

   The Intergovernmental Panel on Climate Change (IPCC) is a group of more than 100 scientists and economists from many countries that is investigating the possibility of global warming and proposing ways that the nations of the world should respond. Do some research on the IPCC and what they have reported.

2. **Earth Summit**

   Investigate the 1997 United Nations Earth Summit in New York. What did the world's nations agree to at the Summit? Have the nations stuck to their promises?

EarthComm

Preparing for the Chapter Challenge

Students should be able to find newspaper articles that include information about the causes of global warming. They can also interview community members about their views on global warming. They can incorporate quotes from the newspaper articles and from their interview into their own newspaper article. You may want to review with students how to cite a reference correctly.

For assessment of this section (see **Assessment Rubric for Chapter Challenge on Climate Change** on pages 296 to 298), students would need to include a discussion on:
- causes of climate change, including carbon dioxide levels
- greenhouse gases
- how humans have increased the levels of CO_2 in the atmosphere
- why scientists think increased CO_2 might lead to global warming
- possible effects of global warming, with a community focus

Be sure to review these criteria with students so that they are comfortable with what they are being asked to do.

Inquiring Further

1. Intergovernmental Panel on Climate Change (IPCC)

The Intergovernmental Panel on Climate Change (IPCC) was established in 1988 to assess the scientific, technical, and socioeconomic information related to the risk of human-induced climate change. It does not carry out new research, nor does it monitor climate-related data. Thus far, the findings of the IPCC indicate that the Earth's climate has changed, and some of the changes are attributable to human activities. The group also reported that carbon dioxide, surface temperatures, precipitation, and sea level are all projected to increase globally during the 21st century because of human activities. The panel noted that projected changes in climate will have both beneficial and adverse effects, but predicted that the larger the changes are, the more adverse the effects will be.

2. Earth Summit

The first Earth Summit took place in 1992, in Brazil. At this time, the Commission of Sustainable Development (CSD) was formed and agreed to meet in five years to discuss progress. The CSD reconvened in 1997 in New York City. Issues discussed included poverty, consumption and production patterns, population, forests, fresh water, oceans, climate, energy, hazardous wastes, and more. Students can learn more about the specifics of the meeting by visiting the *EarthComm* web site.

ACTIVITY 6 — HOW MIGHT GLOBAL WARMING AFFECT YOUR COMMUNITY?

Background Information

Global Warming

The issue of global warming is much in the news these days. It is generally agreed that global yearly surface temperatures, averaged around the globe, have increased slightly over the past several decades. The magnitude of this warming is not entirely certain, however, because of difficulties in intercalibrating measurements made with older instruments and measurements made with newer instruments and techniques. Adding to the uncertainty are the various assumptions and corrections needed to take into account the effects of urbanization (the so-called urban-heat-bubble effect, which is also sometimes referred to as the urban-heat-island effect) on temperature records from the great many weather stations located in urban areas. There is also much indirect evidence that points to global warming, especially the general retreat of glaciers in most parts of the world and the thinning of the pack ice in the Arctic Ocean.

Much of the controversy about global warming hinges upon how much of the observed warming is caused by natural effects and how much by human-induced (so-called "anthropogenic") effects. On the one hand, it is known from the climate record of the past few millennia that global temperatures can change rapidly, by magnitudes comparable to the increase in recent decades, without human intervention. On the other hand, it is generally (although not universally) agreed that the great increase in greenhouse gases—especially carbon dioxide—is likely to cause an increase in global temperature.

Climatologists are working to develop ever more sophisticated computer models of the Earth's climate in order to simulate, and therefore predict, future global temperature and rainfall. It is not yet possible to take all of the important physical effects into account. However, the computer models are in the process of continual refinement. There will never be certainty in prediction of future climate. In the view of many scientists, the concern about global warming comes down to a choice society has to make. We must balance the desirability of trying to forestall the deleterious effects of global climate change that seem likely on the basis of what climatologists know now, against the potentially serious economic disruptions that would arise from such action.

Over the past several thousand years of Earth's history, there has been a very strong positive correlation between global temperature and atmospheric carbon dioxide concentrations: when temperature has been high, carbon dioxide concentrations have been correspondingly high. Many nonscientists assume that this is an indication that the increase in anthropogenic carbon dioxide concentrations will lead to global warming. Be on your guard about such thinking, however. It's a truism in science that correlation does not prove cause and effect.

The correlation between global temperature and carbon dioxide concentration could be interpreted in three ways:
• higher carbon dioxide causes higher temperatures

- higher temperatures cause higher carbon dioxide
- both higher temperatures and higher carbon dioxide are caused by some third factor

Almost all climatologists agree that climate change in recent geologic time has been caused by the astronomical effects described in the sections discussing Milankovitch cycles and variations in the Earth's orbital parameters. If that is so, then the warming must somehow have caused the increase in carbon dioxide! The observed correlation then becomes irrelevant to the extremely important problem of the extent to which the present anthropogenic increase in carbon dioxide will lead to global warming.

Humankind is engaged in an unprecedented experiment on global climate. Unfortunately, the past gives us little basis for predicting the outcome of that experiment.

More Information – on the Web
Visit the *EarthComm* web site www.agiweb.org/earthcomm/ to access a variety of links to web sites that will help you deepen your understanding of content and prepare you to teach this activity. Many of the sites also contain images that you can download.

Chapter 2

Goals and Assessment

Clarify that the goals indicate what students should understand and be able to do as a result of the activity. Make sure students understand that Chapter Assessments are based upon these goals.

Goal	Location in Activity	Assessment Opportunity
Brainstorm the ways that global warming might influence the Earth's spheres.	Investigate	List is comprehensive; items on list are explained.
Make a list of the ways that global warming might affect your community.	Investigate; **Understanding and Applying What You Have Learned** Question 2	List is reasonable and comprehensive.
Design an experiment on paper to test your ideas.	Investigate; **Understanding and Applying What You Have Learned** Question 3	Experimental design is useful and is applicable to problem.
Explain some of the effects of global warming that computer models of global climate have predicted.	**Digging Deeper; Check Your Understanding** Question 2	Question is answered correctly; response closely matches Teacher's Edition.
Understand positive and negative feedback loops and their relationship to climate change.	**Digging Deeper; Check Your Understanding** Question 1	Question is answered correctly; response closely matches Teacher's Edition.
Evaluate and understand the limitations of models in studying climate change through time.	**Digging Deeper; Check Your Understanding** Question 3	Question is answered correctly; response closely matches Teacher's Edition.

NOTES

_____

_____

_____

_____

_____

_____

_____

_____

_____

_____

_____

_____

_____

_____

_____

_____

_____

_____

_____

_____

_____

_____

_____

_____

Chapter 2

Earth System Evolution Climate Change

Activity 6

How Might Global Warming Affect Your Community?

Goals

In this activity you will:

• Brainstorm the ways that global warming might influence the Earth.

• List ways that global warming might affect your community.

• Design an experiment on paper to test your ideas.

• Explain some of the effects of global warming that computer models of global climate have predicted.

• Understand positive and negative feedback loops and their relationship to climate change.

• Evaluate and understand the limitations of models in studying climate change through time.

Think about It

Some scientists think that the average global temperature may increase by several degrees Fahrenheit by the end of the 21st century.

• How do you think global warming could affect your community?

What do you think? List several ideas about this question in your *EarthComm* notebook. Be prepared to discuss your ideas with your small group and the class.

Activity Overview

Students work with their groups to brainstorm all the possible causes of global warming. From the list they have generated, they decide which causes are reasonable. They prepare a poster to display their list, emphasizing how each item affects the different spheres of the Earth system and how each item would affect their community. Students then choose one item from their list and design an experiment on paper to test their idea. **Digging Deeper** introduces some of the problems associated with trying to predict changes in climate, and reviews the findings of computer models thus far.

Preparation and Materials Needed

Preparation

No advance preparation is required to complete this activity.

Materials

• Poster board

Think about It

Student Conceptions

Encourage students to think beyond the obvious response that "it will get warmer." Have them look at the image of the Earth Systems on page Exii in their text and consider how an increase in temperature would affect each of the spheres. This will help prepare them for the investigation.

Answer for the Teacher Only

There is a diversity of possible answers, depending in part upon the location of the community. The most important factors to be considered include:
  • sea-level rise affecting coastal areas
  • increased or decreased average rainfall
  • longer growing seasons
  • more severe weather events
  • local economic dislocations arising from effects of global warming in other parts of the world.

> ### Assessment Tool
>
> **Think about It Evaluation Sheet**
> Use this evaluation sheet to help students understand and internalize the basic expectations for the warm-up activity.

Chapter 2

Investigate

1. In small groups, brainstorm as many effects of higher global temperatures (a few degrees Fahrenheit) as you can. Each time you come up with a possible result of global warming, ask yourselves what effect that result might have. For example, if you think glaciers will recede, ask yourself what the implications of that would be. At this point, do not edit yourself or criticize the contributions of others. Try to generate as many ideas as possible. Here are a few ideas to get your discussion going.

   How might higher temperatures affect the following processes?

   • evaporation

   • precipitation

   • glacial activity

   • ocean circulation

   • plant life

   • animal life.

   a) List all the ideas generated.

2. As a group, review your list and cross off those that everyone in the group agrees are probably incorrect or too far-fetched. The ideas that remain are those that the group agrees are possible (not necessarily proven or even likely, but possible). It's okay if some of the ideas are contradictory. Example: More cloud cover might block out more solar radiation (a cooling effect) vs. More cloud cover might increase the greenhouse effect (a warming effect).

   a) Make a poster listing the ideas that remain. Organize your ideas on the poster using the following headings:

   • geosphere

   • hydrosphere

   • atmosphere

   • cryosphere

   • biosphere.

   b) On a separate piece of paper, write down how each of the possible results might affect your community.

3. Imagine that your group is a group of scientists who are going to write a proposal asking for grant money to do an experiment. Pick one of the ideas on your poster that you would like to investigate.

   a) On paper, design an experiment or project to test the idea. Choose ONE of the following:

   • Design an experiment that you could do in a laboratory that would model the process. Draw a diagram illustrating the model. Tell what materials you would need and how the model would work. Describe what the results would mean. Tell which parts of the experiment would be difficult to design or run, and explain why.

 If you plan to perform your experiments do so only under careful supervision by a knowledgeable adult.

Investigate

1. **a)** Answers will vary. Encourage students to think beyond the examples given in the text.

2. **a)** Posters will vary.

    b) Answers will vary.

3. **a)** Student experiments will vary. Be sure to approve student designs before they begin to conduct their experiments.

Assessment Tools

EarthComm Notebook Entry-Checklist
Use this checklist as a quick guide for student self-assessment and/or an opportunity to quickly score student work. Add further criteria specific to your classroom needs or to this particular investigation.

Investigate Notebook Entry-Evaluation Sheet
Point out the criteria listed on this evaluation sheet that are relevant to this particular investigation. Encourage students to internalize the criteria by making them part of your "assessment conversations" as you circulate around the classroom.

Teaching Tip

You may want to make sure that each student group is working on a different effect of global warming, so that a larger breadth of topics is covered. Note that students are asked to write about the effects of global warming on rainfall patterns, extreme weather events, sea level, ocean currents, and agriculture in the **Preparing for the Chapter Challenge** section. If each group researches one of these topics now, it will reduce the amount of outside research students will need to complete later.

Earth System Evolution Climate Change

• Design a project in which you would gather data from the real world. Include a diagram or sketches illustrating how you would gather data. Tell what kind of data you would gather, how you would get it, how frequently and how long you would collect it, and how you would analyze it. Tell which parts of the project would be difficult to design or carry out, and explain why.

4. Present your poster and your proposal for an experiment or project to the rest of the class.

Reflecting on the Activity and the Challenge

In this activity, you brainstormed ways in which an increase in global temperatures might affect the geosphere, the hydrosphere, the atmosphere, the cryosphere, the biosphere, and your community. Then you designed an experiment or a project for how you might test one of your ideas. This process modeled the way in which scientists begin to think about how to investigate an idea. This will help you to explain which possible effects of global warming would have the greatest impact on your community, and also why it is difficult for scientists to accurately predict climate change.

Geo Words

urban heat-island effect: the observed condition that urban areas tend to be warmer than surrounding rural areas.

Digging Deeper

EFFECTS OF GLOBAL WARMING IN YOUR COMMUNITY

Problems with Making Predictions

Many scientists believe that the world's climate is becoming warmer as a result of the greenhouse gases (carbon dioxide, methane, and nitrogen-oxide compounds) that humans are adding to the atmosphere. Because the world's climate naturally experiences warmer years and colder years, it is hard to say for sure whether global average temperature has been increasing. Nowadays, remote sensing of the land and ocean surface by satellites makes it easy to obtain a good estimate of global temperature. The problem is that such techniques didn't exist in the past. Therefore, climatologists have to rely on conventional weather records from weather stations. That involves several problems. Thermometers change. The locations of the weather stations themselves often have to be changed. The move is usually away from city centers. As urban areas have been developed, they become warmer because of the addition of pavement and the removal of cooling vegetation. That is called the **urban heat-island effect**. Climatologists try to make

Teaching Tip

Encourage students to take notes on their classmates' presentations. Students will find that they can use information given in the presentations when they complete the **Preparing for the Chapter Challenge** section on the possible effects of global warming.

Reflecting on the Activity and the Challenge

This activity required students to pull together all of the knowledge about climate change that they had obtained thus far. This is a good point to review the chapter and make sure that students are prepared to write their final newspaper articles. Ask them if they have any additional questions about climate change.

Digging Deeper

Assign the reading for homework, along with the questions in **Check Your Understanding** if desired.

Assessment Opportunity

Reword or restructure the questions in **Check Your Understanding** for a brief quiz. Use the quiz (or a class discussion of the questions in the textbook) to assess your students' understanding of the main ideas in the reading and the activity.

Teaching Tip

Use **Blackline Master Climate Change 6.1, Relative Global Temperature** of *Figure 1* on page E139 to make an overhead transparency. Incorporate the overhead into a discussion of how global climate has changed over time, and how these changes may or may not have affected people. Do students think that the current increase in the rate of climate change is unusual, given the longer-term trends? If so, how and why?

corrections for these effects. The consensus is that global average temperature really is increasing. The questions then become: how much of the warming is caused by humankind, and how much is natural? It is known that there have been large variations in global temperature on scales of decades, centuries, and millennia long before humankind was releasing large quantities of carbon dioxide into the atmosphere (*Figure 1*).

Geo Words

feedback loops: the processes in which the output of a system causes positive or negative changes to some measured component of the system.

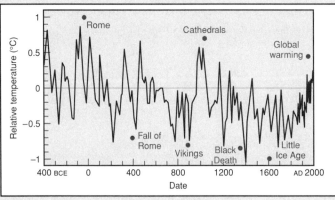

Figure 1 Relative global temperature from 400 BCE to the present.

Scientists who study global warming use very complicated computer models to try to predict what might happen. These computer models do not simply look at carbon dioxide and global temperature. They also do calculations based on many other factors that might be involved in climate change—everything from how much moisture is held in the Earth's soils to the rate at which plants transpire (give off water vapor). The physics of clouds is an especially important but also especially uncertain factor. The workings of the Earth's atmosphere are still much more complex than any computer model. Climatologists are hard at work trying to improve their models.

Drawbacks to the Computer Models

As you have seen, many factors influence the climate on Earth: carbon dioxide and other greenhouse gases, Milankovitch cycles, ocean currents, the positions of continents, weathering of rocks, and volcanic activity. Many of these factors interact with each other in ways that scientists do not fully understand. This makes it hard to make accurate predictions about how the atmosphere will respond to any particular change. The ways that different factors interact in global climate change are called **feedback loops**.

E 139

Blackline Master Climate Change 6.1
Relative Global Temperature

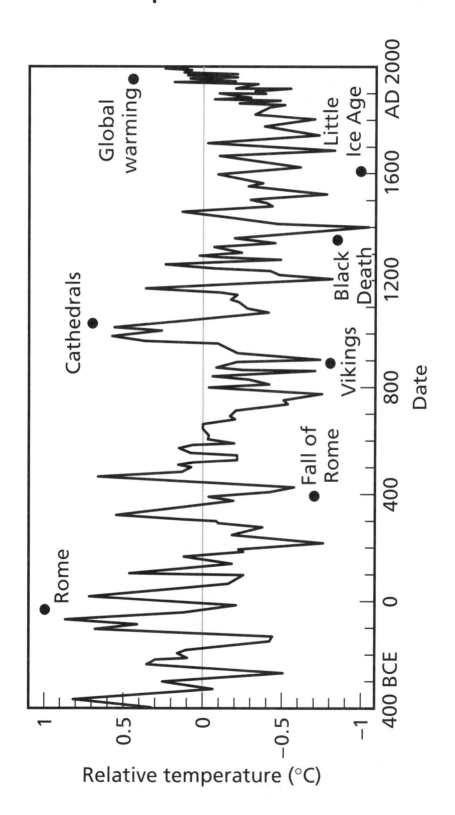

Chapter 2

Earth System Evolution Climate Change

Feedback may be positive or negative. Positive feedback occurs when two factors operate together and their effects add up. For example, as the climate cools, ice sheets grow larger. Ice reflects a greater proportion of the Sun's radiation, thereby causing the Earth to absorb less heat. This results in the Earth becoming cooler, which leads to more ice forming. Ice cover and global cooling have a positive feedback relationship.

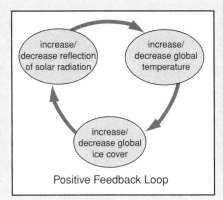

Positive Feedback Loop

Figure 2 An example of a positive feedback loop.

In a negative feedback relationship, two variables operate in opposition with each other. Each tends to counteract the effects of the other. Weathering and carbon dioxide are one such negative feedback pair. Weathering uses up carbon dioxide, which causes the temperature to drop. When the temperature drops, weathering rates slow down, using up less carbon dioxide, and slowing the rate of temperature decrease. In this sense, weathering acts as a negative feedback for global cooling. The Earth's global climate involves many feedback loops like these. Understanding of such feedback loops is extremely important in understanding how the Earth's physical environment changes through time.

How many feedback loops are there? How do they work? These are questions that scientists are working on every day. The uncertainty about feedback loops is one thing that makes it hard to predict how the Earth's climate might respond to an increase in carbon dioxide. Another major unknown in global warming is clouds. With warmer temperatures, there will be more evaporation and therefore more clouds. Clouds reflect incoming solar radiation into space—a cooling effect. But clouds also act like a blanket to hold in heat—a warming effect. Which effect predominates? Scientists aren't yet sure.

Figure 3 How can clouds influence climate?

Teaching Tip

Figure 2 shows an example of a positive feedback loop. Use **Blackline Master Climate Change 6.2, Positive Feedback Loop** to make an overhead of this illustration. Using the overhead as a guide, discuss with students whether or not any of the factors they examined in the investigation were positive or negative feedback loops. Can they think of any additional examples?

Teaching Tip

The role of clouds (see *Figure 3* on page E140) in global warming is perhaps the most troublesome element in the development of computer models of global climate change. Climatologists generally agree that our understanding of cloud physics and its role in climate change is still inadequate. Ask students to hypothesize on what role clouds could play in climate change.

Teaching Tip

It might seem paradoxical to your students that global warming might give rise to bigger snowstorms (see *Figure 4* on page E141). In many areas of the United States, winter snowfall is limited to a greater extent by availability of atmospheric moisture than by temperature. As global temperatures increase, in many areas there will likely be greater delivery of moisture-laden air from lower-latitude regions to feed winter snowstorms.

Chapter 2

What Do the Computer Models Say?

Scientists continue working on their computer models of global climate. They learn as they go and make improvements all the time. Because some of the models have been used for years, scientists can test some of the predictions of past years against weather data collected recently. This helps them make changes to the computer models to make them work better.

Computer climate models have come up with some possible scenarios that may result from the increased concentration of greenhouse gases in the atmosphere. Remember, however, that these scenarios are theoretical outcomes, not certainties.

Changes in Precipitation

Warmer temperatures lead to more clouds. More clouds lead to more rain. Some models predict more rain with global warming. Others predict a change in rainfall patterns—more precipitation in the winter and less in the summer, for example. Some areas of the world would receive more rain, others less. In that respect, some countries would be winners, and others, losers. With the increase in evaporation brought on by warmer temperatures, an increase in extreme events (stronger hurricanes and winter snowstorms) might be likely.

Figure 4 Global warming may cause an increase in the number of extreme winter snowstorms.

Blackline Master Climate Change 6.2
Positive Feedback Loop

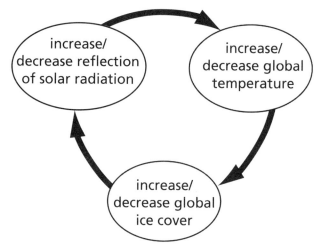

Positive Feedback Loop

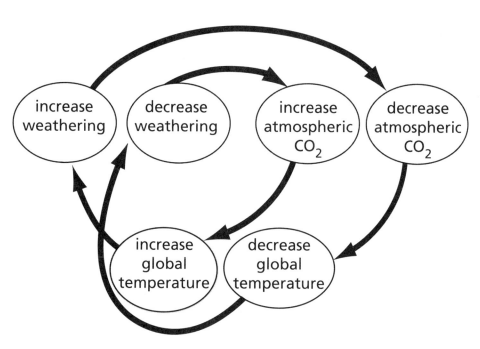

Negative Feedback Loop

Earth System Evolution Climate Change

Changes in Sea Level

Glaciers around the world have been shrinking in recent years. If the Earth's climate continues to warm, more and more glacier ice sheets will melt. Meltwater is returned to the ocean. This would result in a worldwide rise in sea level. Some models predict a sea-level rise of as much as a meter by 2100.

Changes in Agriculture

In the Northern Hemisphere, where most of the world's cropland is located, warmer temperatures would cause a northward shift of the regions where certain crops are grown. Agriculture would also be affected by changes in rainfall patterns. Some regions might become too dry to support present crops. Other places might become too wet to support present crops. Many areas might continue growing traditional crops but experience declines in productivity. In other words, farmers might still grow corn in Iowa but produce fewer bushels per year. An example of a change that might decrease crop productivity is an increase in nighttime temperatures. Corn and some other grain crops do best when the temperature drops below 70°F at night. Another change that could reduce productivity (or increase costs) is a switch to wetter winters and drier summers.

Changes in Ocean Circulation

The addition of fresh meltwater from glaciers into the North Atlantic could disturb the production of North Atlantic Deep Water. The same thing happened 12,000 years ago (see **Activity 4**). The circulation of North Atlantic Deep Water helps distribute heat from solar radiation evenly around the globe. If this flow is disturbed, there might be far-reaching effects on global climate.

Check Your Understanding

1. Explain how ice cover and global cooling work as a positive feedback loop.

2. How might global warming lead to increased precipitation?

3. Why is it hard to predict how the global climate might react to an increase of carbon dioxide in the atmosphere?

Understanding and Applying What You Have Learned

1. Using the information in **Digging Deeper**, add to the poster you made in the **Investigate** section.

2. For each new item you added to your poster, hypothesize how your community would be affected by

that outcome.

3. Using what you learned in **Digging Deeper**, modify the experiment or model you proposed in the **Investigate** section OR design another experiment or model.

Check Your Understanding

1. As the climate cools, ice sheets grow larger, and the ice reflects a greater proportion of the Sun's radiation. Increased reflection of radiation causes the Earth to absorb less heat, which results in the Earth becoming cooler. A cooler climate leads in turn to more ice formation.

2. Warmer temperatures lead to greater evaporation and atmospheric moisture. This leads to more clouds and increased precipitation.

3. The world's climate naturally experiences warmer years and colder years through time, making it difficult to assess the effect increased CO_2 levels have on global temperatures. Additionally, there are many factors that could affect the fluxes of carbon between the difference reservoirs in the carbon cycle. Also, there are several factors beyond increased CO_2 that influence global climate.

Assessment Tool

Check Your Understanding Notebook Entry-Evaluation Sheet
Use this sheet to evaluate the extent to which students understand the key concepts explored in **Activity 6** and explained in the **Digging Deeper** reading section, and to evaluate the students' clarity of expression.

Understanding and Applying What You Have Learned

1. Additions will vary.

2. Hypotheses will vary according to changes made in **Understanding and Applying What You Have Learned, Step 1**.

3. Modifications will vary.

Preparing for the Chapter Challenge

1. Using a style of writing appropriate for a newspaper, write a paragraph on each of the following possible effects of global warming. Make sure you make it clear that these are only possible scenarios, not certainties:

   • changes in rainfall patterns
   • increase in extreme events
   • changes in sea level
   • changes in ocean currents
   • changes in agriculture.

2. Write an editorial about how you think your community should respond to global warming. Should your community wait for further research? Should your community take action? What kind of action should be taken? Would these actions benefit your community in other ways, in addition to slowing global warming?

Inquiring Further

1. **Community energy use**

   Make a plan for calculating how much energy your school uses for heating, air conditioning, lights, and other electrical uses. Make a plan for calculating the energy used in gasoline for students, teachers, staff, and administrators to travel to and from school each day. How could you test your estimates to see how accurate they are? What are some ways your school could reduce its energy use? How can a reduction in energy use influence climate?

2. **"CO$_2$-free" energy sources**

   Investigate some sources of energy that do not produce carbon dioxide, like solar and wind power.

3. **Climate change and crops**

   Call your state's cooperative extension service and find out what are the top three crops grown in your state. Visit the *EarthComm* web site to determine if your state's cooperative extension service has a web site. What are the optimal climatic conditions for maximizing productivity of these crops? How might climate changes due to global warming affect farmers who grow these crops in your state?

Preparing for the Chapter Challenge

Students may need to do some additional research to write their articles. They may also find it useful to rely on other student groups for information.

For assessment of this section (see **Assessment Rubric for Chapter Challenge on Climate Change** on pages 296 to 298), remind students that their article should contain information about:
- possible effects of global warming, with a community focus
- difficulties of predicting climate change
- whether or not your community should be concerned about global warming, and what they can do in response to global warming

Inquiring Further

1. Community energy use

Student plans for calculating energy use will vary, as will the results of the plan. Ways that a school can reduce energy consumption include:
- using more energy-efficient appliances
- turning off lights when rooms are empty
- turning down the heat slightly in the cold season

Students can visit the *EarthComm* web site to learn more about energy conservation. At this point, they should recognize that the burning of fossil fuels to produce energy can release CO_2 into the environment, which can contribute to global warming. Therefore, a reduction in energy use can help slow the rate of global warming.

2. "CO_2-free" energy sources

Solar power involves the use of the Sun's energy for heat, light, hot water, electricity, and even cooling. Photovoltaic (PV) cells convert sunlight directly into electricity. Solar collectors are used to heat water and buildings. Buildings can also be heated through passive solar heating, which involves designing a structure in such a way that it absorbs and retains heat from the Sun.

Wind energy takes advantage of the energy in the wind for such uses as generating electricity, charging batteries, pumping water, or grinding grain. Wind power involves the use of wind turbines with propeller-like blades. The wind turbines sit on top of towers and capture the wind's energy. When the wind blows, it causes the blades of the turbine to spin. The turbine then spins a generator to make electricity. Large, modern wind turbines operate together in wind farms to produce electricity for utilities. Small turbines are sometimes used by homeowners and farmers to help meet energy needs.

3. Climate change and crops

Answers will vary depending upon the state in which your community is located.

Earth Science at Work

ATMOSPHERE: *Plant Manager*
Some companies are providing their workers with retraining to use new equipment that has been designed to reduce the emission of greenhouse gases and control global warming.

BIOSPHERE: *Farmer*
Agriculture is an area of the economy that is very vulnerable to climate change. Climate change that disturbs agriculture can affect all countries in the world. However, there are also steps that farmers can take to reduce the amount of carbon dioxide and other greenhouse gases.

CRYOSPHERE: *Mountaineering Guide*
Regions where water is found in solid form are among the most sensitive to temperature change. Ice and snow exist relatively close to their melting point and frequently change phase from solid to liquid and back again. This can cause snow in mountainous areas to become unstable and dangerous.

GEOSPHERE: *Volcanologist*
Volcanoes can emit huge amounts of carbon dioxide gas as well as sulfur dioxide gas into the atmosphere with each eruption. Both of these substances can have adverse effects on the atmosphere, including global warming as a possible result.

HYDROSPHERE: *Shipping Lines*
Even under "normal" climate conditions, ocean circulation can vary. A changing climate could result in major changes in ocean currents. Changes in the patterns of ocean currents and storm patterns will have important consequences to shipping routes.

How is each person's work related to the Earth system, and to Climate Change?

E 144

NOTES

Chapter 2

Climate Change and Your Community: End-of-Chapter Assessment

Use the topographic map provided to answer questions 1-3.

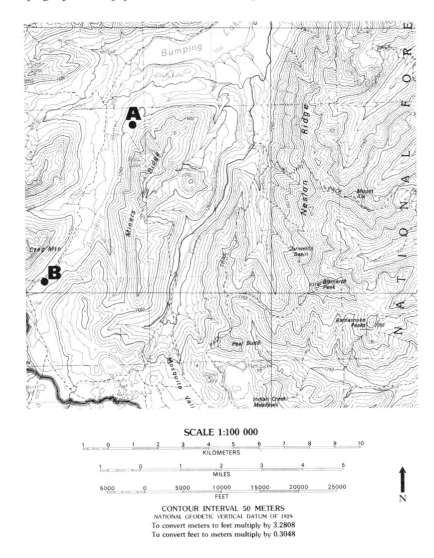

SCALE 1:100 000

KILOMETERS

MILES

FEET

CONTOUR INTERVAL 50 METERS
NATIONAL GEODETIC VERTICAL DATUM OF 1929
To convert meters to feet multiply by 3.2808
To convert feet to meters multiply by 0.3048

N

1. What is the elevation of point A on the map?
    a) 1200 meters
    b) 1200 feet
    c) 1300 meters
    d) 1300 feet

2. What physiographic feature (landform) is found at location B?
    a) a cliff
    b) a valley
    c) a hill
    d) a plain

3. On this map, in which direction is north?
   a) towards the top of the page
   b) towards the bottom of the page
   c) to the right
   d) to the left

4. A high coastal mountain range is on the western coast of a continent. Warm moist air blows from west to east over the mountains. From this information you would expect the western side of the mountain range to have_____ compared to the eastern side of the mountain range.
   a) more annual precipitation
   b) less annual precipitation
   c) hotter summers
   d) cooler winters
   e) the same climate

5. Two cities are located at approximately the same latitude. City A is located on an island in the ocean, and city B is located in the middle of a continent. Both cities are at the same elevation. All other aspects about their physical environments are identical. On the basis of these observations, which of the following is a valid conclusion?
   a) City A would tend to have warmer summers and cooler winters than city B.
   b) City A would tend to have cooler summers and warmer winters than city B.
   c) The two climates would be the same because the cities are at the same latitude.
   d) City A would have a shorter winter and longer summer compared to city B.

6. Which of the following factors or concepts distinguishes climate from weather?
   a) Accuracy of prediction
   b) Kinds of measurements
   c) Time scale of observations
   d) Size of an area

7. Which of the following would not be considered a climate proxy?
   a) historical climate data
   b) isotopic data from ice cores
   c) data on tree-ring thickness data
   d) foraminifera data from deep-sea sediments

8. Ice-core data provide climate records up to _____ of years before present.
   a) several thousands
   b) several tens of thousands
   c) several hundreds of thousands
   d) several millions

9. Which of the following statements is true?
   a) Global climate has been warming steadily for the past hundred thousand years.
   b) Global climate has been cooling steadily for the past hundred thousand years.
   c) Global climate has only started changing in the past few hundred years.
   d) Global climate has cooled and warmed many times over the past few hundred thousand years.

10. One way scientists use ice cores to understand climate change is by measuring the _____ different ice layers
    a) temperature of
    b) density of
    c) composition of
    d) thickness of

Use the following diagram of a lake bottom sediment core and description below to answer questions 11-13.

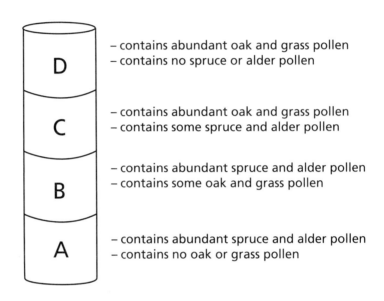

D
– contains abundant oak and grass pollen
– contains no spruce or alder pollen

C
– contains abundant oak and grass pollen
– contains some spruce and alder pollen

B
– contains abundant spruce and alder pollen
– contains some oak and grass pollen

A
– contains abundant spruce and alder pollen
– contains no oak or grass pollen

11. Which layer is composed of the oldest sediment?
    a) A
    b) B
    c) C
    d) D

12. Which layer represents the warmest climate?
    a) A
    b) B
    c) C
    d) D

13. Which statement is the best interpretation of the sediment core?
    a) As time passed, the climate near the lake warmed.
    b) As time passed, the climate near the lake cooled.
    c) As time passed, global climate warmed.
    d) As time passed, global climate cooled.

14. If the tilt of the Earth's rotational axis was less than it is today, which of the
following statements would be true (all other things being equal)?
a) Northern and Southern Hemisphere winters would be warmer.
b) Only the Northern Hemisphere winter would be warmer.
c) Northern Hemisphere winters would be colder.
d) Summers would be hotter in high latitude regions.

15. The Northern Hemisphere summer always occurs:
a) When the Earth is closest to the Sun.
b) When the Earth is farthest from the Sun.
c) When the north end of the Earth's rotation axis is pointed towards the Sun.
d) When the south end of the Earth's axis is pointed towards the Sun.

16. During the vernal (spring) and fall (autumnal) equinoxes, the Earth's axis of
rotation is:
a) Tilted towards the Sun.
b) Tilted away from the Sun.
c) Not tilted (oriented straight up-and-down).
d) Tilted neither toward nor away from the Sun.

17. The combined effects of the axial and orbital precession cycle affect:
a) The degree of tilt of the Earth's rotation axis.
b) The Earth-Sun distance during the different seasons.
c) The shape of the Earth's orbit about the Sun.
d) The shape of the Moon's orbit about the Earth.

18. All other things being equal, a planetary body that is 100,000,000 miles from
the Sun receives _____ solar energy that a planet 50,000,000 miles from
the Sun receives.
a) the same
b) twice the
c) one-half of the
d) one-quarter of the

19. The most appropriate timescale to observe climate changes caused by changes in
the positions of continents is
a) hundreds of years
b) thousands of years
c) millions of years
d) billions of years

20. Movements of the South American, African, and Australian continents since the
Eocene have resulted in the movement of the continents away from Antarctica.
The movements affected global climate mostly because:
a) The continents drifted to higher latitudes.
b) Antarctic ocean currents stopped flowing northward.
c) Isolation of Antarctica melted Antarctic glaciers.
d) Changes in continental positions affected Earth's axial tilt.

Chapter 2

21. From the relationships observed between atmospheric carbon dioxide (CO_2) concentration and global temperature, scientists can confidently conclude that:
    a) Atmospheric CO_2 concentration and global temperature are correlated to one anther.
    b) Atmospheric CO_2 concentration controls global temperature.
    c) Global temperature controls atmospheric CO_2 concentration.
    d) There is no correlation between atmospheric CO_2 concentration and global temperature.

22. Data indicate that atmospheric carbon dioxide concentration:
    a) Has never been as great as it is today.
    b) Has been constant through most of geologic time.
    c) Has fluctuated significantly during the course of geologic history.
    d) Only changed slightly until the Industrial Revolution.

23. Which of the following is NOT a significant greenhouse gas?
    a) water
    b) carbon dioxide
    c) hydrogen sulfide
    d) methane

24. Increased levels of greenhouse gases in the atmosphere result in:
    a) Greater absorption of ultraviolet radiation.
    b) Greater absorption of infrared radiation.
    c) Destruction of the thermosphere.
    d) Depletion of the ozone layer.

25. An increase in the birthrate (number of births per day) causes an increase in population. This increase in population results in an increase in birthrate.

    The statements above describe which of the following?
    a) a positive feedback loop
    b) a negative feedback loop
    c) logarithmic growth
    d) infinite progression

26. Which of the following is NOT a concern currently identified by models of the effects of global warming?
    a) changes in weather patterns
    b) depletion of oxygen in the atmosphere
    c) increases on sea level
    d) changes in ocean circulation patterns

Answer Key

1. c
2. b
3. a
4. a
5. b
6. c
7. a
8. c
9. d
10. c
11. a
12 d
13. a
14. a
15. c
16. d
17. b
18. d
19. c
20. b
21. a
22. c
23. c
24. b
25. a
26. b

Teacher Review

Use this section to reflect on and review the investigation. Keep in mind that your notes here are likely to be especially helpful when you teach this investigation again. Questions listed here are examples only.

Student Achievement

What evidence do you have that all students have met the science content objectives?

_____

Are there any students who need more help in reaching these objectives? If so, how can you provide this?_____

_____

What evidence do you have that all students have demonstrated their understanding of the inquiry processes?_____

_____

Which of these inquiry objectives do your students need to improve upon in future investigations? _____

_____

What evidence do the journal entries contain about what your students learned from this investigation? _____

_____

Planning

How well did this investigation fit into your class time?_____

_____

What changes can you make to improve your planning next time? _____

_____

Guiding and Facilitating Learning

How well did you focus and support inquiry while interacting with students?

_____

What changes can you make to improve classroom management for the next investigation or the next time you teach this investigation? _____

_____

How successful were you in encouraging all students to participate fully in science learning?_____

_____

How did you encourage and model the skills values, and attitudes of scientific inquiry? _____

_____

How did you nurture collaboration among students?_____

_____

Materials and Resources

What challenges did you encounter obtaining or using materials and/or resources needed for the activity? _____

_____

What changes can you make to better obtain and better manage materials and resources next time? _____

_____

Student Evaluation

Describe how you evaluated student progress. What worked well? What needs to be improved? _____

_____

How will you adapt your evaluation methods for next time?_____

_____

Describe how you guided students in self-assessment. _____

_____

Self Evaluation

How would you rate your teaching of this investigation? _____

_____

What advice would you give to a colleague who is planning to teach this investigation? _____

_____

Chapter 2

3

Changing Life
...and Your Community

EARTH SYSTEM EVOLUTION
CHAPTER 3

CHANGING LIFE...
AND YOUR COMMUNITY

Chapter Overview

In **Chapter 3**, students are challenged to produce a display that illustrates the biological changes experienced by their community over several scales of geologic time. Students begin the chapter by exploring the process of fossilization and determining which organisms in their community are most likely to be preserved in the fossil record. They look at how climate influences the kinds of organisms found in a specific place. They use this information to hypothesize about how organisms in their community might have differed 20,000 years ago, when glaciers covered large areas of North America. They explore extinction events and consider how they fit into the evolutionary scheme. Finally, students investigate how their community has changed through geologic time, and what effect this might have had on the kinds of organisms found in their community.

Chapter Goals for Students

- Understand how biological changes are related to all of the Earth systems.

- Participate in scientific inquiry and construct logical conclusions based on evidence.

- Appreciate the value of Earth science information in improving the quality of life, globally and in the community.

Chapter Timeline

Chapter 3 takes about three weeks to complete, assuming one 45-minute period per day, five days per week. Adjust this guide to suit your school's schedule and standards. Build flexibility into your schedule by manipulating homework and class activities to meet your students' needs.

A sample outline for presenting the chapter is shown on the following page. This plan assumes that you assign homework at least three nights a week, and that you assign **Understanding and Applying What You Have Learned** and **Preparing for the Chapter Challenge** as group work to be completed during class. This outline also assumes that **Inquiring Further** sections are reserved as additional, out-of-class activities.

This outline is only a sample, not a suggested or recommended method of working through the chapter. Adjust your daily and weekly plans to meet the needs of your students and your school.

Day	Activity	Homework
1	Getting Started; Scenario; Chapter Challenge; Assessment Criteria	
2	Activity 1 – Investigate, Parts A and B	Digging Deeper; Check Your Understanding
3	Activity 1 – Investigate Part A continued; Review; Understanding and Applying; Preparing for the Chapter Challenge	
4	Activity 2 – Investigate	Digging Deeper; Check Your Understanding
5	Activity 2 – Review; Understanding and Applying; Preparing for the Chapter Challenge	
6	Activity 3 – Investigate, Parts A and B	Digging Deeper; Check Your Understanding
7	Activity 3 – Review; Understanding and Applying; Preparing for the Chapter Challenge	
8	Activity 4 – Investigate, Part A	
9	Activity 4 – Investigate, Part B	Digging Deeper; Check Your Understanding
10	Activity 5 – Activity 4 – Review; Understanding and Applying; Preparing for the Chapter Challenge	

Chapter 3

Day	Activity	Homework
11	Activity 5 – Investigate	Digging Deeper; Check Your Understanding
12	Activity 5 – Investigate continued	
13	Activity 5 – Review; Understanding and Applying; Preparing for the Chapter Challenge	
14	Complete Chapter Report	Finalize Chapter Report
15	Present Chapter Report	

National Science Education Standards

Preparing a display to educate the community about biological changes in the area sets the stage for the **Chapter Challenge**. Students learn about the evidence for changing life and the factors that can cause biological changes over time. Through a series of activities, they begin to develop the content understandings outlined below.

CONTENT STANDARDS

Unifying Concepts and Processes
- Systems, order, and organization
- Evidence, models, and explanation
- Constancy, change, and measurement
- Evolution and equilibrium
- Form and function

Science as Inquiry
- Identify questions and concepts that guide scientific investigations
- Design and conduct scientific investigations
- Use technology and mathematics to improve investigations
- Formulate and revise scientific explanations and models using logic and evidence
- Communicate and defend a scientific argument
- Understand scientific inquiry

Earth and Space Science
- Energy in the Earth system
- Origin and evolution of the Earth system

Science and Technology
- Communicate the problem, process, and solution

Science in Personal and Social Perspectives
- Population growth

History and Nature of Science
- Science as a human endeavor
- Nature of scientific knowledge
- Historical Perspectives

Chapter 3

Key Science Concepts and Skills

Activities Summaries	Earth Science Principles
Activity 1: The Fossil Record and Your Community Students use plaster and a clamshell to understand what fossils are and how they form. They examine the different levels of the ecosystem in their community and identify which organisms are more or less likely to be preserved in the fossil record.	• Food chains and food webs • Fossils • Fossilization • Geologic time scale • Fossiliferous rocks
Activity 2: North American Biomes Students look at photographs of the major biomes of North America and write definitions of each one. They decide which biome their own community is found in, and they compare this biome to the other North American biomes. Students think about the factors that govern which plants and animals are commonly found in their community. They consider how the organisms in the community might change if the physical and chemical conditions were to change.	• Climate • Biomes
Activity 3: Your Community and the Last Glacial Maximum Students examine maps that show the distribution of pollen and spores over time for two different kinds of trees. On the basis of the changing distribution of the trees, students write a description of how climate has changed in the United States from 20,000 years ago to the present. Students then look at online data to determine how plant and animal distributions have changed in their community from the Pleistocene to the present.	• Paleoclimate • Glacial/interglacial periods • Response of biomes to climate change
Activity 4: The Mesozoic–Cenozoic Boundary Event Students use the Internet to collect data about the paleoclimate before and after the Mesozoic–Cenozoic boundary event. They also search for information about the fossil evidence from these time periods to understand how climate changes can cause extinctions. Students then compare photographs of skulls of organisms from before and after the boundary event.	• Extinction events • Causes of extinction
Activity 5: How Different Is Your Community Today from that of the Very Deep Past? Students examine a geologic map of their community to determine what geologic time periods the rocks in the area represent. They select one of these time periods, and they research the plants and animals that lived in the community at that time. Students then prepare a diorama for display.	• Biodiversity • Extinction and evolution

Equipment List for Chapter Three:

Materials needed for each group per activity.

Activity 1 Part A
- Paper towels
- Petroleum jelly
- Small deli container
- Plaster mix (and water)
- Clamshell (both halves)
- Confetti
- Small ballpeen hammer

Activity 1 Part B
No additional materials needed.

Activity 2
- Internet access (or scientific information about most common plants and animals in your biome)*

Activity 3 Part A
- Copies of the oak and spruce distribution maps for last 20,000 years (**Blackline Master 3.1 Oak and Spruce Distribution Maps**)

Activity 3 Part B
- Internet access (or information about pollen distribution from the Late Pleistocene to the present)*
- Internet access (or information about distribution of animals from the Late Pleistocene to the present)*

Activity 4 Part A
- Internet access (or information about paleoclimate before and after the Mesozoic–Cenozoic boundary)*
- Internet access (or information about organisms living before and after the Mesozoic–Cenozoic boundary)*

Activity 4 Part B
No additional materials needed

Activity 5
- Geologic map of your community*
- Resources like textbooks, encyclopedias, the Internet, and museum displays
- Materials to construct dioramas

Chapter 3

* See the *EarthComm* web site for information about how to obtain these resources at
http://www.agiweb.org/earthcomm

3

Changing Life ...and Your Community

Getting Started

When you travel across the continental United States by land, you can see obvious changes in the plants and animals. For example, you would expect to find cactus in the Southwest but not in Oregon, alligators in Florida but not in Maine, and sea otters in California but not in Michigan. Organisms live in particular areas because they are adapted to a range of climate conditions. If the range is exceeded, that organism can no longer live in that area.

- Do the plants and animals that live in your community also live in other areas of your state or region? If so, over what geographic area?

- What would happen to these plants and animals in your community, state, or region if climate or geological setting changed?

What do you think? In your *EarthComm* notebook write down the factors that affect what plants and animals live in your community. Consider both the climate and the geography. Be prepared to discuss your ideas with your working group and the class.

Scenario

Scientists have recognized that there are short-term and long-term changes in plants and animals during Earth's history. The United Nations is interested in changes in the number of different organisms and where they live. A special task force has been organized to make recommendations on what should be done over different time scales (decades, centuries, millennia) to maintain the ecosystem in your community. Your class has been asked to help the task force by explaining how and why plants

E146

Getting Started

Uncovering students' conceptions about Changing Life and the Earth System

Use **Getting Started** to elicit students' ideas about the main topic. The goal of **Getting Started** is not to seek closure (i.e., the "right" answer) but to provide you, the teacher, with information about the students' starting point and about the diversity of ideas and beliefs in the classroom. By the end of the chapter, your students will have developed a more detailed and accurate understanding of biological changes and the Earth system.

Ask students to work independently or in pairs and to exchange their ideas with others. Avoid labeling answers right or wrong. Accept all responses, and encourage clarity of expression and detail.

The first **Getting Started** question is intended to encourage students to think about biodiversity and the kinds of factors that influence it, and to provide you with important information about the students' knowledge of the plants and animals around them. This question should also lead students to recognize that some organisms are better adapted to thrive over a wide range of conditions, while others are more limited in the conditions under which they can thrive. Students' experiences in noticing the plants and animals around them are likely to be highly variable. Some students may not have really considered the local plants and animals at all, particularly if they have not traveled to others places or to climates different from their own. Many students, however, will recognize that some organisms that live in their community also live many places around the country (or even world), whereas others have a much more limited range.

The second question is intended to lead students to think about the fact that the kinds of plants and animals that inhabit a region, or even the Earth in general, evolve as time passes. Most students will probably realize that climate is an important factor controlling the animals and plants in a community, recognizing that in climates very different from their own (for example, tropical or polar climates) many, if not all, of the plants and animals are different from those in their community. In particular, if students have completed **Chapter 2** of this module they will recognize that the Earth (or at least North America) has many different climate zones, and that climate has changed over time. Students will probably recognize that urbanization plays an important role in floral and faunal assemblage. However, they probably have not considered in detail the role that geography and geologic history plays in establishing the flora and fauna of a region. For example, they may not have considered why or how places in different parts of the world that have similar climates can have very different plants and animals. If they have completed **Chapter 2**, they will recognize that the presence of certain geographic features affects the climate of a region. Students probably won't recognize, however, that the array of different plant and animals in a region relates in other ways to the region's geologic and geographic history.

and animals in your community have changed over time. This information will help the task force to make predictions about how life might change in the future. Can the *EarthComm* team meet this challenge?

Chapter Challenge

You have been asked to create a display that illustrates the biological changes your community has experienced over several scales of geologic time. You will need to address:

- Evidence that there has been change in the life forms (trees, shrubs, herbivores, carnivores, etc.) and biodiversity (numbers of different organisms) of your community over time.

- Short-term and long-term factors that have influenced biological changes in your community.

- The natural processes that have been responsible for the appearance and disappearance of life forms throughout geologic time.

- Suggestions as to what could be done to reduce biodiversity loss caused by natural changes in Earth systems.

Assessment Criteria

Think about what you have been asked to do. Scan ahead through the chapter activities to see how they might help you to meet the challenge. Work with your classmates and your teachers to define the criteria for assessing your work. Record all of this information. Make sure that you understand the criteria as well as you can before you begin. Your teacher may provide you with a sample rubric to help you get started.

E147

Scenario and Chapter Challenge

Read (or have a student read) the **Chapter Challenge** aloud to the class. Allow students to discuss what they have been asked to do. Have students meet in teams to begin brainstorming what they would like to include in their **Chapter Challenge** displays. Ask them to summarize briefly in their own words what they have been asked to do. Review the attributes of a high-quality presentation.

Alternatively, lead a class discussion about the challenge and the expectations. Review the titles of the activities in the Table of Contents. To remind students that the content of the activities corresponds to the content expected for the chapter project, ask them to explain how the title of each activity relates to the expectations for the **Chapter Challenge**. Familiarize students with the way activities are structured, pointing out the sections that are common to all activities. Note particularly the section titled **Preparing for the Chapter Challenge**.

Guiding questions for discussion include:
- What do the activities have to do with the expectations of the challenge?
- What have you been asked to do?
- What should a good display contain?

Assessment Criteria

A sample rubric for assessing the **Chapter Challenge** is shown on the following page. Copy and distribute the rubric as is, or use it as a baseline for developing scoring guidelines and expectations that suit your needs. For example:
- You may wish to ensure that core concepts and abilities derived from your local or state science frameworks also appear on the rubric.
- You may wish to modify the format of the rubric to make it more consistent with your evaluation system.

However you decide to evaluate the Chapter Report, keep in mind that all expectations should be communicated to students and that the expectations should be outlined at the start of their work. Please review **Assessment Criteria** (pages xxiv to xxv of this Teacher's Edition) for a more detailed explanation of the assessment system developed for the *EarthComm* program.

Chapter 3

Assessment Rubric for Chapter Challenge on Changing Life

Meets the standard of excellence. **5**	<u>*Significant*</u> information is presented about <u>*all*</u> of the following: • What evidence has been found to indicate change in the life forms and biodiversity of your community over time? • What short-term and long-term factors have influenced biological changes in your community? • What natural processes have been responsible for the appearance and disappearance of life forms through geologic time? • What could be done to reduce biodiversity loss caused by natural changes in Earth systems? <u>*All*</u> of the information is accurate and appropriate. The writing is clear and interesting.
Approaches the standard of excellence. **4**	<u>*Significant*</u> information is presented about <u>*most*</u> of the following: • What evidence has been found to indicate change in the life forms and biodiversity of your community over time? • Which short-term and long-term factors have influenced biological changes in your community? • Which natural processes have been responsible for the appearance and disappearance of life forms through geologic time? • What could be done to reduce biodiversity loss caused by natural changes in Earth systems? <u>*All*</u> of the information is accurate and appropriate. The writing is clear and interesting.
Meets an acceptable standard. **3**	<u>*Significant*</u> information is presented about <u>*most*</u> of the following: • What evidence has been found to indicate change in the life forms and biodiversity of your community over time? • Which short-term and long-term factors have influenced biological changes in your community? • Which natural processes have been responsible for the appearance and disappearance of life forms through geologic time? • What could be done to reduce biodiversity loss caused by natural changes in Earth systems? <u>*Most*</u> of the information is accurate and appropriate. The writing is clear and interesting.
Below acceptable standard and requires remedial help. **2**	<u>*Limited*</u> information is presented about the following: • What evidence has been found to indicate change in the life forms and biodiversity of your community over time? • Which short-term and long-term factors have influenced biological changes in your community? • Which natural processes have been responsible for the appearance and disappearance of life forms through geologic time? • What could be done to reduce biodiversity loss caused by natural changes in Earth systems? <u>*Most*</u> of the information is accurate and appropriate. Generally, the writing does not hold the reader's attention.
Basic level that requires remedial help or demonstrates a lack of effort. **1**	<u>*Limited*</u> information is presented about the following: • What evidence has been found to indicate change in the life forms and biodiversity of your community over time? • Which short-term and long-term factors have influenced biological changes in your community? • Which natural processes have been responsible for the appearance and disappearance of life forms through geologic time? • What could be done to reduce biodiversity loss caused by natural changes in Earth systems? <u>*Little*</u> of the information is accurate and appropriate. The writing is difficult to follow.

NOTES

Chapter 3

ACTIVITY 1 — THE FOSSIL RECORD AND YOUR COMMUNITY

Background Information

Fossils

A fossil is any evidence of past life. This broad definition is very inclusive, although it probably should not go so far as to cover fresh road kill! When your students hear the word "fossil," they probably think of shells or dinosaur bones. These are indeed good examples of what are called body fossils. Even more abundant than body fossils, however, are trace fossils—physical evidence of the life activities of now-vanished organisms. Tracks, trails, burrows, feeding marks, and resting marks are all trace fossils.

The fossil record is quite extensive: a few hundred thousand fossil species have been discovered and described. The average life span of species, from the time of origination to the time of extinction, is very poorly known. It probably lies somewhere in the range from half a million years to five million years.

It has been estimated that the total number of plant and animal species that exist today is four or five million. Given the enormous duration of geologic time, the total number of species that have ever existed must be far greater then the number of species now living. This means that the fossil record represents a minuscule proportion of the total number of potentially fossilizable species. The relatively small fossil record also underscores that body fossilization is an extremely rare event. Moreover, it is generally observed that a great many sedimentary rocks are loaded with trace fossils but contain few, if any, body fossils representing the organisms that generated the traces.

It is understandable that there are great differences in the potential for fossilization among the various groups of plant and animal species. It should seem almost like common sense to your students that something as durable as a clamshell has a much greater potential for fossilization than, say, a worm. In fact, the percentage of all species of robustly shelled organisms—like trilobites, mollusks, and brachiopods—that are represented in the fossil record is fairly high. The fossil record of worms, on the other hand, is very scanty, even though there is good reason to believe that worm species have been abundant throughout the latter part of geologic time.

Fossilization

For the body of an organism to become preserved as a fossil, it must escape destruction—at least in part—both before and after it is buried with sediments. Destruction before burial might result from chemical and/or biological decomposition, or from mechanical effects like abrasion and/or breakage during transport by wind or water currents, or a combination of both.

Most subaerial environments (that is, those exposed to the open air rather than being underwater) are fully oxygenated. Therefore, the soft tissues of dead organisms, whether plants or animals, are susceptible to decay. Microorganisms like bacteria are especially important in facilitating such decomposition.

Many—if not most—subaqueous (underwater) environments are also oxygenated, owing to the ability of water to dissolve the oxygen of the atmosphere. For organisms to escape decay, burial must be extremely rapid, or the depositional environment must be anoxic (that is, without the presence of oxygen). Some of the best-preserved soft-body fossils have been found in deposits that are interpreted to have formed in marine basins in which there is little or no vertical exchange of water. In this case, the bottom waters are stagnant and thus anoxic. At the same time, there is a rain of organic matter from the near-surface waters. Free-floating organisms that fall to the bottom in such a water body have an excellent chance of preservation. Probably the best modern example of such an environment is the Black Sea.

Hard skeletal materials, like bones and shells, have a far higher probability of preservation than soft tissues. For a bone or shell to be preserved, it must only survive breakage and abrasion before burial, and chemical dissolution of its constituent mineral material before and after deposition. Even if the object is dissolved after deposition, it is likely to be represented by a cavity. This cavity serves the paleontologist almost as well as the entire preserved object. Except in the youngest sedimentary rocks, imprints of the shells of marine invertebrates are just as common as the shells themselves, and usually even more so.

Fossiliferous Rocks

Almost all fossils are contained in sedimentary rocks. They are virtually nonexistent in igneous rocks, and they are extremely uncommon in metamorphic rocks, although certain robust body fossils can survive a certain degree of metamorphism. Finding a fossil in a metamorphic rock is a significant and exciting event for a geologist, because it is extremely difficult otherwise to

date the time of deposition of the sedimentary precursors of now-metamorphosed rocks.

Limestones are generally the most fossiliferous of sedimentary rocks. Most coarse-grained limestones, and many fine-grained limestones as well, consist mostly of whole shells or fragments of shells. Such fragments, although recognizably derived from whole organisms, are not usually the subject of special study by paleontologists. They are not sufficiently intact to carry detailed information about the nature of the organisms from which they were derived (although technically they are nevertheless fossils).

Shales, which are derived from freshly deposited mud, are often rich in trace fossils. Shales are less likely to contain body fossils except when the chemical conditions during deposition were conducive to preservation rather than decomposition. The best representatives of soft-bodied organisms are from shales, although—frustratingly for paleontologists—instances of such preservation are uncommon.

Many sandstones are fossiliferous as well, although the body fossils in sandstones are usually relatively robust shelly materials, which are not highly susceptible to chemical decomposition.

Conglomerates are the least fossiliferous of sedimentary rocks.

More Information – on the Web
Visit the *EarthComm* web site www.agiweb.org/earthcomm to access a variety of links to web sites that will help you deepen your understanding of content and prepare you to teach this activity. Many of the sites also contain images that you can download.

Chapter 3

Goals and Assessment

Clarify that the goals indicate what students should understand and be able to do as a result of the activity. Make sure students understand that Chapter Assessments are based upon these goals.

Goal	Location in Activity	Assessment Opportunity
Understand the process of fossilization.	**Investigate** Part A **Digging Deeper; Check Your Understanding** Questions 2 – 3, and 5 **Understanding and Applying What You Have Learned** Questions 1 – 5	Observations are correct. Questions are answered in a reasonable way, on the basis of observations and reading.
Determine which plant and animal parts have the highest and lowest potential for becoming fossilized, and understand why this is the case.	**Investigate** Part B **Digging Deeper; Check Your Understanding** Question 3 **Understanding and Applying What You Have Learned** Question 4	Reasoning behind likelihood for fossilization is logical and accurate. Questions are answered correctly.
Determine which organisms in your community are most likely to become preserved in the fossil record.	**Investigate** Part B **Understanding and Applying What You Have Learned** Question 2	List of organisms found in the community is complete. Reasoning behind likelihood for fossilization is logical and accurate.
Determine where fossils may be forming within your community.	**Understanding and Applying What You Have Learned** Question 1	Rocks that may contain fossils in community are correctly identified.
Understand the hierarchy of a food chain and how this affects the likelihood that an organism will be preserved in the fossil record.	**Investigate** Part B **Digging Deeper; Check Your Understanding** Question 1	Question is answered correctly.

NOTES

Chapter 3

Earth System Evolution Changing Life

Activity 1   The Fossil Record and Your Community

Goals

In this activity you will:

- Understand the process of fossilization.

- Determine which plant and animal parts have the highest and lowest potential for becoming fossilized and understand why this is the case.

- Determine which organisms in your community are most likely to become preserved in the fossil record.

- Determine where fossils may be forming within your community.

- Understand the hierarchy of a food chain and how this affects the likelihood that an organism will be preserved in the fossil record.

Think about It

Imagine that it is hundreds of thousands of years into the future. A geologist has just discovered your community and is planning an excavation.

- What evidence would the geologist find to know that life had existed in your community?

What do you think? Record your ideas about this question in your *EarthComm* notebook. Be prepared to discuss your response with your small group and the class.

Activity Overview

Students start **Activity 1** by using a clamshell and plaster to model how fossils form. They brainstorm about the most common organisms in their community, and place these organisms within the appropriate trophic levels. Students then consider which of these organisms are more or less likely to be preserved in the fossil record. **Digging Deeper** explains what food chains are and why understanding them is important in understanding the fossil record. The reading also defines the different types of fossils and reviews the process of fossilization.

Preparation and Materials Needed

Beyond collecting the necessary materials, no advanced preparation is required for this activity. The degree of success of **Part A** of this investigation is somewhat dependent on good technique, so you may want to try out this part of the investigation beforehand.

(Note: For **Question 1** of **Understanding and Applying What You Have Learned**, students will need a geologic map of the community.)

Materials
Part A
- Paper towels
- Petroleum jelly
- Small deli container
- Plaster mix (and water)
- Clamshell (both halves)
- Confetti
- Small ballpeen hammer

Part B
No additional materials needed.

Think about It

Student Conceptions

Responses will vary. Students are likely to say that evidence such as buildings and tools would suggest the presence of life. They may also say that bones or fossils could be used as evidence of past life in their community.

Chapter 3

Answer for the Teacher Only

Fossils are of two basic kinds: body fossils and trace fossils. The most likely body fossils would be human skeletons, because of the great care we take in sealing corpses in caskets and burying them in cemeteries. The bones of other animals might be preserved as well, if the animals died and were buried in areas like river beds and lake beds where deposition rather than erosion was occurring. Plant fossils, in the form of impressions of leaves and bark, might be preserved as well. Trace fossils produced by the life activities of humans would be likely as well; think of all the ways we shape and modify our physical environment, in the form of building foundations, roadcuts, mines, and the like. In a real sense, these artifacts are fossils!

Assessment Tool

Think about It Evaluation Sheet
Use this evaluation sheet to help students understand and internalize the basic expectations for the warm-up activity.

NOTES

Investigate

Part A: How Fossils Form

1. With a paper towel, smear petroleum jelly all over the inside of a container.

2. Following the directions on the package, mix plaster in a mixing bowl. Complete **Steps 3** through **6** immediately after this, so that the plaster does not set while you are still preparing the "rock."

3. Fill the container half full of the plaster.

4. With a paper towel, smear a thin coating of petroleum jelly on both surfaces of a clamshell. Place the clamshell in the middle of the container and press it gently into the plaster.

5. Sprinkle some confetti onto the rest of the surface of the plaster, enough to cover about 50% of the surface.

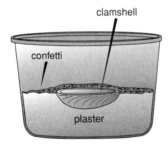

clamshell

confetti

plaster

6. Fill the rest of the container with more of the plaster.

7. Let the plaster harden overnight. In the next class, remove the plaster "rock" from the container by turning the container upside down and banging it down against the floor or tabletop.

 Wear goggles throughout Part A.

8. Set the plaster on its edge on a hard surface. Hit it gently with the hammer, at about the level where the confetti was sprinkled. It should split along the plane where the clamshell was placed.

Cover the plaster cast with a towel before hitting it with the hammer.

a) Why did the "rock" break along the plane where the fossil is located rather than somewhere else?

b) Do you think that a rock would usually break through the locations of fossils? Why or why not?

c) Clams have two parts to their shell. The parts are called valves. Each valve has an inner surface and an outer surface. How many different kinds of imprints might be seen when the valves of a clam are buried and fossilized as in this investigation?

d) If you had not seen the original shell that went into the plaster and did not see it when the "rock" was split open, how might you reconstruct what the shell looked like, just from studying the fossil evidence?

Investigate

Assessment Tools

EarthComm Notebook Entry-Checklist
Use this checklist as a quick guide for student self-assessment and/or an opportunity to quickly score student work. Add further criteria specific to your classroom needs or to this particular investigation.

Investigate Notebook Entry-Evaluation Sheet
Point out the criteria listed on this evaluation sheet that are relevant to this particular investigation. Encourage students to internalize the criteria by making them part of your "assessment conversations" as you circulate around the classroom. For example, while students are working, ask them criteria-driven questions such as:
  • Is your work thorough and complete?
  • Are all of you participating in the activity?
  • Do you each have a role to play in solving this problem? And so on.

Part A: How Fossils Form

1. – 7. Success in this part of the investigation depends on good technique and good timing. It might be wise for you to try out this procedure before class.

8. a) The "rock" breaks along the plane where the fossil is located because the weakest part of the "rock" is along the lubricated surfaces of the clamshell.

    b) In the real world of rocks and fossils, sometimes the rock breaks through the fossil and sometimes not. This often depends just on slight differences in strength.

    c) Imprints can be made of the inside of each of the valves and of the outside of each of the valves. Each of these four kinds of imprints can be seen in either "negative" form (as molds) or "positive" form (as casts), so eight different views are possible.

    d) Answers will vary. Students might respond that they could sketch the imprint to get an idea of how the shell looked. Or, they might say they could produce a replica of the shell by pouring a setting material, like plaster, into the void left by the shell, and allowing it to harden.

Earth System Evolution Changing Life

Part B: Fossils in Your Community

1. Natural ecosystems have different energy levels called trophic levels. Plants belong to the first level because the chemical energy they store comes directly from solar energy. They are primary producers. Organisms that eat only plants are at a higher level. They are called primary consumers or herbivores. Organisms that eat only other animals are higher-order consumers or carnivores. (Omnivores eat both plants and animals.)

   a) Identify the organisms shown at each of these levels in the diagram.

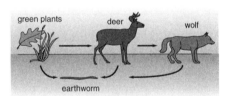

   b) Identify the most common organisms at each of these levels within your community. Try to list at least two or three different organisms for each level.

   c) Draw a diagram that illustrates the connections between each of these organisms. You will have essentially constructed a food web for your community.

2. Think about the organisms and their parts in your community.

   a) Which parts of each organism are the least resistant to decay and decomposition?

   b) Which parts of each organism are the most resistant to decay and decomposition?

3. Compare your list of resistant parts that you have identified for each plant and animal group with others in your group.

   a) Which type of organism has the highest probability of leaving some sort of fossil record?

   b) Why do some plants or animals have a low probability of leaving some sort of fossil record?

   c) How do you think that an organism's position on the food chain affects its likelihood of being fossilized?

Reflecting on the Activity and the Challenge

In this activity you learned how some types of fossils are formed. You also came to understand that not every organism has the same potential for becoming a fossil. Only certain types of parts have a high potential for becoming part of the fossil record. You also realized that the greater the number of parts an organism has, the higher the probability that one individual part may become fossilized at some time if the environmental conditions are right. You are now in a better position to evaluate the kinds of fossil evidence that may be found in your community.

E 150

Part B: Fossils in Your Community

1. a) In the diagram, the green plants are the primary producers, the deer is the primary consumer, the wolf is the higher-order consumer, and the earthworm is the decomposer. (Note: Students may have difficulty identifying the role of the earthworm as a decomposer because decomposers are not specifically addressed until the **Digging Deeper** reading section.)

   b) Answers will vary by community.

   c) Diagrams will vary.

2. a) – b) Answers will vary. For example, the different parts of a fish might include its scales, eyes, teeth, bones, and inner organs. Of these parts, the eyes and inner organs are least resistant to decay and decomposition; the teeth and bones are most resistant to decay and decomposition.

3. a) – b) Answers will vary. Students should recognize that organisms with the most parts that are resistant to decay (teeth, bones, etc.) are the most likely to be preserved in the fossil record.

   c) This question is not directly addressed in the investigation or the **Digging Deeper** reading section, so students' responses will likely vary. An organism's position on the food chain strongly affects the likelihood of that organism being preyed upon by a consumer. At first consideration it may seem that this would relate strongly to the potential for fossil preservation, but a low likelihood of being preyed upon does not necessarily translate to a greater potential for fossilization. Many factors contribute to the likelihood of fossilization. Generally speaking, the lower a species is in the food web, the more individuals of that species exist. In some cases, the primary producers may outnumber the top-tier consumers by billions of individuals. Greater numbers of individuals improves a species' chance of preservation in the fossil record. Size also plays an important role. Large organisms, like many top-tier consumers, are not easily buried by sediments, often times leaving them exposed to decay processes longer than a small organism in a similar circumstance.

Reflecting on the Activity and the Challenge

Review this section with your students and make sure that they have a firm understanding of how an organism becomes fossilized. Being able to understand what determines whether or not an organism is preserved in the fossil record will help students as they begin to interpret the fossil record of their own community.

Chapter 3

Digging Deeper

FOSSILS

Food Chains and Food Webs

Plants use energy from the Sun to make food, through the process of photosynthesis. Organisms that make their own food are called producers. Other organisms, like animals, are not able to make their own food. They must eat plants, or other animals that eat plants, to obtain energy. Such organisms that rely on plants for food are called consumers. Scientists use a kind of flowchart, called a food chain, to show how organisms are connected to each other by the food they eat. It is a convenient way to show how energy and matter are transferred from producers to the next levels of consumers. In most ecosystems, consumers rely on more than one source of food. Therefore, it is more realistic to show the relationships in the form of a food web. *Figure I* shows a sample food web.

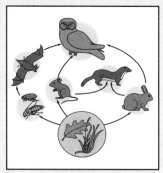

Figure I Can you identify the different trophic levels of this food web?

Geo Words

fossil: any remains, trace, or imprint of a plant or animal that has been preserved in the Earth's crust since some past geologic or prehistoric time.

body fossil: any remains or imprint of actual organic material from a creature or plant that has been preserved in the geologic record (like a bone).

trace fossil: a fossilized track, trail, burrow, tube, boring, tunnel or other remnant resulting from the life activities of an animal.

bias: a purposeful or accidental distortion of observations, data, or calculations in a systematic or nonrandom manner.

All living things die. Also, consumers generate waste materials from the food they eat. A special group of consumers, called decomposers, obtain the matter and energy they need from wastes and dead plants and animals. Decomposers play an essential role in food webs.

What is a Fossil?

A **fossil** is any evidence of a past plant or animal contained in a sediment or rock. There are two kinds of fossils: **body fossils** and **trace fossils**. Body fossils, like those shown in *Figure 2*, are the actual organisms or some part of them. They may also be the imprint of the organism or some part of it. Bones, teeth, shells, and other hard body parts are relatively easily preserved as fossils. However, even they may become broken, worn, or even dissolved before they might be buried by sediment. The soft bodies of organisms are relatively difficult to preserve. Special conditions of burial are needed to preserve delicate organisms like jellyfish. Sometimes, such organisms fall into a muddy sea bottom in quiet water. There they might be buried rapidly by more mud. Only in circumstances like these can such organisms be fossilized. For that reason, the fossil record of soft-bodied organisms is far less well known. There is a strong **bias** in the fossil record. Some organisms rarely have the chance of becoming fossilized. Under very specific circumstances, however, even these can become part of the fossil record.

Figure 2 Dinosaur bones collected in Dinosaur National Monument in Utah.

Digging Deeper

As students read the **Digging Deeper** section, the relevance of the concepts investigated in **Activity 1** will become clearer to them. Assign the reading for homework. The questions in **Check Your Understanding** can be provided as a homework assignment.

Assessment Opportunity

Reword or restructure the questions in **Check Your Understanding** for a brief quiz. Use the quiz (or a class discussion of the questions in the textbook) to assess your students' understanding of the main ideas in the reading and the activity.

Teaching Tips

Use **Blackline Master Changing Life 1.1, Food Web** to make an overhead of the food web shown in *Figure 1* on page E151. Discuss with students the different trophic levels shown in the diagram. They should be able to identify the green plants as the primary producers; the insects, mouse, and rabbit as the primary consumers; and the bat, weasel, and owl as the higher-order consumers. This diagram can be used as part of a discussion regarding why the food web shown in *Figure 1* is a more realistic representation than a food chain diagram, like the one shown on page E150.

Dinosaur bones, like those shown in *Figure 2*, are examples of body fossils. These particular dinosaur bones were found in Dinosaur National Monument. The national park, which is located near Vernal, Utah, and Dinosaur, Colorado, takes its name from the deposit of fossil bones found in the sands of an ancient riverbed. The deposit, known as Dinosaur Quarry, has yielded thousands of bones, including several nearly complete skeletons.

Teaching Tips

Figure 3 on page E152 shows an image of trace fossils. Trace fossils include any preserved evidence of the life activities of an organism. Trace fossils include things like footprints, trails, coprolites (fossil feces), eggs, borings, and burrows. The study of trace fossils is known as ichnology. Burrows, like those shown in *Figure 3*, are a kind of trace fossil found commonly in sedimentary rocks. They represent the life activities of an organism on the surface of the sediment, as the deposit is being formed, or within the sediment, after the sediment has been deposited.

Figure 4 on page E152 shows a fallen tree on the forest floor. Have students look at the image and identify the different trophic levels represented in the photo. Ask them to consider the role played in the process of fossilization by the fungus that is decomposing the tree. Would they expect the tree or flowers in the photograph to eventually be fossilized?

Earth System Evolution Changing Life

Trace fossils, like those shown in *Figure 3*, are the record of life activities of organisms rather than the organisms themselves. Tracks, trails, burrows, feeding marks, and resting marks are all trace fossils. Trace fossils are useful for geologists and **paleontologists** because certain kinds of organisms, which live in specific environmental conditions, make distinctive traces.

Figure 3 Feeding trails and burrows are kinds of trace fossils. Note the penny for scale.

In relatively young **sediments** and rocks, the actual body parts of an organism are often preserved. You modeled this in **Part A** of the investigation. In older rocks, the body parts are usually dissolved away. They may also be recrystallized or replaced by another kind of mineral. Even so, the imprints of the organisms are still preserved. They can be studied if the rock splits apart in the right place and the right orientation to reveal the imprint. Paleontologists usually collect large numbers of rock pieces. They then split the rock in the laboratory with special mechanical splitting devices to try to find at least a few fossils.

Geo Words

paleontologist: a scientist who studies the fossilized remains of animals and/or plants.

sediments: solid fragmental material that originates from weathering of rocks and is transported or deposited by air, water, or ice, or that accumulates by other natural agents, such as chemical precipitation from solution or secretion by organisms.

Fossilization

Figure 4 Fungi attack a fallen log in the woods.

As you saw when you looked at food chains and food webs, only a very small part of what once lived is spared being a meal for some other organism. There is a very high probability that any organism on Earth will be either consumed by another organism or decomposed by microorganisms following death. Decay affects not only soft body parts but also some of the harder, more resistant body parts. Think of a forest floor like the one in *Figure 4*. Each plant and animal that lives in the forest eventually ends up on the forest floor in some form. Soft tissues of animals, leaves, and flowers are used by decomposers. They decay within several weeks or are used by some other organism as a food source. The most resistant body parts include insect exoskeletons, vertebrate bones, wood, leaf cuticle, seeds, pollen, and spores. They may remain on the forest floor for many years or even centuries. This

Blackline Master Changing Life 1.1
Food Web

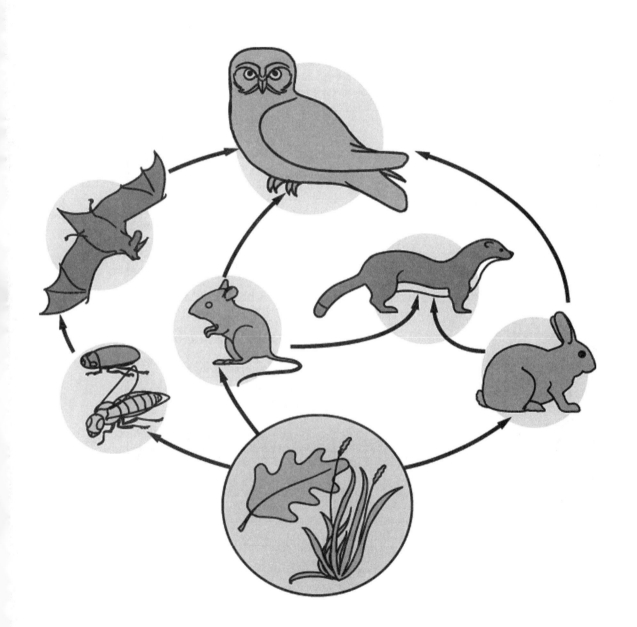

depends upon the physical and chemical conditions of the soil. A similar situation exists on the ocean floor.

For an organism or body part to become a fossil, it must either live within or be moved to a place where it can be buried. Burial alone does not guarantee that fossilization will occur. Under normal burial conditions all organisms undergo a scientifically predictable decay trend. Hence, there must be other factors operating during or immediately following burial to slow or stop decay. The conditions necessary for fossilization do not exist everywhere all of the time. In fact, they exist in only a few places and for only a tiny fraction of the time.

Despite this, enormous numbers of organisms have become fossilized. The reason lies in the extent of geologic time. It is difficult to imagine how long a million years is. Yet physical, chemical, and biological processes have been operating on Earth not just millions of years but billions of years. *Figure 5* shows a simple geologic time scale that indicates when various kinds of organisms are first seen in the fossil record.

Important Evolutionary Events of Geologic Time
(boundaries in millions of years before present)

Era	Period	Event	
Cenozoic	Quaternary	modern humans	
Cenozoic	Tertiary	abundant mammals	1.8
Mesozoic	Cretaceous	flowering plants; dinosaur and ammonoid extinctions	65
Mesozoic	Jurassic	first birds and mammals; abundant dinosaurs	145
Mesozoic	Triassic	abundant coniferous trees	213
Paleozoic	Permian	extinction of trilobites and other marine animals	248
Paleozoic	Pennsylvanian	fern forests; abundant insects; first reptiles	286
Paleozoic	Mississippian	sharks; large primitive trees	325
Paleozoic	Devonian	amphibians and ammonoids	360
Paleozoic	Silurian	early plants and animals on land	410
Paleozoic	Ordovician	first fish	440
Paleozoic	Cambrian	abundant marine invertebrates; trilobites dominant	505
Proterozoic		primitive aquatic plants	544
Archean		oldest fossils; bacteria and algae	2500

Figure 5 Important evolutionary events of geologic time. Note that the divisions of geologic time are not drawn to scale.

Blackline Master Changing Life 1.2
Major Divisions of Geologic Time

Major Divisions of Geologic Time (boundaries in millions of years before present)			
Era	Period	Event	
Cenozoic	Quaternary	modern humans	
			1.8
	Tertiary	abundant mammals	
			65
Mesozoic	Cretaceous	flowering plants; dinosaur and ammonoid extinctions	
			145
	Jurassic	first birds and mammals; abundant dinosaurs	
			213
	Triassic	abundant coniferous trees	
			248
Paleozoic	Permian	extinction of trilobites and other marine animals	
			286
	Pennsylvanian	fern forests; abundant insects; first reptiles	
			325
	Mississippian	sharks; large primitive trees	
			360
	Devonian	early plants and animals on land	
			410
	Silurian	amphibians and ammonoids	
			440
	Ordovician	first fish	
			505
	Cambrian	abundant marine invertebrates; trilobites dominant	
			544
Proterozoic		primitive aquatic plants	
			2500
Archean		oldest fossils; bacteria and algae	

Chapter 3

Earth System Evolution Changing Life

Fossiliferous Rocks

A rock, like the one shown in *Figure 6*, that contains fossils is said to be **fossiliferous**. Not all **sedimentary rocks** contain fossils. If you parachuted out of an airplane and landed on sedimentary rock, the chance of your finding a fossil would be rather small.

Geo Words

fossiliferous rock: a rock containing fossils.

sedimentary rock: a rock resulting from the consolidation of accumulated sediments.

Figure 6 A fossiliferous limestone composed mostly of brachiopod fossils.

Some kinds of sedimentary rocks contain more fossils than others. Limestones are the most fossiliferous sedimentary rocks. That should not be surprising, because most limestones consist in part, or even entirely, of the body parts of shelly marine organisms. Some shales are fossiliferous as well, because certain organisms like to live on muddy sea floors. Sandstones are usually much less fossiliferous than limestones. Fewer kinds of organisms can tolerate the strong currents and shifting sand beds that are typical of areas where sand is being deposited. For the same reason, conglomerates are the least fossiliferous of sedimentary rocks.

Check Your Understanding

1. What is a food chain, and what role does it play in fossilization?

2. How does the rate of burial relate to the likelihood of fossilization?

3. What proportion of living organisms has the likelihood of becoming fossils?

4. What is the difference between a body fossil and a trace fossil?

5. Which kinds of sedimentary rocks tend to be the most fossiliferous, and which tend to be the least fossiliferous?

Teaching Tip

Figure 5 on page E153 shows the geologic time scale with major biological events. Geologic time can be described in two ways: relative and absolute.

"Relative time" refers to age relationships based on a specific order, usually the position of layers with a sequence of rocks. Relative ages are usually determined on the basis of fossils. Geologists and paleontologists have developed the scale of relative geologic time over the past 200 years. The various subdivisions of relative time have been established by convention and have been given official names, the Cambrian Period being just one example.

"Absolute time" refers to numerical ages in thousands, millions, or billions of years. These ages are most commonly obtained by isotopic dating methods, which are based on the known rates of decay of radioactive isotopes that are fixed in minerals at the time of their crystallization. Measurements of absolute time can be used to calibrate the relative time scale, refining age estimates and putting numbers on the divisions of relative time.

The geologic time scale continues to be revised and refined, as new measurements of absolute time become available. Therefore, the time scales given in the student text may be slightly different from time scales you find elsewhere.

You may find it useful to use **Blackline Master Changing Life 1.2, Important Evolutionary Events of Geologic Time** and **Blackline Master Changing Life 4.1, Major Divisions of Geologic Time** (a more detailed account of the geologic time scale) to make copies of the time scale for students to insert into their notebooks for quick reference.

Teaching Tip

The photograph in *Figure 6* is a fossiliferous limestone with abundant brachiopod fossils. Brachiopods are marine organisms with two shells, called valves, which are hinged on one side and open at the opposite side, much like clams. However, these two kinds of organisms are only very distantly related: brachiopods belong to the phylum Brachiopoda, whereas clams belong to the phylum Mollusca. There are about 300 living species of brachiopods in the oceans today; they are a minor part of the marine invertebrate fauna. During the Paleozoic Era, however, brachiopods were one of most prolific kinds of marine animals.

Chapter 3

Check Your Understanding

1. A food chain is a diagram that shows how organisms are connected to each other by the food they eat. (In reality, the food chain is more than just the diagram; it is the nature of the relationships involved.) In most ecosystems, however, consumers rely on more than one source of food. For this reason, the more complex food web diagrams are considered to portray a more realistic representation of how organisms are connected to one another by the food that they eat. In general, organisms that are on a higher level of the food web are less likely to be preyed upon than those that are at a lower level of the food web. The role of the food chain in the likelihood of fossilization is not directly addressed, so student responses to this part of the question will likely vary. Refer to the answer given for **Part B, Step 3c** of this investigation for further detail on the relationship between trophic levels and the likelihood of fossilization.

2. The more quickly an organism is buried after death, the more likely it is to be fossilized.

3. Given the specialized conditions required for fossilization, a relatively small proportion of all organisms are actually fossilized.

4. A body fossil is the preserved remains of the actual organism, some part of the organism, or an imprint of the organism. Trace fossils are the preserved records of life activities of organisms, like tracks, trails, or burrows.

5. Limestones are the most fossiliferous sedimentary rocks, and conglomerates are the least fossiliferous sedimentary rocks.

Assessment Tool

Check Your Understanding Notebook Entry-Evaluation Sheet
Use this sheet to evaluate the extent to which students understand the key concepts explored in **Activity 1** and explained in **Digging Deeper**, and to evaluate the students' clarity of expression.

NOTES

Chapter 3

Understanding and Applying What You Have Learned

1. A geologic map shows the distribution of bedrock at the Earth's surface. Every geologic map has a legend that shows the kinds of bedrock that are present in the map area. The legend also shows the rock bodies or rock units that these rocks belong to, and their geologic age. Look at a geologic map of your community.

   a) What kinds of rocks are found in your community?
   b) Are fossils likely to be found in these rocks? Why or why not?

2. Revisit the **Think about It** question at the start of the activity.

   a) What evidence do you think the paleontologist would find in your community as proof of past life?
   b) Where would the paleontologist look to find this evidence?

   c) Do the organisms living in your community provide a biological signal that is unique to your area? Explain.

3. Must an organism die as a requirement to be represented in the fossil record? Explain your answer.

4. How are the physical and chemical processes responsible for preservation of plants and soft-bodied animals different from those for organisms that have hard skeletal parts?

5. Why would you expect that organisms living in ponds, lakes, or oceans have a greater chance of becoming part of the fossil record than organisms that live on land?

Preparing for the Chapter Challenge

Write a background summary that introduces fossils and fossilization. Be sure to identify those plant and animal parts in your community that you think are likely to leave some kind of fossil record, and discuss the reasons why you would expect one organism part to be more common than another. Also indicate where you would expect these fossils to be found within your community.

Inquiring Further

1. **Taphonomy**

   Taphonomy is the subdiscipline of paleontology that is concerned with the study of fossilization. The field of forensic science applies these principles to police and detective work. Investigate these subjects further.

2. **Common geological settings for preservation**

   Where on land or in the ocean might an organism be buried "dead or alive?" What are the conditions necessary to preserve soft tissues in these settings? What examples can you find in the fossil record of soft-tissue preservation?

Understanding and Applying What You Have Learned

1. **a)** Answers will vary.

    b) Answers will vary. Fossils will be found in sedimentary rocks. Limestones and shales are the most likely rocks to contain fossils.

2. **a) – c)** Answers will vary according to the nature of the communities considered. Certainly, the amount of evidence that human civilization is leaving to mark its existence is unprecedented in the history of the Earth. The modern practices that humans use to honor their dead to some extent protect their remains from the processes of fossilization, and the range of things that could be preserved as trace fossils is enormous. Encourage students to consider not only body fossils but also the plethora of trace fossils that could potentially survive the ravages of geologic time.

3. The death of an organism is not required for it to be represented in the fossil record. Trace fossils are the record of life activities of organisms and are therefore left by organisms that had not yet died.

4. In order to be preserved in the fossil record, soft-bodied animals and plants must fall to a muddy sea bottom in quiet water, where they are buried rapidly by more mud. Conditions that are harsher, physically or chemically, are not conducive to the preservation of soft-bodied organisms. Organisms with hard skeletal parts, on the other hand, are more resistant to harsher physical or chemical conditions.

5. In ponds, lakes, or oceans, organisms are more likely to settle to the bottom and be buried rapidly with little disturbance. Organisms that live on land are much less likely to be buried by sediments. This leaves their remains much more susceptible to the activities of other creatures and the processes of decay.

Preparing for the Chapter Challenge

This section gives students an opportunity to apply what they have learned about fossils and the conditions necessary for fossilization to the **Chapter Challenge**. They can work on this as a homework assignment or during class time within groups. Understanding the concepts they have been introduced to in this activity will help students understand why the fossil record in their community appears the way it does. This knowledge will also help them interpret biases in the fossil record.

For assessment of this activity you may wish to consider how this activity relates to the goals and Assessment Criteria for the overall **Chapter Challenge** (see **Assessment Rubric for Chapter Challenge on Changing Life** on page 520). This activity provides the framework for students to understand the process of fossilization, and how that process affects the evidence that gets preserved in the geologic record. Finally, students consider how all of this relates to which organisms in their community are likely to have evidence of their existence geologically

Chapter 3

preserved. As such, this activity is relevant to the question of "What evidence has been found to indicate change in the life forms and biodiversity of your community over time?" as presented in the **Assessment Rubric for Chapter Challenge on Changing Life** on page 520.

Inquiring Further

1. Taphonomy

Taphonomy involves the study of how an organism is fossilized and also how the processes of fossilization might influence information obtained from a fossil. More specifically, taphonomy examines:

- events that affected the organisms while they were alive, like climate changes, food availability, etc.
- how the organism came to be preserved in the fossil record
- the chemical and physical interactions that may have affected the organism from the time that it was buried until its discovery as a fossil

Forensic scientists are concerned with many of the same issues. For example, they look for evidence of action taken on a body, like cremation or fracture. They want to know what events affected persons while they were alive, and how they came to be deposited where their bodies were discovered. These scientists also are concerned with the biological and chemical interactions that may have altered the remains, including things like temperature, soil chemistry and composition, and the action of bacteria and insects.

2. Common geological settings for preservation

In order for a soft-bodied organism to be preserved in the rock record, it must be buried very rapidly, so as not to be exposed to further transport or decomposition. It must also remain in a chemical environment that prevents further decay. An excellent example of soft-tissue preservation is the fossil assemblage preserved in the Burgess Shale, in the Canadian Rockies, during the Cambrian Period.

NOTES

Chapter 3

ACTIVITY 2 — NORTH AMERICAN BIOMES
Background Information

A biome is a large-scale community of animals and plants that characterize an extensive area, sometimes thousands of miles in length or width, with relatively uniform climatic conditions. Because biomes are strongly related to both climate and their physical environment, the background materials presented below discuss both elements of climate and ecosystems. There are a number of different biomes that occur on the North American continent, and descriptions of each biome are detailed below in the comments that address **Step 1(a)** of the **Investigate** section of this activity.

Elements of Climate

Climate is more than just the long-term average of temperature and precipitation in a region. Daily, weekly, and yearly range in temperature is also an important determinant of climate. In addition to yearly averages of precipitation, frequency is important: does rain fall often in small amounts, or does it tend to fall less often but in large amounts?

The relative timing of yearly temperature and precipitation is very important for growth of trees and shrubs. In areas with large seasonal variations in precipitation, the wet season might coincide with the warm growing season. Or, the wet season might coincide with the cold season, in which trees and shrubs are dormant. Arid or semiarid regions with the latter conditions can support much more woody vegetation, given the same level of annual precipitation, than regions characterized by the former condition. Several other factors, like wind speed, wind direction, cloud cover (which is correlated with amount of precipitation, but not strongly in some areas), relative humidity, and frequency of occurrence of unusually strong storms like tropical cyclones (hurricanes and typhoons) are also important.

The most sensitive indicator of climate is vegetation. In fact, the connection between vegetation and climate is so strong that climatologists have to a great extent been guided by vegetation type in choosing criteria by which to classify climates.

For more details about climate, see the **Background Information** for **Chapter 2, Climate Change and Your Community,** in this Teacher's Edition. Note particularly the **Background Information** sections for Activities 1, 3, 5, and 6 on pages 300, 368, 440, and 478.

Ecosystems

The term ecology, which tends to be used too loosely in the popular media these days, is the study of the interrelationships between organisms and their environment. Similarly, paleoecology is the study of the ecology of fossil species. Paleoecologists labor under the disadvantage that they cannot observe, sample, and measure ancient environments directly, in the way that ecologists can study modern environments and their interactions with modern species of animals and plants.

The largest entity studied in ecology is the ecosystem, which consists of some chosen part of the physical and chemical environment and all of the organisms within it. An ecosystem involves all of the physical, chemical, and biological processes and interactions that operate within the given part of the environment. An ecosystem can be as large as the entire biosphere or as small as a puddle of water at your feet.

Your students should be familiar with some of the elements of an ecosystem. *Habitat* is the local environment in which a given organism lives. Ecosystems usually involve a number of different and distinctive habitats. A related concept is that of an *ecological niche*, which is the position of the organism in its habitat, including its way of life and its role in the ecosystem. Most habitats are occupied by several species, each with its own ecological niche. Each species in an ecosystem is represented by a number (usually large, but not always) of individual organisms, called a *population*. A population of two or more species that occupy a given habitat is called a *community*. There may be more than one community in a given ecosystem.

Within any ecosystem, there are typically many kinds of interactions among constituents, both living and nonliving. These interactions can be viewed in terms of flows of matter and energy through the ecosystem.

Organic compounds are synthesized from the environment by producers, which in all but the most specialized ecosystems are photosynthesizing plants. The producers are in turn consumed by plant-eating animals, called herbivores. Some of the herbivores are in turn consumed by carnivores (or by omnivores, which eat both plants and animals).

Other elements of the ecosystem are parasites, which feed on living organisms without killing them, and scavengers, which feed on dead organisms. The sequence of species ranging from the producers, at one end, to carnivores that no other carnivores eat, at the other end, is called a food chain. Often, the term food web is more appropriate, because the actual situation is more complex than a simple linear arrangement of species. The tissues of producers, herbivores, and carnivores that are not consumed by species higher in the food chain are broken down by organisms called decomposers, which are usually bacteria.

More Information – on the Web

Visit the *EarthComm* web site www.agiweb.org/earthcomm to access a variety of links to web sites that will help you deepen your understanding of content and prepare you to teach this activity. Many of the sites also contain images that you can download.

Chapter 3

Goals and Assessment

Clarify that the goals indicate what students should understand and be able to do as a result of the activity. Make sure students understand that Chapter Assessments are based upon these goals.

Goal	Location in Activity	Assessment Opportunity
Define the major biomes of North America and identify your community's biome.	Investigate	Description of biomes is accurate. Community biome is correctly identified.
Understand that organisms on land and in the ocean have physical and chemical limits to where they live.	Investigate; Digging Deeper; Check Your Understanding Question 3 Understanding and Applying What You Have Learned Questions 1, 3 – 4	Responses to questions are reasonable.
Recognize the most common plants and animals in your community.	Investigate; Understanding and Applying What You Have Learned Questions 1 – 2	Responses to questions are reasonable.
Explore how a change in physical and chemical conditions in your community could alter your community's biome.	Investigate; Digging Deeper; Check Your Understanding Question 4 Understanding and Applying What You Have Learned Question 5	Responses to questions are reasonable.
Understand that because certain organisms have physical and chemical limits to where they live, there are predictable relationships in the locations of the different biomes in North America.	Investigate; Digging Deeper; Check Your Understanding Question 3 Understanding and Applying What You Have Learned Questions 3 – 4	Responses to questions are reasonable.

NOTES

Chapter 3

Earth System Evolution Changing Life

Activity 2 North American Biomes

Goals

In this activity you will:

• Define the major biomes of North America and identify your community's biome.

• Understand that organisms on land and in the ocean have physical and chemical limits to where they live.

• Recognize the most common plants and animals in your community.

• Explore how a change in physical and chemical conditions within your community could alter your community's biome.

• Understand that there are predictable relationships between where the different biomes occur in North America.

Think about It

Humans have adapted to a wide variety of climates. Most plants and animals, too, have a range of climatic tolerances in which they live.

• Why don't the same plants and animals live all over the United States?

What do you think? Record your ideas about this question in your *EarthComm* notebook. Be prepared to discuss your response with your small group and the class.

Activity Overview

Students examine photographs of the major biomes of North America and write descriptions of each biome. They identify the biome in which their community is found. Students list the most common plants and animals found in the community and compare this list to information about what scientists have found to be the most common plants and animals in that biome. **Digging Deeper** defines the term biome and considers the factors that affect the distribution of organisms on Earth.

Preparation and Materials Needed

Preparation

You will need to provide students with a list of the most common plants and animals found in your biome. Alternatively, you need to provide students with time and resources to research this information themselves. Some information regarding some of the plants and animals common to each biome is provided below in the **Investigate** section. Additional information can be found on the *EarthComm* web site.

Materials
- Internet access (or scientific information about the most common plants and animals in your biome)*

Think about It

Student Conceptions

Students are likely to recognize that certain plants and animals cannot live in certain parts of the United States because of differences in climate. You may want to give your students some examples to direct their thinking. For example, if you live in Florida, you might ask your students whether a polar bear could live in your community, and why or why not.

Answer for the Teacher Only

Although many kinds of plants and animals do have a range of climatic tolerances, others do not. For example, there are many plants and animals that are not well suited to survive in regions with long, cold winters. Even for those organisms with broad climatic tolerances, in any given climate some species will tend to be more successful than others. Many organisms have evolved to have adaptations that make them particularly well suited to thrive in specific climatic and environmental conditions. For these reasons, the makeup of major biologic communities (or biomes) is strongly influenced by climatic and other environmental factors. Differences in climate and the physical environment give each kind of biome a unique makeup and character.

* See the *EarthComm* web site for information about how to obtain these resources at
http://www.agiweb.org/earthcomm

Assessment Tool

Think about It Evaluation Sheet
Use this evaluation sheet to help students understand and internalize the basic expectations for the warm-up activity.

NOTES

_____

_____

_____

_____

_____

_____

_____

_____

_____

_____

_____

_____

_____

_____

_____

_____

_____

_____

_____

Chapter 3

Investigate

1. A biome is a major biologic community. It is classified according to the main type of vegetation present. The organisms that live in each biome are characteristic of that area. Examine the photographs of the major biomes of North America that have been identified by ecologists.

tundra

taiga

chaparral

desert

grassland

mountain zones

E 157

Investigate

1. a) **Tundra** occupies the northernmost areas of North America and circumpolar regions in the Northern Hemisphere. The climate is dry (14–36 cm of precipitation per year) and cold (freezing temperatures can occur any day of the year, and temperatures rarely exceed 15°C even during the warmest months). Trees are lacking. Vegetation consists of low shrubs and herbs, lichens, sedges, rushes and grasses, and willows. Animals include migratory birds, wolf, Arctic fox, weasel, owl, bear, caribou, musk ox, polar bear, hare, vole, and lemming.

Teaching Tip

The photos on pages E157 and E158 are examples taken within their respective biomes. It is important to recognize that some degree of variation exists within every biome. The photos are not necessarily demonstrative of the entire biome. Go to the *EarthComm* web site to see more photos that show the diversity of plants and animals within each of the different biomes.

Chaparral occupies coastal southern California and northern Baja California, Mexico. The climate is cool and wet in winter and hot and dry otherwise. Precipitation occurs as rain, mainly in late November to early April (35 to 75 cm). Plants include shrubs like chamise, California buckwheat, oak and pine trees, and many kinds of wildflowers. Animals include a wide variety of birds, especially ground birds, as well as deer and small mammals.

Grassland occurs west of the Mississippi and extends to the foothills of the Rocky Mountains. Rainfall ranges from 25 to 100 cm annually, and temperatures typically range from 50°C in the summer to –45°C in the winter. Plants include the tall grasses, mixed grasses, and short grasses. Animals include bison, pronghorn antelope, and prairie dog.

Taiga covers circumpolar regions in the Northern Hemisphere. Climate is dry (43–51 cm of precipitation per year) and cold (freezing temperatures, extreme cold; –70°C is not uncommon). The temperature rarely exceeds 30°C, even during the warmest months of the year. Coniferous vegetation includes pine, fir, spruce, and tamarack. Hardwood trees include paper birch and quaking aspen. Other plants include shrubs, grasses, lichens, and bryophytes. Animals include elk, caribou, deer, bear (brown, grizzly), wolf, lynx, wolverine, fox, and migratory and some resident birds, especially songbirds.

Desert regions are associated with a lack of water, usually less than 10 cm of precipitation per year. Extreme daily variations in temperature are characteristic. Plants include those with deep roots to reach water below the Earth's surface. Plants can deal with drought by shedding their leaves during times of stress and being in the seed condition during times of drought, and, in the case of plants like cacti, by the ability to store water. Examples include sagebrush, creosote bush, and cacti. Animals include kangaroo rats, lizards, and snakes.

Mountain zones occur at the higher elevations in the Rocky Mountains and the Sierra Nevada. The forests are very similar to the taiga biome and include Douglas firs, coastal redwoods, and ponderosa pines. Water is abundant, owing to fog and rain on the windward sides of the mountains.

Tropical rainforest occurs from the Yucatan Peninsula south through Central America to Panama. Rainfall levels are high, ranging from 200 to 400 cm annually. Temperatures range between 25 and 32°C, and relative humidity seldom falls below 80%. There are no seasons based on temperature. Broadleaf evergreen trees dominate, with canopy contiguous.

Temperate evergreen forest begins, in general, at the 2750 m level and ends at the tree line, at about 3700 m. Rainfall ranges from 25 to 125 cm per year. Temperatures range from −30 to 20°C. Plants include evergreen trees like spruce, fir, pine, hemlock, and cedar.

Temperate deciduous forest occurs in the eastern United States, extending west to about the Mississippi valley. This region receives 75 to 250 cm of rainfall annually. Climate is temperate with distinct seasons; summers are warm and winters are cold. Plants are characterized by having leaves that fall in the autumn of the year, and include beech, maple, basswood, hemlock, oak, and hickory.

Polar regions are usually ice covered. The climate in these regions is cold and dry, with temperatures typically below 10°C; precipitation, almost exclusively as snow, is sparse. Plants are mainly lichens and moss on exposed rock and soil. Animals include polar bears, caribou, penguins, and foxes.

Blackline Master Changing Life 2.1
Map of North American Biomes

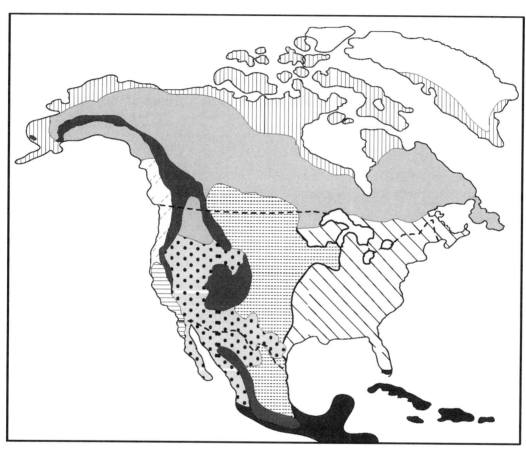

⦀ tundra		■ mountain zones	
▤ chaparral		■ tropical rainforest	
▦ grassland		◹ temperate evergreen forest	
▨ taiga		◺ temperate deciduous forest	
⦂ desert		☐ polar ice	

Chapter 3

Earth System Evolution Changing Life

tropical rainforest

temperate deciduous forest

temperate evergreen forest

polar ice

a) Use the photographs above and on the previous page and the *EarthComm* web site to write a short description for each biome, identifying the predominant plants and animals found in the biome.

b) Which biome most closely resembles your community?

c) How were you able to recognize your biome? What characteristic landscape did you use to pick the most representative photograph?

2. Look at the map of North American biomes. Match the photographs of the biomes with their locations shown on the map.

a) How do you think that the climate of your community's biome differs from the climates of other biomes in North America? Consider the temperature and precipitation of the area as well as the temperature and precipitation changes from season to season. Refer also to the map of the climatic regions of the United States on E85.

E 158

b) Answers will vary depending upon where your community is located.

c) Answers will vary depending upon where your community is located.

Teaching Tip

Use **Blackline Master Changing Life 2.1, Map of North American Biomes** to make an overhead of the map of the different biomes in North America. Use this map as part of a discussion about biomes and the major factors that control their distribution.

2. a) Answers will vary depending upon where your community is located.

Assessment Tools

EarthComm Notebook Entry-Checklist
Use this checklist as a quick guide for student self-assessment and/or an opportunity to quickly score student work. Add further criteria specific to your classroom needs or to this particular investigation.

Investigate Notebook Entry-Evaluation Sheet
Point out the criteria listed on this evaluation sheet that are relevant to this particular investigation. Encourage students to internalize the criteria by making them part of your "assessment conversations" as you circulate around the classroom.

Chapter 3

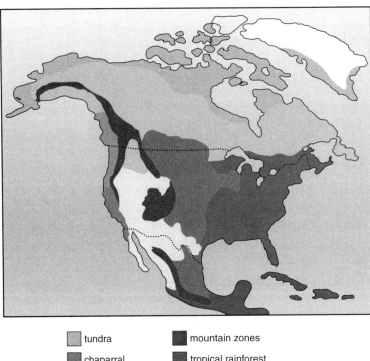

tundra

chaparral

grassland

taiga

desert

mountain zones

tropical rainforest

temperate evergreen forest

temperate deciduous forest

polar ice

3. As a group, think of the common plants and animals that occur naturally in your community. Use any available resources.

   a) Make a list of the most common naturally occurring plants. Identify at least five plants.

   b) Make a list of the most common naturally occurring animals. Identify at least five animals.

4. Share your lists of common plants and animals with the rest of your class using a class discussion format. To make comparisons between groups easier, each group may want to write their lists on the chalkboard.

   a) From the compiled data, develop a single class list of the 10 most common plants and animals in your community. Rank the plants and animals in order from most common to least common.

3. **a) – b)** Answers will vary depending upon where your community is located. You may want to supply students with photographs of different areas of your community to help them brainstorm about the plants and animals found in the community.

4. **a)** Answers will vary depending upon where your community is located. Students should attempt to assemble a list of 10 each of the most common plants and animals.

Chapter 3

Earth System Evolution Changing Life

5. Compare the class list with what scientists have found to be the most common plants and animals in your biome.

   a) How does your class list and ranking compare with the scientific data?

   b) Which plants and animals that you thought were common on the basis of your experiences were the same as those identified by scientists?

   c) What could account for any differences between your observations and the scientific data?

6. Compare the class list with what scientists have found to be the most common plants and animals in a North American biome other than the one in which you live.

   a) How do the plants and animals living in these areas differ?

   b) What are the physical and climatic factors that control the kinds of organisms that live in these two different biomes?

   c) What might happen to the plant and animal life in your community if the physical and chemical conditions suddenly changed to those found in the other biome?

Reflecting on the Activity and the Challenge

You have now seen the interactions between plants and animals in your community and compared these same relationships with another part of North America. You have also seen that different plants and animals characterize other parts of North America because their requirements for life are met by the physical and chemical conditions presently in those regions. You should now be able to begin to explain how biome boundaries can change in response to changing physical and chemical conditions on Earth over time.

5. **a) – c)** Answers will vary depending upon where your community is located. Students can visit the *EarthComm* web site to view more extensive lists of plants and animals commonly found in each biome.

6. **a) – c)** Answers to these questions will vary. Encourage students to select a biome that is distinctively different from your own. For example, if your community is located in the temperate deciduous forest, students might look at the desert biome. They should recognize that if the chemical and physical conditions in their biome were to change, organisms would adapt to the new conditions, leave the area, or die out.

Reflecting on the Activity and the Challenge

Have students read this brief passage and share their thoughts about the main point of the activity in their own words. Hold a class discussion about how the investigation relates to what is being asked of the students in the **Chapter Challenge**. Ask them to consider why they think it is important to understand the boundaries that define where an organism lives when trying to interpret the fossil record. Students who have completed **Chapter 2, Climate Change and Your Community**, will probably have a sense that because the Earth's climate changes through time, so does the distribution of organisms on the Earth's surface.

Chapter 3

Digging Deeper

CLIMATE AND BIOMES

An **ecosystem** is a community of plants and animals together with the physical and chemical environment in which that community exists. In the broadest sense, there are two major kinds of ecosystems on Earth today: aquatic and terrestrial ecosystems. Both of these kinds of ecosystems can be traced far back in geologic time. Each can be split into smaller subdivisions, using different criteria. For example, aquatic ecosystems can be subdivided into two categories using water chemistry: freshwater ecosystems and saltwater ecosystems. Each of these can be further subdivided. The aquatic ecosystem shown in *Figure 1* is a freshwater ecosystem. The terrestrial ecosystem can be divided into many categories, using the tallest plants in the area.

Figure 1 An example of a freshwater ecosystem.

Biological subdivisions on land are called **biomes**. Many present-day biomes parallel the lines of **latitude**. Climate conditions across the globe at these latitudes tend to be similar. In other words, there is a relationship between the physical and chemical processes that operate within latitude belts and the plants and animals that are adapted to these conditions. There is a slight difference in the case of mountains, because temperature decreases with elevation, as shown in *Figure 2*. Often, lowland biomes at high latitudes

Figure 2 How do you think the organisms found at the base of a mountain differ from those found near the top of the mountain?

Geo Words

ecosystem: a unit in ecology consisting of the environment with its living elements, plus the nonliving factors that exist in it and affect it.

biome: a recognizable assemblage of plants and animals that characterizes a large geographic area on the Earth; a number of different biomes have been recognized, and the distribution of the biomes is controlled mainly by climate.

latitude: a north-south measurement of position on the Earth. It is defined by the angle measured from the Earth's equatorial plane.

E 161

Digging Deeper

Assign the reading for homework, along with the questions in **Check Your Understanding** if desired.

Assessment Opportunity

Reword or restructure the questions in **Check Your Understanding** for a brief quiz. Use the quiz (or a class discussion of the questions in the textbook) to assess your students' understanding of the main ideas in the reading and the activity.

Teaching Tips

The photograph of ducklings living on a small pond (*Figure 1*) illustrates one of several different kinds of freshwater ecosystems. Other freshwater ecosystems include lakes, rivers, and different kinds of wetlands. Ask the student to identify some of the important roles that the freshwater ecosystems in your community play in both sustaining the environment and the community.

Discuss the photograph shown in *Figure 2* and the question raised by the caption. This photo should help students understand the role of elevation in determining which organisms are found in a specific area.

Chapter 3

Earth System Evolution Changing Life

Geo Words

climate: the characteristic weather of a region, particularly as regards temperature and precipitation, averaged over some significant interval of time.

weather: the condition of the Earth's atmosphere, specifically, its temperature, barometric pressure, wind velocity, humidity, clouds, and precipitation.

(closer to the poles) are very similar to highland biomes at lower latitudes (closer to the Equator). For example, plant communities on the rocky, windswept peaks in the Appalachians are similar to those in many places in subarctic Canada.

Climate involves the long-term characteristics of the **weather** in a given region of the Earth. The most important factors in climate are temperature and precipitation. Both the average values and the deviations from the average are important. Everybody knows that some places are hot and some are cold. Some places are wet and others dry. What is also important is how much temperature and precipitation change from season to season. The average rainfall in an area may be high, but there may also be a long dry season. Plants need adaptations for surviving those dry periods before renewing their growth during the wet season. In areas with rainy seasons and dry seasons, it is also important for plant communities whether the wet season coincides with summer or with winter. You can easily see why climate plays the most important role in the distribution of biomes.

What links common animals with common plants is the dietary requirements of the plant eaters (herbivores). Similarly, the dietary requirements and taste preferences of meat eaters (carnivores) are linked to the specific herbivores that live in a biome. The most common plants, like the ferns, mosses, and trees shown in *Figure 3*, provide food as well as

Figure 3 Temperate evergreen forest, Hoh Rainforest, Washington.

E 162

Teaching Tip

The Hoh Rainforest (pictured in *Figure 3*) is located on the Olympic Peninsula of Washington State. This temperate evergreen forest receives enough annual precipitation to be classified as one of the few temperate rainforests in the world. Temperate rainforests are found on the western edge of North and South America, where moist air from the Pacific Ocean drops between 140 cm and 500 cm of rain a year. One way that these rainforests differ from the widely publicized tropical rainforests is that temperate rainforests have significant seasonal climatic variation. The temperate rainforests of the United States exist in localized areas within the pink zone of the map of North American biomes on page E159 (the temperate evergreen forest biome). Ask the students to consider whether there are any tropical rainforests in North America and whether there are other rainforests within the United States. (A Hawaiian rainforest is pictured on page E158.)

Figure 4 How do you think an organism like an owl responds to climate change?

Geo Words

global climate: mean climatic conditions over the surface of the Earth as determined by the averaging of a large number of observations spatially distributed throughout the entire region of the globe.

shelter for many animals in the biome. When plants and animals die, decomposers break down the dead and discarded organic matter, recycling it to the soils to be used again by plants. The rate of decay and the release of nutrients into the soil are also affected by the climate.

Global climate change is in the news nowadays. There is abundant evidence, however, that climate has changed continually in the past, long before the development of modern human society. These changes have taken place on time scales that range from as short as decades to as long as hundreds of millions of years. **Chapter 2, Climate Change and Your Community**, investigates climate change in greater depth.

How do biomes respond to climate change? Generally, animals in a biome can migrate quite rapidly in response to climate change. Plants migrate much more slowly than animals. Plants are not able to respond immediately when the climate conditions change and the plant's tolerances are exceeded! Plants can shift their range only by dispersal of seeds and spores by the wind or by animals. Keep in mind that all animals depend, directly or indirectly, on plants as their food source. If climate change is slow, biomes can respond without much disruption as they shift in position. If climate change is more rapid, there can be great changes in the makeup of the plant and animal communities in a given area.

Check Your Understanding

1. What are the two major kinds of ecosystems found on Earth?

2. Define, in your own words, the term biome.

3. What factor plays the greatest role in determining the distribution of biomes on Earth?

4. How do biomes respond to changes in climate?

EarthComm

Teaching Tip

Like all organisms, owls like the one pictured in *Figure 4* are adapted to a particular set of environmental conditions. Owls are predators. If a variation in climate changes the kind or abundance of the animals on which they prey, the owls may have a greater or smaller food supply. Also, most owls live in trees. If rapid climate change causes the disappearance of tree species used by owls for living quarters, the owls must adapt to different trees, shift their geographic range, or become extinct.

Check Your Understanding

1. The two main ecosystems found on Earth today are the aquatic and terrestrial ecosystems.

2. A biome is a biological subdivision on land, classified according to the predominant vegetation and characterized by adaptations of organisms to that particular environment.

3. Latitude plays the greatest role in determining the distribution of biomes, because climate conditions are controlled largely by latitude.

4. In response to climate change, most animals can migrate rapidly. Plants, on the other hand, can shift their range only slowly, by dispersing seeds and spores.

Assessment Tool

Check Your Understanding Notebook Entry-Evaluation Sheet
Use this sheet to evaluate the extent to which students understand the key concepts explored in **Activity 2** and explained in the **Digging Deeper** reading section, and to evaluate the students' clarity of expression.

Chapter 3

Earth System Evolution Changing Life

Understanding and Applying What You Have Learned

1. Why might different teams in your class have developed different rankings of the common plants and animals in your community? What role might the location of your home have in what you determined to be the most commonly occurring organisms?

2. Determine the date when your town or city was incorporated.

   a) What plants and animals were common before that date?
   b) How have the common organisms in your community changed since that date?
   c) How do you explain these changes?

3. What are the main climate factors that restrict the distribution of plants and animals within your biome?

4. Refer back to the biome map of North America.

   a) What proportion of the United States lives in the same biome as your community?
   b) What is the relationship between biome distribution and latitude?
   c) What is the relationship between biome distribution and elevation?

5. What changes in the animals and plants might you expect if the climate of your community became colder? Became warmer?

Preparing for the Chapter Challenge

Write a short paper in which you describe the present biome in your community. Indicate which physical and chemical factors determine the limits of your biome, and indicate how variations in these factors could cause changes in the distribution of organisms in your community. For example, suppose that your community is now located in a mild climate. How might a decrease in the yearly average temperature affect the organisms in your community?

Inquiring Further

1. **Herbivores and carnivores in your community**

   What relationships do herbivores and carnivores have with plants in your biome? Do all consumers have the same preference for food, or do different groups of animals use different food sources?

2. **Animal adaptations to climate**

   Animals in Alaska must survive the extreme cold of the arctic and subarctic climate. What adaptations have Alaskan animals developed to survive? How do you think these adaptations might affect how these organisms would be preserved in the fossil record?

E 164

Understanding and Applying What You Have Learned

1. Answers will vary depending upon your community. If you live in a large, diverse community, it is likely that the location of students' homes will affect how they describe which organisms are found most commonly in the community.

2. Answers will vary. Most likely, organisms have changed since the incorporation of your town. This will be more obvious in urban areas than in small rural communities. Information about the history for most towns and cities can be found at the local Town Hall or on the Internet. (An Internet search using the town or city and state names often quickly leads to historical information about the town.) Information regarding changes in the common biota of the community can be researched using local libraries or the Internet, or by interviewing some of the more elderly members of the community.

3. Answers will vary.

4. a) Answers will vary.

    b) Biome distribution is generally controlled strongly by latitude, because latitude is the most important control on climate. In the central portion of North America, however, the presence of several mountain ranges greatly affects the biome distribution.

    c) Biome distribution can vary with elevation because temperatures at higher elevations are cooler than those at lower elevations. Additionally, the presence of large mountain ranges can strongly affect precipitation and biome distribution.

5. Answers will vary. Students can draw upon work they completed earlier in the investigation—comparing their community biome to a different biome.

Preparing for the Chapter Challenge

In **Activity 2**, students learned about the chemical and physical conditions that limit the geographical distribution of an organism. Students should reexamine the **Chapter Challenge** to consider how the concepts studied in this activity will help them as they begin to construct the display of the changing biology of their community through time.

For assessment of this section (see **Assessment Rubric for Chapter Challenge on Changing Life** on page 520), students should begin to discuss:

- what short-term and long-term factors [may] have influenced biological changes in your community

Chapter 3

Inquiring Further

1. Herbivores and carnivores in your community

Answers will vary depending upon the organisms found in your community. Students will most likely find that not all consumers have the same food preference and that different groups of animals rely on different food sources.

2. Animal adaptations to climate

Plants and animals in the Arctic tundra must withstand nine months of winter, with temperatures as low as –50°C, high winds, and several months of the year when the Sun does not rise. To survive, organisms have adapted traits to withstand the cold, the wind, and often a limited food supply. Many animals have developed long fur coats or thick feathers to withstand cold temperatures. Others are capable of finding shelter; some migrate from the Arctic for the winter. Many tundra birds and mammals have a larger body size and small appendages, which allows them to retain heat better. Plants in the Arctic tend to grow closer to the ground to shelter themselves from the cold and wind.

NOTES

ACTIVITY 3 — YOUR COMMUNITY AND THE LAST GLACIAL MAXIMUM

Background Information

Pleistocene Glaciations

There have been four distinct periods of glaciation in North America in the last 1.6 million years, with briefer interglacial periods. The most recent glacial period ended about 10,000 years ago. The time between 1.6 million years ago and 10,000 years ago is called the Pleistocene Epoch (one of the standard time divisions of Earth history). Ice covered approximately 22 million km^2 of the Earth's surface during the height of the Pleistocene glaciations. In contrast, the areas of the present ice sheets in Greenland and Antarctica are 2,175,600 and 14,200,000 km^2, respectively. The Pleistocene Epoch is particularly interesting because the features of many present landscapes reflect the work of Pleistocene glaciers. Before the Pleistocene, there seem to have been at least three other periods of glacial activity, at about 2 billion, 600 million, and 250 million years ago.

Each of the major glaciations developed slowly, as global climate cooled. Continental ice sheets began to develop in favorable areas in the interiors of the Northern Hemisphere continents of North America and Eurasia. In North America, the largest continental ice sheet, called the Laurentide ice sheet, was centered on east-central Canada. Presumably, the combination of high land, fairly abundant winter snowfall, and cool summers led to the development of the nucleus of the Laurentide ice sheet.

The Laurentide ice sheet expanded to eventually cover a substantial portion of northeastern and north–central North America. The southern limit of advance of the ice is well defined: it is marked by long ridges of glacial sediments formed as the southward-moving ice delivered its load of sediment to an ice margin that was almost stationary for many thousands of years. Interestingly, the southern limit of each of the four major Pleistocene glaciations lies in approximately the same position across the northeastern and central United States, stretching from Massachusetts across New Jersey and Pennsylvania, and continuing across southern Ohio, Indiana, and Illinois to northern Missouri and Iowa, and then northward into central Canada.

In time, the margin of the ice sheet retreated toward its original nucleus in Labrador. The retreat and disappearance of each ice sheet was much more rapid than the advance: from its maximum stand at about 20,000 years before the present, the most recent ice sheet melted away completely by 6000 years before the present.

Climates south of the ice sheet, in southern and south–central United States, and even in areas far removed from glaciation, as in the American Southwest, were very different during the glacial periods than they are now. Conditions were generally wetter and cooler south of the ice sheet than in interglacial times. Plant communities, and the attendant animal communities as well, were shifted far southward.

It is not easy to imagine the great differences in climate between glacial and interglacial times in the areas of the United States south of the ice sheets. Try to conjure up in your mind the great differences between a rigorous tundra climate and the verdant deciduous forests of temperate latitudes—a change that was completed in the brief space of about 10,000 years!

The Causes of Ice Ages

Many theories for the causes of the ice ages have been proposed. Clearly, glaciers form because the Earth's climate changes in such a way as to make possible the development of continent-scale ice sheets. But that just pushes the question back further: what causes the climate to change in the first place? Keep in mind that it's not just a matter of the climate becoming colder. To be conducive to glaciation, the climate must be such that the

conditions for winter snowfall are enhanced and/or the conditions for summer melting are lessened. Temperature is a major factor, but precipitation is also important. A warmer climate means more summer melting, but it is also conducive to greater winter snowfall. See the **Background Information** section accompanying **Activity 3** of **Chapter 2** on page 368) for more details about the causes of the ice ages.

More Information – on the Web

Visit the *EarthComm* web site www.agiweb.org/earthcomm to access a variety of links to web sites that will help you deepen your understanding of content and prepare you to teach this activity. Many of the sites also contain images that you can download.

Chapter 3

Goals and Assessment

Clarify that the goals indicate what students should understand and be able to do as a result of the activity. Make sure students understand that Chapter Assessments are based upon these goals.

Goal	Location in Activity	Assessment Opportunity
Investigate how changes in climate are linked to shifts in the distribution of oak and spruce trees in North America, 20 ka (20,000 years ago) to the present.	Investigate Part A	Maps are correctly contoured. Answers to questions are reasonable and comprehensive.
Understand how scientists can use data collected from pollen and spores to reconstruct past environments.	Investigate Parts A and B **Digging Deeper;** **Check Your Understanding** Question 2	Answers to questions are reasonable and comprehensive. Research is well conducted.
Understand how the climate of your community has changed over the past 18,000 years.	Investigate Part B **Digging Deeper;** **Check Your Understanding** Question 3 **Understanding and Applying What You Have Learned** Question 1	Answers to questions are reasonable and comprehensive.

NOTES

Activity 3 Your Community and the Last Glacial Maximum

Goals

In this activity you will:

- Investigate how changes in climate are linked to shifts in the distribution of oak and spruce trees in North America, 20 ka (20,000 years ago) to the present.

- Understand how scientists can use data collected from pollen and spores to reconstruct past environments.

- Understand how the climate of your community has changed over the past 18,000 years.

- Understand how this climate change has affected the plants and animals of your community.

Think about It

The Earth's climate has varied significantly over geologic time. As recently as 18,000 years ago much of North America was covered by thick glacial ice.

- What plants and animals (if any) lived in your community when large parts of North America were covered by ice?
- As the continental glaciers retreated northward, what changes in the plants and animals happened over time?

What do you think? Record your ideas about these questions in your *EarthComm* notebook. Be prepared to discuss your response with your small group and the class.

Activity Overview

Students begin the investigation by examining maps that show the distribution of oak and spruce trees in the United States over the past 20,000 years. They describe the shifts in the distribution of the trees and correlate these shifts to changes in climate. Students then use the information from this exercise to form hypotheses about how the climate in their own community might have differed during the Pleistocene, and how plants and animals might have differed during this time. Students then complete Internet research to determine what fossil evidence suggests about life in their community 18,000 years ago. The **Digging Deeper** reading section reviews the use of pollen and spores as indicators of climate change, defines the term paleoclimate, and explains how biomes respond to changes in climate.

Preparation and Materials Needed

Preparation

Part A
Use **Blackline Master Changing Life 3.1, Oak and Spruce Distribution Maps** to make photocopies of the oak and spruce distribution maps for students to contour.

Part B
Students will need Internet access to complete this part of the investigation, so you will need to plan ahead and make any necessary arrangements. If you cannot provide students with Internet access in the classroom, you can assign this part of the investigation as a homework exercise, or you can provide students with printouts of the necessary information.

Materials

Part A
- Copies of the oak and spruce distribution maps for last 20,000 years (**Blackline Master Changing Life 3.1, Oak and Spruce Distribution Maps**)

Part B
- Internet access (or information about pollen distribution from the Late Pleistocene to the present)*
- Internet access (or information about distribution of animals from the Late Pleistocene to the present)*

* See the *EarthComm* web site for information about how to obtain these resources at http://www.agiweb.org/earthcomm

Think about It

Student Conceptions

Answers will vary. Students are likely to recognize that when glaciers are present in an area, the climate of that area will be cold and inhospitable to many plants and animals. They will likely observe that as the glaciers retreated, the warmer climate influenced what kinds of plants and animals could thrive.

Answer for the Teacher Only

In most areas of the United States, both the fauna and the flora have changed greatly from the time of the last glacial maximum to the present. The main reason for that is the northward shift in climate zones as the glaciers retreated and temperatures increased. In the process, some species of plants and animals were able to shift their region of occurrence quickly enough to keep pace with the shifts in climate; others were not able to keep pace, and thus became extinct.

There is still controversy among the experts about the role of early humans in the extinction of the large North American mammals, like mammoths. Did these species become extinct because of overhunting by the recently arrived humans, or was it entirely a matter of rapid climate change? Perhaps both of these factors played a part.

Blackline Master Changing Life 3.1
Oak and Spruce Distribution Maps

Oak and Spruce Distribution Maps for Last 20,000 Years.
oak (*Quercus*)   spruce (*Picea*)

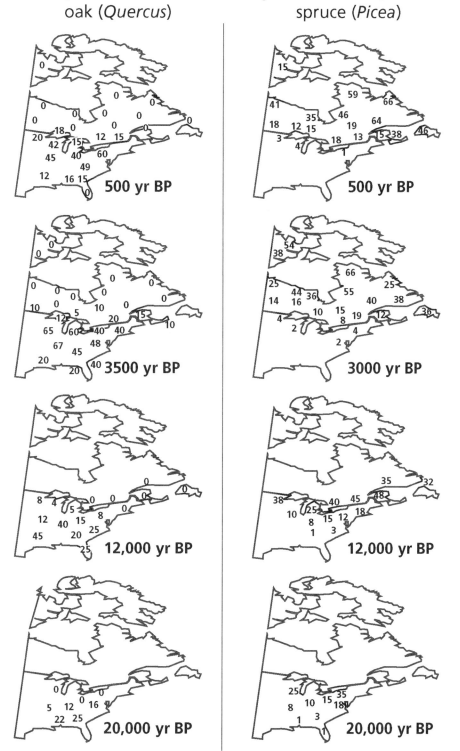

Chapter 3

Earth System Evolution Changing Life

Investigate

Part A: Can You Recognize a Forest without the Trees?

1. Examine the maps. Pollen and spore records were recovered from lakes, ponds, and bogs across North America. Coupled with carbon-14 radio-isotopic dating, scientists have reconstructed the changes in plant types over space and time. The maps show sample sites across North America for approximately the last 20,000 years. The percentages of spruce and oak pollen are plotted on each map set.

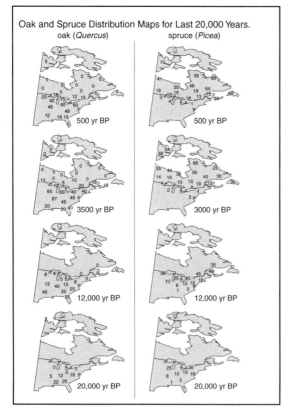

Oak and Spruce Distribution Maps for Last 20,000 Years.
oak (*Quercus*)   spruce (*Picea*)

a) Obtain copies of the maps. Using a different color for each plant, draw the 5% and 20% contour lines neatly across each map.

b) Where is the maximum percentage of spruce found 500, 3000, 12,000 and 20,000 years ago?

c) Where is the maximum percentage of oak found 500, 3500, 12,000 and 20,000 years ago?

2. Oak trees are characteristic of deciduous forest, and spruce trees are representative of the taiga.

a) Do the general trends in oak parallel the changes in spruce? Explain.

b) What other plants and animals might you expect to find in the fossil records in the collection sites 3000 and 20,000 years ago?

Part B: Your Community Biome: Pleistocene to Present

1. Develop a hypothesis concerning the plants and animals that may have lived in your community when North America experienced the transition from a glacial period to the present day (nonglacial period).

a) Record your prediction and the reason for your prediction in your *EarthComm* notebook.

2. Visit the *EarthComm* web site to find a list of databases and animations regarding pollen distribution from the Late Pleistocene to the present.

a) Make a list of plants that have been found in your community from approximately 18,000 years ago to the present.

EarthComm

Investigate

Part A: Can You Recognize a Forest without the Trees?

1. a) Drawing contour lines is not always straightforward. The idea is to draw smooth curves that fall between two adjacent values that bracket the given contour value. For example, the "10" contour line must fall between two neighboring values, one of which is "6" and the other "13."

   Also, the contour line is placed in a position that is proportional to the two values. In the case above, the "10" contour line must lie slightly closer to the "13" value than to the "6" value, because the difference between 13 and 10 is less than the difference between 10 and 6. (i.e., 10 is closer in value to 13 than it is to 6.)

   One troublesome problem with contouring is that the point values usually show some scatter: they are not perfectly representative of the ideal values. The person doing the contouring has to draw the contours to minimize the inevitable inconsistencies. Contouring is usually an imperfect activity, and it depends quite a lot on the judgment of the person doing the contouring.

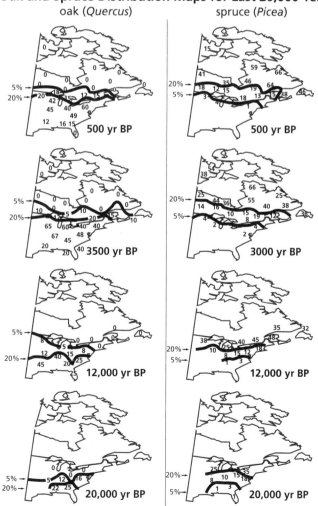

Oak and Spruce Distribution Maps for Last 20,000 Years.
oak (*Quercus*)   spruce (*Picea*)

b) At 20,000 years before the present, spruce occupied much of the eastern United States, from Florida to the Great Lakes. At that time, the highest percentage of spruce was located around the Great Lakes region. With time (and the receding of the glaciers), the distribution of spruce shifted gradually northward. At 3000 years before the present, the maximum percentage of spruce was centered around the northernmost regions of Canada. At 500 years before the present the maximum percentage of spruce was still located in northern Canada, although the distribution was changed somewhat from that of 3000 years before the present.

c) At 20,000 years before the present, oak occupied the central and southern regions of the United States. At 12,000 years BP, the distribution of oak spread farther north and east, with the maximum percentage shown occurring in the Midwest region of the United States. At 3500 years BP, the distribution of oak was relatively uniform across the United States, with small percentages in Canada. At this time, the percentages of oak pollen reach their greatest levels. By 500 years BP, the oak pollen maximum had shifted slightly east of where it was at 3500 years BP, and the maximum percentages of oak pollen had decreased slightly.

2. a) Overall, the trends are similar, with both trees showing a shift northward. However, the shift observed in the distribution of spruce trees appears much more dramatic. This is because as time progresses, the spruce pollen disappears completely from the southern United States whereas the oak pollen does not.

b) Students may expect to find fossils characteristic of the taiga biome where the spruce pollen is dominant. The preservation of other conifers like pine, fir, and tamarack, as well as deciduous trees such as paper birch and quaking aspen, is also possible. Shrubs, grasses, lichens, and bryophytes are additional plants likely to be found in areas associated with a high percentage of spruce pollen. Fauna that inhabit the taiga biome and could be preserved as fossils include elk, caribou, deer, bear (brown and grizzly), wolf, lynx, wolverine, and fox. Migratory birds and some resident birds, especially songbirds, also live in this biome.

Students will likely characterize areas where oak pollen was dominant as temperate deciduous forests. In addition to oak, flora preserved in the fossil record may also include such species as hemlock, hickory, beech, elm, maple, basswood, cottonwood, willow, and flowering herbs. Animals that inhabit this biome and could be preserved as fossils include squirrels, rabbits, deer, mountain lions, bobcats, black bear, and foxes.

Part B: Your Community Biome: Pleistocene to Present

1. **a)** Answers will vary depending upon your community.

2. **a)** Answers will vary depending upon your community.

    b) Most likely, the two lists will be different.

    c) Most likely, the plants found in your community 18,000 years ago are representative of a much colder climate than exists in your community today.

b) How does this list compare to the list of plants commonly found in your community, which you developed in **Activity 2**?

c) How do you account for the differences between the plants found in your community today and the plants that were found in your community in the past?

3. Databases also exist that document the fossil animals found across North America through time. Visit the *EarthComm* web site to find a list of databases that show animal distribution in North America.

a) Why are fossil animal records not as comprehensive as pollen records?

b) Make a list of animals that have been found in your community from approximately 18,000 years ago to the present.

c) How does this list compare to the list of animals commonly found in your community, which you developed in **Activity 2**?

d) How do you account for the differences between the animals found in your community today and the animals that were found in your community in the past?

4. Evaluate the hypothesis you generated at the beginning of **Part B** of the investigation.

a) Do the data you have collected verify or refute your hypothesis?

b) What modification(s) to your hypothesis must be made, in light of the data you found? Revise your hypothesis according to your limited data set.

Reflecting on the Activity and the Challenge

You have learned that the plants and animals that once lived in your community were probably very different than today. As changes in climate accompanied the retreat of the glaciers, there were changes in the composition of the ecosystem. Compiling information on past communities helped you to think about the potential change in plants and animals in your state, region, and community under future changes in climate.

3. a) Students should recognize from previous activities that the conditions required to fossilize animals are very demanding and do not occur frequently. Pollen grains, however, are relatively resistant to decay and are easily buried in sediments.

   b) Answers will vary depending upon your community.

   c) Most likely, the two lists will be different.

   d) Most likely, the animals found in your community 18,000 years ago are representative of a much colder climate than exists in your community today, and some of the pleistocene animals may now be extinct.

4. Answers will vary depending upon the hypotheses originally developed by students.

Reflecting on the Activity and the Challenge

Students examined the distribution of two kinds of trees through time and used shifts in distribution to make assumptions about the climate during this time period. They also learned how plants and animals that inhabited their community 18,000 years ago differed from the plants and animals of today. These concepts will be explored further in the **Digging Deeper** reading section. Take some time to review the main points of the investigation, and make sure that students understand that the organisms were changing or moving in response to changes in climate.

Chapter 3

Earth System Evolution Changing Life

Geo Words

spore: a typically unicellular reproductive structure capable of developing independently into an adult organism either directly if asexual or after union with another spore if sexual.

pollen: a collective term for pollen grains, which are microspores containing the several-celled microgametophyte (male gametophyte) of seed plants.

Digging Deeper

ORGANISM RESPONSE TO CLIMATE CHANGE

Pollen and Spores

The reproduction in plants is different from that in animals. As part of the plant life cycle, **spores** (from mosses, ferns, and horsetails) or **pollen** (from most other kinds of plants) are produced in the hundreds of thousands per individual plant each year. Spores and pollen can be dispersed in a number of ways. The most common way is by wind. This is made easier by the very small size of individual pollen or spore grains, generally less than 50 μm (0.05 mm). *Figure 1* shows a 15 μm pollen grain magnified 1600 times its actual size.

Figure 1A Transmitted-light image of a pollen grain (of the genus *Cupuliferoipollenites*), 15 μm in diameter.

Figure 1B Scanning electron microscope image of the same pollen grain.

All species are genetically unique. The genes of an organism control its specific growth and development. When plants reproduce, each plant species produces a unique spore or pollen grain. Spores and pollen grains differ greatly in their shape, size, and surface features. Some plants produce spores that are triangular, and others produce spores that are football shaped, as shown in *Figure 2*. Some spores are covered with short spines or club-shaped structures, and

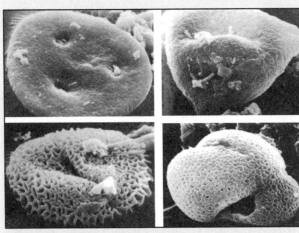

Figure 2 Scanning electron microscope image of four different pollen grains, ranging in diameter from 18 to 38 μm.

Digging Deeper

You may wish to assign the reading for homework. The questions in **Check Your Understanding** can be provided as a homework assignment.

Assessment Opportunity

Reword or restructure the questions in **Check Your Understanding** for a brief quiz. Use the quiz (or a class discussion of the questions in the textbook) to assess your students' understanding of the main ideas in the reading and the activity.

Assessment Tool

Check Your Understanding Notebook Entry-Evaluation Sheet
Use this sheet to evaluate the extent to which students understand the key concepts explored in **Activity 3** and explained in **Digging Deeper**, and to evaluate the students' clarity of expression.

Teaching Tip

Although the **Digging Deeper** reading section states that the very small size of pollen (like those pictured in *Figures 1* and *2*) and spores makes them easily dispersed by wind, the actual size of a micrometer (also called a micron and abbreviated μm) is a concept that is unfamiliar to most people. Accordingly, it might be useful to have students relate the size of one of these pollen grains to a measure that they are more familiar. Students can calculate how many of these pollen grains could be lined up in the space of one millimeter and use a metric ruler to then visualize the size of a 15 μm diameter pollen grain.

Chapter 3

others are perfectly smooth. All plants of the same species produce the same type of pollen each year throughout each generation. The pollen grains are transported to ponds, lakes, and rivers where they settle with the sediment and are buried.

Pollen and spores are very resistant to decay. Pollen and spores can be preserved for hundreds of millions of years under the right burial conditions. This resistance to decay is the main reason why spores and pollen can be extracted not only from recently deposited sediments but also from ancient sedimentary rocks from the geologic past.

Paleoclimate

Paleoclimatology is the study of the Earth's past climates. Climatologists know from several kinds of evidence that the Earth's climate has changed continually, and substantially, through geologic time. Over the past million years (a short time, geologically!) there have been several colder periods when vast ice sheets, as much as a few kilometers thick, expanded to cover much of northern and central North America and Eurasia. In between these **glacial periods**, there were shorter periods, called **interglacials**, when global temperature was much warmer and ice sheets disappeared from North America and Eurasia. *Figure 3* shows temperature changes over the past 420,000 years as interpreted from data recovered from the Vostok ice core (Antarctica).

Geo Words

paleoclimatology: the scientific study of the Earth's climate during the past.

glacial period: an interval in time that is marked by one or more major advances of glacier ice. Note that the time interval is not necessarily of the same magnitude as the "Period" rank of the geologic time scale.

interglacial period: the period of time during an ice age when glaciers retreated because of milder temperatures.

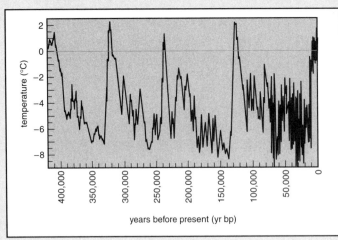

Figure 3 Temperature variation over the past 420,000 years relative to the modern surface temperature at Vostok (−55.5°C).

Blackline Master Changing Life 3.2

Temperature Variation over the Past 420,000 Years

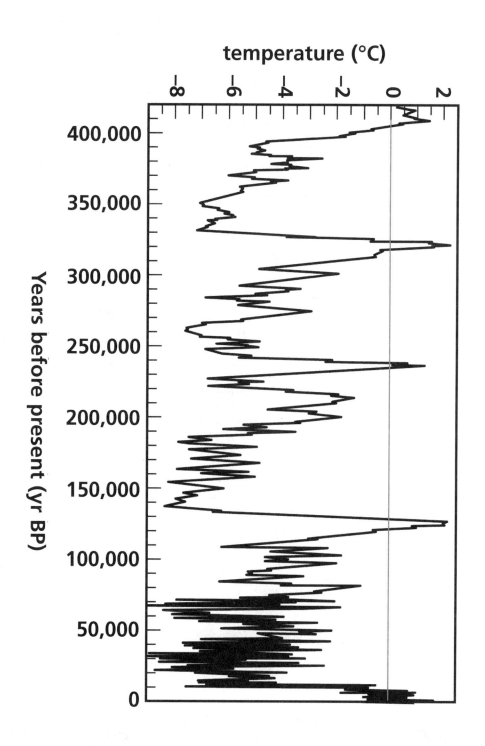

Earth System Evolution Changing Life

Geo Words

steppe: an extensive, treeless grassland found in semiarid mid-latitude regions. Steppes are typically considered to be drier than the prairie.

Glacial periods last longer than interglacials. Ice sheets build up slowly over several tens of thousands of years, and then retreat rapidly from their maximum advance over a brief period of several thousand years. Interglacials last only 20,000 to 25,000 years. The last glacial maximum was about 20,000 years ago. By about 7000 years ago, almost all of the Northern Hemisphere continental ice sheets, except for the Greenland ice sheet, had melted away. Since then, global climate has been relatively warm, although with significant fluctuations over decades, centuries, and millennia.

The Response of Biomes to Climate Change

Each glacial advance modified the climate across North America. In response to these climatic changes, there was a change in biome boundaries. Each plant species has evolved to be adapted to a particular range of climatic conditions. Outside of that range of conditions, the climate is too cold or warm, or too wet or too dry, for that species to survive over the long term. As climatic belts shift during times of global climate change, plant species shift accordingly in their range of distribution. If the change is too rapid, they become extinct. The animals that rely upon specific kinds of plants for food or shelter are similarly affected. Some of the organisms that lived during the most recent glacial and interglacial periods continue to exist today in North America. Some of the organisms that lived in the past are now extinct.

The distribution of biomes during the last glacial period, as determined from the fossil record across North America, was very different from the present distribution. For example, much of the southeastern United States was covered in an open woodland or forest that graded westward into forest **steppe** and open, dry treeless steppe at the glacial maximum. Fossil evidence indicates that parts of the Florida peninsula may have been desertlike. At the same time a variety of forest types persisted in the northernmost part of the state. To the east of the Appalachians, cool-climate pine forests and prairie herbs existed. The central Rocky Mountains were covered in cold-tolerant conifer woodlands. Farther south there was a mixture of semidesert scrub and sparse conifer woodland. Where there are dry desert conditions in the southwest, temperate open woodlands or grasslands covered the region during glacial maximum.

Not only was biome distribution different in response to the shift in climate belts across North America, but many of the organisms within them were also different. Fossil evidence indicates that some associations between plants and animals, particularly among ice-age North American mammals that exist today did not exist in the Pleistocene. For example, the snowshoe hare still found in Minnesota and Wisconsin is known from fossil evidence in

Teaching Tip

The data for the graph shown in *Figure 3* on page E169 were obtained from a long ice core taken from the Antarctic ice sheet. The temperatures were derived from a study of samples of air that were trapped in bubbles in the ice. These temperatures are not a direct measure of global temperature, but they give a good idea of the sense and magnitude of change in global temperature with time. The periodicity of climate change is clear from the graph.

Chapter 3

Missouri, Kentucky, and Virginia when these states were covered in taiga. It's known that extinction of large mammals in North America occurred at the time when the last continental ice sheets melted. Two hypotheses have been proposed to account for this. One hypothesis proposes that ecological changes at the end of the Pleistocene disrupted the biological balance. This hypothesis suggests that at the time of biome readjustment, a large number of species were unable to adapt. The other hypothesis, the "overkill model," proposes that

the expansion of human migrants from Asia into North America resulted in overhunting and depletion of the mammal populations. Some authors suggest that humans killed everything in their path, whereas others hypothesize that humans reduced mammal populations to low numbers. Such low populations, in turn, were unable to maintain birthrates high enough for the species to survive. Whatever the reason (or reasons), North America today is very different from the North America of just 10,000 years ago.

Check Your Understanding

1. What are the two hypotheses to account for the extinction of some animals at the end of the Pleistocene?

2. Why are pollen and spores conducive to long-term preservation?

3. How was the climate of North America different during the last full glacial episode? What impact did this have on organisms in the area?

Understanding and Applying What You Have Learned

1. A scientist is like a detective. Rarely do detectives have all of the information to solve the case. However, they usually get enough evidence to show beyond a reasonable doubt what happened and when it happened. Use evidence that you have gathered to answer the following questions:

   a) How can you demonstrate that the biome in which you now live was different during and after the last glacial maximum?

   b) How many different kinds of fossil data have been found in your community or region?

   c) What do these fossil data tell you about the plants and animals that once lived where you do?

   d) What happened to the Late Pleistocene plants and animals that today are not found in your community?

2. It has been stated that the fossil record is biased. What biases might have affected the Late Pleistocene fossil record of your community's recent past?

3. How good is the Pleistocene fossil record? How much data have been collected and used to characterize the North American Pleistocene?

Check Your Understanding

1. Two hypotheses to account for the extinction of some animals at the end of the Pleistocene suggest:

    - that ecological changes at the end of the Pleistocene disrupted the biological balance

    - that the expansion of human migrants from Asia into North America resulted in overhunting and depletion of mammal populations

2. Pollen and spores are very resistant to decay, which makes them conducive to long-term preservation.

3. The climate of North America was much colder during the last full glacial episode. Organisms living in the area during this time had to be adapted to much colder and harsher conditions.

Understanding and Applying What You Have Learned

1. a) Answers will vary. Students may say that they could look for fossil evidence of plants and animals that may have lived in their community and are suggestive of a colder climate.

    b) Answers will vary.

    c) Answers will vary.

    d) Answers will vary. The plants and animals have either evolved to survive the current climate conditions, migrated to a different region, or become extinct.

2. Answers will vary. Many factors come to bear on the likelihood of an organism being fossilized and preserved in the geologic record. Many of these are discussed in the **Digging Deeper** reading section of **Activity 1**, and student answers will likely reflect the content of that section. Some factors that influence the likelihood of fossilization are: the environmental and geographic conditions of the area and the nature of the organism itself (i.e., does the organism have hard parts that are easily preserved, did the organism live in an environment conducive to fossilization, etc.). Because not all organisms have an equal potential for fossilization, the fossil record by itself does not portray a complete representation of floral and faunal assemblages.

3. The fossil record of the Pleistocene is better than the record of fossils at almost any other time in Earth history. This should not be surprising, because generally the older the deposits, the less chance fossils have to survive. The fossil record of land plants, preserved as pollen and spores, is especially good. In many cases, the depositional environments in which these materials were preserved, like ponds and swamps, have not yet been removed by erosion.

Earth System Evolution Changing Life

4. a) For the Pleistocene plants and animals that no longer live in your community and survived the Pleistocene extinction, what part(s) of the food web do they occupy? Why might they no longer be found in your community naturally?

b) For the Pleistocene plants and animals that lived in your community but didn't survive the Pleistocene extinction, what part(s) of the food web did they occupy? Why might they have become extinct?

Preparing for the Chapter Challenge

Think about what you have learned about the plants and animals in the present and past biome of your area. Prepare an illustration that shows the relationships between the organisms of the past and present. Can you document a progression from the organisms of the past to the organisms of the present? Include a narrative in which you explain why the organisms changed through time.

Inquiring Further

1. **Carbon-14 dating techniques**

   Search your library and the web for information about how it is possible to determine the carbon-14 age of organic matter. Prepare a report for the class on the physical and chemical principles upon which the technique is based, and how advances in analytical equipment have reduced the uncertainty in the age estimates.

2. **Late Pleistocene extinctions**

   Approximately 11,000 years ago, a variety of animals across North America became extinct. These were mostly large mammals (over 50 kg), like saber-toothed cats, mammoths, mastodons, and giant beavers. What mechanisms have been proposed to have caused the extinctions in North America? Did Late Pleistocene extinctions occur in places other than North America? If so, where and what types of organisms became extinct?

4. a) – b) Answers to these questions will vary depending upon your community. For **Part A**, one possible explanation is that urbanization has changed their habitat and driven them away from the region. One possible explanation for **Part B** is that the organism was unable to adequately adapt to the changes in its environment that resulted from climatic changes that have occurred since the Pleistocene.

Preparing for the Chapter Challenge

This exercise will help students as they begin to develop their final displays. They should be able to see a progression from the organisms that were found in the community during the Pleistocene and the organisms found in the community today. Students who are very creative will particularly enjoy creating the illustration. Remind them that they should include a written explanation of their illustration.

Criteria for assessment (see **Assessment Rubric for Chapter Challenge on Changing Life** on page 520) include discussions of:

- evidence that there has been change in the life forms and biodiversity of your community over time
- short-term and long-term factors that have influenced biological changes in your community

Review these criteria with your students so that they can be certain to include the appropriate information in their illustrations and short papers.

Inquiring Further

1. Carbon-14 dating techniques
Carbon-14 dating is a way to determine the age of organic material. Carbon-14 (conventionally written ^{14}C) is a radioactive isotope of carbon, with a half-life of about 5700 years. ^{14}C is produced in the upper atmosphere when high-energy cosmic rays interact with the atmosphere. The interaction of cosmic rays with the atmosphere results in the release of neutrons. Some of the slow-moving neutrons in turn strike nitrogen-14 (^{14}N) atoms. The collision causes the nitrogen atom to capture the neutron and emit a proton, thus changing it to a ^{14}C atom (nitrogen has 7 protons and carbon has 6). The ^{14}C combines with oxygen to form a molecule of CO_2, which is then mixed with the CO_2 of the atmosphere.

In contrast to the other isotopes of carbon, ^{12}C and ^{13}C, which are stable, ^{14}C is radioactive: eventually the atom of ^{14}C decays back to ^{14}N. The CO_2 is incorporated into plants by photosynthesis and into animals when they eat plants. The ratio of ^{14}C to the stable isotopes of carbon is almost exactly the same in every living organism at any given time.

When an organism dies, it stops taking in new carbon. The ratio of ^{14}C to ^{12}C at the time of death is the same as that of every living thing. If the organic matter is sealed away from the atmosphere, so that no new ^{14}C is introduced, the ratio of ^{14}C to ^{12}C slowly decreases. Scientists can measure the $^{14}C/^{12}C$ ratio in a suitable sample and then use the known rate of ^{14}C decay to calculate the age of the sample.

Because the half-life of ^{14}C is only about 5700 years, however, the carbon-dating technique is useful only for samples up to about 60,000 to 70,000 years old. The technique is especially useful for samples of woody plant material and for shells consisting of calcium carbonate minerals.

2. Late Pleistocene extinctions

Nearly 70 species, approximately 95% of the megafauna present in North America during the Pleistocene, became extinct at about 11,000 ka. This included large mammals like mammoths, mastodons, horses, camels, ground sloths, giant beavers, giant jaguars, and sabre-tooth cats.

The Pleistocene extinctions in North America have been linked to environmental changes, including:

- a relatively rapid temperature increase
- changes in rainfall patterns
- glacial melting

Some organisms could not adapt or relocate quickly enough to survive.

Another hypothesis suggests that humans, newly arrived from eastern Asia across the Bering Strait land bridge, hunted many of the mammals of North America to extinction. This is supported by the fact that the extinction of these mammals coincides very closely with the arrival of humans to the North American continent. Cave paintings in North America from that time period depict hunting scenes.

Perhaps there is an element of truth in both hypotheses: the animals might have already been stressed by climate change when the hunting began on a large scale.

NOTES

ACTIVITY 4 — THE MESOZOIC-CENOZOIC BOUNDARY EVENT

Background Information

The Geologic Time Scale

The development of ideas about geologic time had to await the first great revolution in geology, which occurred at the end of the 18th century. This revolution was brought about by James Hutton's ideas about the development of the geological record by the action of processes we can observe and study, like erosion, sedimentation, volcanism, and tectonism. Two pioneers in the early years of the 19th century—William Smith in England and Georges Cuvier in France—laid the groundwork for dating sedimentary rocks by their fossil content. These two scientists were the first to perceive that there is a definite and nonrepeating succession of fossil species in the sedimentary rocks of the world, largely independent of the particular rock types that contain the fossils.

As geologists in the first part of the 19th century began their systematic study of the record of successions of ancient sedimentary rocks around the world, they were led to develop a chronology based on the fossil content of the rocks. For the sake of communication, it became necessary to divide geologic time into manageable, named units. The boundaries between units were generally drawn where there were significant worldwide changes in the faunas contained in the rocks. By the late 1800s, the result of these efforts—the relative geologic time scale—was largely in place. The adjective "relative" is used for this time scale because absolute ages, in years, are not involved.

The Relative Geologic Time Scale

The relative geologic time scale involves a hierarchy of units, ranging from the longest spans of geologic time to the shortest (see the chart on page E174 of the student text):

- eons
- eras
- periods
- epochs

There are further, even finer, subdivisions, but they are not generally recognizable worldwide.

Each of the Paleozoic and Mesozoic periods are subdivided into epochs. The epochs are further subdivided into units called ages.

There is another set of units, which exactly parallel the time units, called time-stratigraphic units. These are defined as consisting of all the rocks that were deposited during the corresponding time unit. For example, the time-stratigraphic equivalents to periods are called systems, and the time-stratigraphic equivalents to epochs are called stages. Your students might be confronted with these differently named units when they use geologic maps.

The Absolute Geologic Time Scale

When the relative time scale was developed, there was still no way to assign absolute ages to the rocks. That was to come much later, in the middle of the 20th century, with the advent of radioisotopic dating methods. Soon after the discovery of natural radioactivity in the final years of the 19th century, geologists realized that the rate of decay of radioactive isotopes like uranium could be used to date rocks. This idea could not be exploited fully, however, until the middle of the 20th century, with the advent of sophisticated devices called mass spectrometers that allow precise separation of the various isotopes involved.

Generally, only igneous rocks are suitable for radioisotopic dating. The sedimentary record must therefore be dated either directly or indirectly. The direct method involves dating beds of volcanic ash that in many places are interbedded with sedimentary rocks. The indirect method uses a "bracketing" technique whereby a date is obtained for two igneous intrusive bodies; one of the intrusive bodies is overlain by sedimentary rocks, and the other cuts through those same sedimentary rocks.

By use of methods like radioisotope dating, absolute dates have gradually been associated with the relative time scale. The process continues today: only recently, the age of the base of the Cambrian System (that is, the time for the beginning of the Cambrian Period) was changed from about 570 million years to 543 million years!

Mass Extinctions

Species originate, and they eventually become extinct. The average duration of existence of species is not easy to estimate, but it probably lies in the range of half a million years to a few million years. The lifespans of individual species, however, range widely, from just several tens of thousands of years to tens of millions of years. Species that become adapted to environments that change little through geologic time obviously last the longest.

The general course of fairly steady rates of origination and extinction of species is known as background extinction. In contrast, at certain times in Earth's history the rate of

extinction of species has experienced sudden peaks. At these times, substantial percentages of all known fossil species became extinct. These events are known as mass extinctions.

The mass extinction that is best known outside the circle of geoscience specialists is the one that happened at the end of the Cretaceous. (It's worth pointing out here that its association with the boundary between the Mesozoic and the Cenozoic is just because those time periods were originally *defined* on the basis of the mass extinction! The early geologists who contributed to the development of the geologic time scale were perceptive in being guided in their placement of boundaries by recognizing sudden and substantial changes in faunas and floras, which we now recognize as mass extinctions.) The greatest mass extinction of all time, however, was at the end of the Paleozoic. Over 90% of all known animal species became extinct.

Cause(s) of Mass Extinction

The cause (or causes) of mass extinctions continues to be controversial. In the case of the end-Cretaceous mass extinction, it is now generally accepted that the cause was collision of an asteroid-size object with the Earth. In many sedimentary successions of that age around the world, there is a spike in concentration of the chemical element iridium. This element has typically very low concentrations in terrestrial rocks but is known to be abundant in certain meteorites. More recently, the site of the impact has been tentatively identified in the Yucatan Peninsula of Mexico.

The cause of the earlier end-Permian extinction remains controversial. Some believe that there is evidence of a catastrophic collision, but another cause (or causes) seems

Chapter 3

to most geoscientists more likely. Coincident with the mass extinction was the great outpouring of basaltic lava to form what are called the Siberian traps. The volume of the outpouring is staggering: millions of cubic kilometers of lava, all within the (geologically) extremely short time span of about a million years. Some believe that such an outgassing of enormous amounts of carbon dioxide and other greenhouse gases would have altered global climate abruptly and fundamentally, thus exerting extreme stress on both marine and terrestrial organisms.

Another hypothesis for the cause of the end-Permian extinction is that global cooling and development of glacial conditions at high latitudes might have triggered a massive outgassing of carbon dioxide and methane from the oceans. There is good evidence that at times of warm and equable global temperatures, the deep-ocean circulation stagnates from the lack of production of cold, dense seawater at high latitudes. This stagnation in turn leads to buildup of gases like carbon dioxide by decay of organic matter that rains down from the surface waters. Geologically sudden release of the dissolved gases upon invigorated circulation might have caused climate change that would have stressed biotas severely.

No consensus has yet developed. The jury is still out on the basic causes of mass extinctions.

More Information – on the Web

Visit the *EarthComm* web site www.agiweb.org/earthcomm to access a variety of links to web sites that will help you deepen your understanding of content and prepare you to teach this activity. Many of the sites also contain images that you can download.

Goals and Assessment

Clarify that the goals indicate what students should understand and be able to do as a result of the activity. Make sure students understand that Chapter Assessments are based upon these goals.

Goal	Location in Activity	Assessment Opportunity
Understand how changes in the Earth's climate have affected organisms throughout geologic time.	**Investigate** Part A **Digging Deeper**	Questions are answered in a reasonable manner; sufficient data is collected to support responses.
Understand that the organisms that dominate the continents today differ from the organisms that dominated the Earth in the deep geologic past.	**Investigate** Part A and Part B **Digging Deeper; Check Your Understanding** Questions 1 – 3 **Understanding and Applying What You Have Learned** Question 1	Answers to questions are accurate and closely match those given in the Teacher's Edition.
Understand that severe ecological disruptions alter the history of life, resulting in extinctions followed by the evolution and appearance of new organisms.	**Investigate** Part A **Digging Deeper; Check Your Understanding** Questions 2, 4 – 5	Answers to questions are accurate and closely match those given in the Teacher's Edition.
Understand that newly evolved organisms develop similar body features as organisms that became extinct, allowing them to use the same resources.	**Investigate** Part A **Digging Deeper; Check Your Understanding** Question 3	Answer to question is accurate and closely matches that given in the Teacher's Edition.

Chapter 3

Activity 4 The Mesozoic—Cenozoic Boundary Event

Goals

In this activity you will:

- Understand how changes in the Earth's climate have affected organisms throughout geologic time.

- Understand that the organisms that dominate the continents today differ from the organisms that dominated the Earth in the deep geologic past.

- Understand that severe ecological disruptions alter the history of life, resulting in extinction followed by the evolution and appearance of new organisms.

- Understand that newly evolved organisms develop similar body features that allow them to use the same resources as those organisms that became extinct.

Think about It

Perhaps no other interval of geologic time has captured the attention of popular culture as much as the Mesozoic Era. This was the time when dinosaurs roamed the Earth.

- If the dinosaurs were such successful creatures, what happened to them, and why?

What do you think? Record your ideas about this question in your *EarthComm* notebook. Be prepared to discuss your ideas with your working group and the class.

E 173

Activity Overview

Students begin the activity by visiting the *EarthComm* web site to collect data on the climate before and after the Mesozoic–Cenozoic boundary. Students then collect data on the fossil record of organisms before and after the boundary. They make connections between the appearance and disappearance of organisms and the change in climate across the boundary. Students then examine photographs of modern carnivore and herbivore skulls and compare them to skulls from the Mesozoic to understand how extinctions can open niches for new organisms. The **Digging Deeper** reading section defines extinction, reviews the causes of extinction, and takes a closer look at the causes of the extinction event that occurred across the Mesozoic–Cenozoic boundary.

Preparation and Materials Needed

Preparation

Part A

Students will need Internet access to complete this part of the investigation, so you will need to plan ahead and make any necessary arrangements. If you cannot provide students with Internet access in the classroom, assign this part of the investigation as a homework exercise, or provide students with printouts of the necessary information.

Part B

No advance preparation is required to complete this activity.

Materials

Part A

- Internet access (or information about paleoclimate before and after the Mesozoic–Cenozoic boundary)*
- Internet access (or information about organisms living before and after the Mesozoic–Cenozoic boundary*

Part B

- No additional materials needed

Think about It

Student Conceptions

Most students are probably familiar with the notion that dinosaurs became extinct suddenly. They may or may not be familiar with the theories about the reasons behind the extinction.

* See the *EarthComm* web site for information about how to obtain these resources at http://www.agiweb.org/earthcomm

Answer for the Teacher Only

The sudden demise of the dinosaurs at the end of the Cretaceous Period is only part of the story: the great mass extinction at the end of the Cretaceous eliminated about one quarter of all known families of animals. The extinction affected both land and sea animals, but it was especially severe for land animals. There has been a long-standing controversy about what caused the great extinction.

Most geoscientists accept the hypothesis that the extinction was caused by the impact of an asteroid-size object with the Earth, which would have shrouded the Earth with dust and soot. The evidence for such an impact is clear. One problem, however, is that not all of the extinctions were "overnight." Many groups died out over a span of as much as a million years.

An alternative hypothesis is that climate cooling and sea-level fall near the end of the Cretaceous, for which there is good evidence, set the stage for extinction of a large part of the Earth's fauna and flora by the meteorite impact. The dinosaurs, in particular, seem already to have been much less successful by the end of the Cretaceous than they had been in earlier times during the Mesozoic.

Assessment Tool

Think about It Evaluation Sheet
Use this evaluation sheet to help students understand and internalize the basic expectations for the warm-up activity.

NOTES

Earth System Evolution Changing Life

Investigate

Part A: Changes in Climate and Life at the End of the Mesozoic Era

1. Look at the geologic time scale. Familiarize yourself with the terms used to name different geologic time intervals. The boundary between the Mesozoic Era and the Cenozoic Era, about 65 million years ago, represents one of the most catastrophic extinction events in Earth history. In this part of the investigation you will focus on the changes in climate and organisms between the end of the Cretaceous Period (the end of the Mesozoic), and the beginning of the Tertiary Period (the beginning of the Cenozoic). You may sometimes see different names used to refer to the Cenozoic Periods. That is because there are two alternative ways that the Cenozoic Era can be subdivided into periods. In the older terminology, the Cenozoic is subdivided into the Tertiary and Quaternary Periods.

Major Divisions of Geologic Time
(boundaries in millions of years before present)

	Epoch	Period		Era	Eon
.01	Holocene	Neogene	Quat.	Cenozoic	Phanerozoic
1.8	Pleistocene				
5.3	Pliocene				
24	Miocene		Tertiary		
34	Oligocene	Paleogene			
56	Eocene				
65	Paleocene				
145		Cretaceous		Mesozoic	
213		Jurassic			
248		Triassic			
286		Permian		Paleozoic	
325		Pennsylvanian			
360		Mississippian			
410		Devonian			
440		Silurian			
505		Ordovician			
544		Cambrian			
2500				Proterozoic	Cryptozoic (Precambrian)
				Archean	

Note that the divisions of geologic time are not drawn to scale. Quat. = Quaternary

E 174

EarthComm

Investigate

Part A: Changes in Climate and Life at the End of the Mesozoic Era

Teaching Tip

Use **Blackline Master Changing Life 4.1, Major Divisions of Geologic Time** to make copies of the figure on page E174 for students to insert into their notebooks for quick reference. This figure gives additional information that was not provided in the earlier figure that detailed the major evolutionary events of geologic time (page E153). Specifically, this figure provides greater detail on the subdivision of time in the Cenozoic Era.

1. The conventional geologic time scale has evolved since geologists first began to study the sedimentary record of the Earth almost 200 years ago. The relative geologic time scale was mostly in place by the late 1800s.

    The boundaries between the major divisions of geologic time (the eras and the periods) are mostly reflective of major changes in fauna and flora. These boundaries are thus natural rather than arbitrary. In particular, the Permian–Triassic boundary and the Cretaceous–Paleogene boundary correspond to the two greatest episodes of mass extinction in Earth history. It was only in the mid-1900s, with the advent of absolute dating techniques based on decay of radioisotopes, that numbers (i.e., dates) were placed in the various boundaries. These numbers continue to be adjusted, as new dates become available and accuracy in dating improves.

Chapter 3

The newer terminology, however, subdivides the Cenozoic Era into the Paleogene and Neogene Periods. It is important to note, as is shown in the diagram on p. 174, that the Period boundaries are not in the same place for the two different terminologies.

2. In your group, visit the *EarthComm* web site to collect data on paleoclimate from the pre-boundary time interval (the Late Cretaceous) and the post-boundary time interval (the earliest Tertiary, also known as the Paleogene Period).

   a) How did the climate change at the end of the Mesozoic?

   b) What evidence is there on how fast the climate changed?

   c) From your findings, what do you think are the dominant plants and animals in North America before the boundary?

   d) From your findings, what do you think are the dominant plants and animals in North America after the boundary?

3. In your group, visit the *EarthComm* web site to collect data on the fossil record of organisms from the pre-boundary time interval (the Late Cretaceous) and the post-boundary time interval (the Paleogene).

   a) How accurate were your predictions on the dominant plants and animals in North America before and after the boundary? Explain any differences.

   b) How do you think that the interactions between organisms differed before and after the boundary? Why?

   c) Which organisms from the fossil record are still living (extant) and which ones are no longer known to be alive on Earth today (extinct)?

   d) Why do you think that some organisms survived beyond the boundary, but others did not?

   e) How do you think that the dominant plants and animals in your community differed before and after the boundary? Why?

Part B: Consumers of the Mesozoic and Cenozoic

The adaptions of organisms that have evolved to function within a given ecosystem appear to have several common features that allow success. You have now seen that plants before and after the extinction between the Cretaceous and Tertiary Periods (the K/T extinction) were very similar but that the dominant animals were very different. Or were they?

1. Examine the photographs of the skull and jaw structure of both a hare (rabbit) and a wolf (shown on the next page.

Hare skull.

2. **a)** The climate cooled at the end of the Mesozoic. The Paleocene (the beginning of the Cenozoic) is thought to have had a cooler climate than the Cretaceous, although still warmer than the climate today. Seasons began to grow more pronounced as the global climate became cooler.

   b) Many scientists have studied the magnitude and rate of climate change during the Late Cretaceous and Early Paleogene. Some of the best data comes from marine sediments and the study of the isotopic composition of fossil foraminifera. These data show a period of rapid global cooling in the latest Cretaceous, around the time interval that surrounds the boundary.

   c) Students will probably indicate that plants and animals adapted to a warm climate were dominant before the boundary. Many may also recognize that dinosaurs existed before the boundary. In greater detail, life in the early part of the Cretaceous Period was much like that of the Jurassic Period that came before it. Forests of ferns, cycads, and conifers occupied the land. Dinosaurs dominated the land fauna, ammonites, belemnites, other mollusks, and fish lived in the waters and were hunted by marine reptiles, and pterosaurs and birds flew in the skies above. Although mammals first appear in the Triassic, in the Cretaceous they were still small, mostly nocturnal animals. The Cretaceous also saw the first appearance of many life forms that would go on to play key roles in the coming Cenozoic world. During the Cretaceous, forests evolved to look similar to present-day forests, with oaks, hickories, and magnolias becoming common in North America by the end of the Cretaceous. One important development was the first appearance of the flowering plants. Flowering plants first appeared about 125 million years ago, and by the close of the Cretaceous a number of forms had evolved that any modern botanist would recognize. In addition to this, many different kinds of insects, including (among others) ants, termites, butterflies, aphids, grasshoppers, and the eusocial bee came into existence during the Cretaceous.

   d) Students will probably indicate that plants and animals more adapted to cooler climates became dominant after the boundary. They may also note that after the end of the Mesozoic, dinosaurs became extinct, triggering a major shift in the dominant land fauna. In greater detail, this situation was the starting point for the great evolutionary success of mammals. Within ten million years, at the end of the Paleocene, mammals had occupied a large part of the vacant ecological niches. The forests supported a great variety of insects, resulting in a great deal of insectivores from this time period. Browsing animals (leafeaters) were also prevalent in the Paleocene, many of which have descendants that can be found in modern tropical and subtropical environments, and carnivorous mammals were stalking their prey. Plesiadapiformes, an order of mammals that appear to be very primate-like, flourished in the Paleocene. Mammals were not the only groups of organisms that continued their existence beyond the Mesozoic into the Cenozoic. Many other groups of organisms, like flowering plants, gastropods and pelecypods (snails and clams), amphibians, lizards and snakes, crocodilians, and mammals survived the Cretaceous – Tertiary boundary, with few or no apparent extinctions.

Chapter 3

3. **a)** Answers will vary. Note the descriptions above for **Steps 2a** and **2b.**

   b) Answers will vary, but may include some discussion of how the extinction of many different organisms (actually orders or organisms) opened up many ecological niches greatly changing the interactions between different species of organisms and the opportunities for those species to thrive.

   c) Many orders of living organisms still survive today. Perhaps one of the best-known examples is the coelacanth. The coelacanth is a kind of fish that appeared 410 million years ago, and there have been hundreds of species in this order. The whole order apparently vanished from the fossil record about 65 million years ago. Everyone thought that this order became extinct, but in 1937 a fisherman caught a live coelacanth. (It was, however, a different species, genus, and family from any fossil coelacanth). Cycads are plants that were abundant in the Mesozoic, and their relatives still exist today. There are several families of cycads from the fossil record that still exist today. Other examples of organisms from the fossil record whose lineage continues to modern times include sharks and crocodilians. The lineage of modern crocodilians extends back to the earliest part of the Mesozoic, and modern sharks first appear in the Jurassic.

   d) Answers will vary, but one possible answer is that some organisms were better equipped to thrive in the environment that was brought about by the event(s) that led to the mass extinction at the end of the Mesozoic.

   e) Answers will vary.

Part B: Consumers of the Mesozoic and Cenozoic

1. **a)** Students should note that the rabbit skull has large, blunt front teeth and broad, flat back teeth. The wolf skull has a tight arrangement of sharp teeth.

   b) The wolf is a carnivore; the rabbit is a herbivore.

   c) The teeth of the rabbit are adapted for eating plants, meaning that it is a primary consumer within the ecosystem. The tooth arrangement of the wolf, on the other hand, is designed for eating flesh, meaning that the wolf is a higher-order consumer.

NOTES

Chapter 3

Earth System Evolution Changing Life

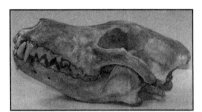

Wolf skull.

a) What are the differences in the organization of teeth in these animals?

b) Which animal is a carnivore and which is a herbivore?

c) How does the tooth arrangement reflect their levels within the ecosystem?

2. Examine the photographs of extinct fossil mammal skulls collected from Paleogene rocks of North America.

a) How are these skulls similar to modern rabbits and wolves?

b) In which level(s) of the ecosystem would you predict that each of these animals existed? That is,

which ones were herbivores and which ones were carnivores? How can you make these interpretations?

3. Examine the photographs of dinosaur skulls collected from Mesozoic rocks of North America.

a) What features do these skulls have that are similar to the modern rabbits and wolves, and to the extinct mammals from the Tertiary Period?

b) Which dinosaurs exhibit skull, jaw, and teeth similar to the herbivores?

c) Which dinosaurs exhibit skull, jaw, and teeth similar to the carnivores?

d) In which level(s) of their ecosystem would you predict that each of these animals existed? How can you make these interpretations?

Archaeohippus (an early horse).

Smilodon *californicus* (saber-toothed tiger).

Diplodocus.

Tyrannosaurus *rex*.

2. a) Students should look at the tooth structure of the skulls, as well as skull shape.

   b) The Archaeohippus and *Smilodon californicus* skulls are from the Paleogene. *Archaeohippus* was a primary consumer, or herbivore. *Smilodon californicus* was a higher-order consumer, or carnivore. Students should be able to assess this by looking at the teeth of the skulls. They can consult **Activity 1** if they have questions about the different levels of the ecosystem.

3. a) The *Diplodocus* and *Tyrannosaurus rex* skulls are from the Mesozoic. Again, in comparing the skulls with the Paleogene and modern skulls, encourage students to look at both tooth structure and skull shape.

   b) The *Diplodocus* skull is similar to that of a herbivore.

   c) The *Tyrannosaurus rex* skull is similar to a carnivore skull.

   d) *Diplodocus* was a primary consumer, or herbivore. *Tyrannosaurus rex* was a higher-order consumer, or carnivore. These interpretations can be made based upon tooth structure and skull shape. Higher-order consumers have eye sockets placed close to the front of the skull, while primary consumers often have eye sockets placed farther back and to the sides of the skull. Having eyes forward in the skull is associated with greater depth perception, an adaptation beneficial to predators. Conversely, having eye sockets set farther back and to the sides of the skull can give a greater ability for peripheral vision, an adaptation useful in avoiding predators.

Assessment Tools

EarthComm Notebook Entry-Checklist
Use this checklist as a quick guide for student self-assessment and/or an opportunity to quickly score student work. Add further criteria specific to your classroom needs or to this particular investigation.

Investigate Notebook Entry-Evaluation Sheet
Point out the criteria listed on this evaluation sheet that are relevant to this particular investigation. Encourage students to internalize the criteria by making them part of your "assessment conversations" as you circulate around the classroom.

Chapter 3

Reflecting on the Activity and the Challenge

This activity gave you the chance to explore the fact that life forms that dominated Earth in the deep past differ in several ways from what exists at the present. However, these extinct life forms occupied all the parts of the same ecological levels that exist today. It also showed you that at certain times in the geologic past, climate and life changed over very short geological times. New organisms evolved to take the ecological place of those that became extinct. How did this activity change your ideas about life on Earth? You will need to explain the appearance of new animals and plants in Paleogenic (post-boundary) rocks that have no fossil record in Cretaceous (pre-boundary) rocks as part of your **Chapter Challenge**.

Digging Deeper

THE EXTINCTION OF SPECIES

The success or failure of a particular species through time is impossible to predict. Many physical and biological factors interact with one another in complicated ways over a long time to determine success or failure. Additionally, the data from the fossil record seem to indicate that different kinds of organisms have different rates of overall success. Paleontologists have documented that some species have persisted for tens of millions of years. Others span only a few million years, or even only a few hundred thousand years.

The Extinction Event at the End of the Mesozoic

You have learned that there was a marked difference in the landscapes of the Mesozoic Era and the Cenozoic Era. (The terms come from the Greek *meso-*, meaning middle, and *kainos-*, meaning new.) The groups of animals that dominated Earth for nearly 130 million years during the Mesozoic in the pre-boundary biosphere mysteriously disappeared. The disappearance was sudden; almost overnight, in terms of geologic time. This extinction affected some plants and many groups of animals that lived on land. It similarly affected much of the food web in the oceans. Groups from lowly **phytoplankton** to top carnivores disappeared from the fossil record, never to be seen again except as fossil material. The post-boundary biosphere, which became established early in the Paleogene was very different in its aspect. It took several million years for the plant and animal groups

Geo Words

phytoplankton: small photosynthetic organisms, mostly algae and bacteria, found inhabiting aquatic ecosystems.

Reflecting on the Activity and the Challenge

Have a student read this section aloud to the class, and discuss the major points raised. Discuss with students the implications of the information they examined in the investigation. Ask them about the relevance of examining modern skulls and Mesozoic skulls. Help them recognize that these organisms occupied similar niches within the ecosystem and that the extinction of the Mesozoic organisms opened up opportunities for the Cenozoic organisms to thrive. This is an important point of the investigation.

Digging Deeper

Assign the reading for homework, along with the questions in **Check Your Understanding** if desired.

Assessment Opportunity

Use a quiz to assess student understanding of the concepts presented in **Activity 4**. Some sample questions are listed below:

Question: The appearance and disappearance of a few species at any time is referred to as a _____, whereas a significant loss in biodiversity over a relatively short period of time is called a

_____.

Answer: Background extinction; mass extinction.

Question: What evidence is used to support the theory that the mass extinction event at the end of the Mesozoic was caused by the impact of an asteroid?

Answer: Geochemical evidence in the form of iridium in sediments of that age worldwide; the existence of the Chicxulub impact crater; and evidence of a gigantic sea wave.

Chapter 3

Earth System Evolution Changing Life

known to exist at present to evolve and fill all of the ecological spaces opened by the end-of-the-Mesozoic extinction event.

Figure 1 Dinosaur tracks provide evidence for these prehistoric life forms.

If you were to examine the post-boundary fossil record in more detail, you would find that changes in the kinds of animals now extinct are related to the appearance and disappearance of their food source(s). The fossil record indicates that when evolution changes the basic composition of plants in a community, those dependent organisms must find a new food source, change how they process food for nutrition, or face extinction. For example, there is no physical evidence in the extensive fossil record, to date, for many grazing animals until the mid-Cenozoic. The first evidence for the evolution and appearance of grasslands appears at about this time. Following this, many new groups of animals whose diets include these plants are found for the first time in the fossil record, not before. In North America such animals include camels, rhinoceroses, horses, and a wealth of other mammals that are now known to be extinct. The extinction of a few species now and then appears to be a normal phenomenon. Scientists refer to the appearance and disappearance of a few species at any time as **background extinction**. When the fossil record documents a significant change in Earth's **biodiversity**, these events are termed **mass extinctions**. There have been five major mass extinctions during the history of life. During these mass extinctions up to 90% of the known biodiversity was lost. The extinction event at the end of the Paleozoic Era, between the Permian and the Triassic Periods, was even more devastating to life on Earth than the extinction at the end of the Mesozoic. The general patterns of evolution and recovery in

Blackline Master Changing Life 4.1
Major Divisions of Geologic Time

Major Divisions of Geologic Time (boundaries in millions of years before present)					
	Epoch	Period		Era	Eon
	Holocene		Quat.	Cenozoic	Phanerozoic
.01	Pleistocene	Neogene			
1.8	Pliocene	Neogene			
5.3	Miocene		Tertiary		
24	Oligocene	Paleogene			
34	Eocene	Paleogene			
56	Paleocene				
65		Cretaceous		Mesozoic	
145		Jurassic			
213		Triassic			
248		Permian		Paleozoic	
286		Pennsylvanian			
325		Mississippian			
360		Devonian			
410		Silurian			
440		Ordovician			
505		Cambrian			
544				Proterozoic	Cryptozoic (Precambrian)
2500				Archean	

Chapter 3

each of the post-extinction worlds are similar, including the appearance of new organisms with similar adaptions for living. Understanding these patterns may provide insight into how the Earth's biosphere responds to severe trauma.

Figure 2 The woolly mammoth became extinct near the end of the Pleistocene Epoch.

Incidentally, are you aware that the subdivision of geologic time into the Paleozoic, Mesozoic, and Cenozoic Eras at the particular places in the span of geologic time was adopted just because of the magnitude and abruptness of the extinctions? To a great extent, the accepted scheme of subdivision of geologic time isn't just arbitrary; it has a natural basis.

The Cause of the Extinction

Geologists and paleontologists have known for more than 150 years about the mass extinction at the end of the Mesozoic. The cause, or causes, of the extinction, however, remained obscure until very recently. In the 1970s a scientist name Luis Alvarez and his coworkers proposed that the extinction was caused by collision of a gigantic asteroid with the Earth. They based that hypothesis on studies of cores and outcrop sections through sedimentary rocks whose deposition spanned the Mesozoic–Cenozoic boundary. In several such sections they found geochemical evidence, in the form of unusually high concentrations of the chemical element iridium, that pointed toward a catastrophic collision. Iridium is known to be introduced into the Earth system during certain meteorite and asteroid impacts. Later, what most geoscientists consider to be the "smoking gun" was found in the Yucatán Peninsula of Mexico: the remnants of a colossal impact structure, called

NOTES

Earth System Evolution Changing Life

the Chicxulub crater. The idea is that the collision placed such quantities of dust and ash into the atmosphere that the climate became dramatically cooler and it is believed that light was unable to reach Earth's surface. The Earth's ecosystems were stressed for long times after the collision, leading to widespread extinction of species of many kinds of organisms. In the Gulf of Mexico, evidence of sediment movement and deposition by a gigantic sea wave, presumably caused by the impact, has strengthened the hypothesis.

Check Your Understanding

1. How long do species last?

2. What happens to animal species when the food sources of the animals change?

3. How can you document that as different groups of organisms evolved they developed similar structures, allowing them to occupy a position in the ecological hierarchy?

4. What is the difference between background extinction and mass extinction?

5. What kinds of evidence can be used to document a major meteorite or asteroid impact in the geologic past?

Figure 3 The Chicxulub Crater, on the Yucatan Peninsula of Mexico, provides evidence for a large impact event at the end of the Cretaceous. Here the crater is magnified using detailed measurements of the Earth's gravitational field and is revealed as the roughly circular feature in the center of this image.

Many aspects of the extinction remain unclear. Extinction of species continued for geologically long times after the collision. One needs to appeal to long-term ecological effects to explain this circumstance. In addition, there is also good evidence for increased volcanic activity around the time of the Mesozoic–Cenozoic boundary. A minority of geoscientists favor an alternative hypothesis for the great extinction, having to do with climate change induced by the volcanism. Evidence is still being gathered to resolve the details of the extinction and its causes.

Check Your Understanding

1. Some species have persisted for tens of millions of years, whereas others span only a few million years, or even only a few hundred thousand years (or less).

2. When food sources in an area change, the animal species in that area also change. The fossil record indicates that when evolution changes the basic composition of plants in a community, those dependent organisms must find a new food source, change how they process food for nutrition, or face extinction.

3. Paleontologists can document changes in organisms through the fossil record. By looking at common traits in a variety of different organisms, paleontologists can document that organisms have evolved to fill niches that are left when either other organisms become extinct or when major changes in the environment take place. For example, the advent of expansive grasslands caused a major change in the environment and ecological hierarchy, and provided a number of different animals the opportunity to evolve to become adapted to that environment. Animals that were once adapted to living in a forested environment would have had to either evolve to be successful in the open environment or go extinct. Many of the adaptations, like teeth that are better suited to eating grass, or bodies that are better suited to sprinting in open terrain, can be common to a variety of different animals.

4. Background extinction refers to the appearance and disappearance of a few species at any time. Mass extinction refers to a significant change in Earth's biodiversity, when up to 90% of known plants and animals are lost.

5. To document a major impact event in the geologic past, scientists can look for:
    - geochemical evidence in the form of iridium
    - remnants of impact structures
    - evidence for gigantic sea waves or sedimentary structures

---

Assessment Tool

Check Your Understanding Notebook Entry-Evaluation Sheet
Use this sheet to evaluate the extent to which students understand the key concepts explored in **Activity 4** and explained in the **Digging Deeper** reading section, and to evaluate the students' clarity of expression.

Understanding and Applying What You Have Learned

1. Think about the Mesozoic and Cenozoic fossil records you have uncovered. How do the organisms of the deep geologic past differ from those in your present community? Why?

2. What is the probability of one or more of the organisms that lived just before or just after the time of the Mesozoic–Cenozoic boundary still existing today somewhere in the world? Explain your answer.

3. If you were to examine the rock record of another planet, how would you identify that a mass extinction occurred when you were unfamiliar with that planet's life forms?

4. How are new data discovered, where are these data reported, and what is the purpose of independent review of a scientist's data and ideas?

Preparing for the Chapter Challenge

With your group, develop a poster presentation that illustrates both the Cretaceous (pre-boundary) and Paleogene (post-boundary) organisms within their ecological context. Indicate on the poster all of the data sources you consulted, the approximate duration of time that is represented by your fossil communities (how many millions of years), and the major differences between pre-boundary and post-boundary biodiversity. Compare these data to the data you collected for the last glacial maximum to the present. Over how many years is it thought that both of these extinction events occurred? Consider which organisms or types of life strategies appear to have been affected most by the extinction events.

Inquiring Further

1. **Hypotheses to explain the Mesozoic–Cenozoic boundary event**

   Using available resources, investigate further the hypotheses that have been proposed to explain the Mesozoic–Cenozoic boundary event. (It would be useful to know that the event has often been called the "Cretaceous–Tertiary boundary event," because in earlier usage, which still exists today, the Cenozoic was divided into the Tertiary and Quaternary periods.) How do the hypotheses differ? What scientific evidence and data exist to support each hypothesis?

2. **Other mass-extinction events**

   Research other mass extinctions that paleontologists have identified in the rock record. When in time did these occur? What groups of organisms were affected?

E 181

EarthComm

Understanding and Applying What You Have Learned

1. Answers will vary. The organisms found in the past are most likely quite different from those found in your community today, in part because of the mass extinction events but in greater part just because of slow background extinction caused by slow changes in environmental conditions, like climate change and sea-level change.

2. The probability that the actual organisms are still alive to day is absolutely zero: no organism lives for 65 million years! The probability that a few species of organisms still survive today is small but nonzero. It is possible for certain species that become extremely well adapted to an environment to change little through geologic time and survive for tens of millions of years. As one moves higher up in the hierarchy of taxonomic units (i.e., to genera, families, and orders) the likelihood of similar kinds of organisms surviving from the beginning of the Cenozoic to today increases.

3. You would first have to establish that the conditions on the planet were favorable for the existence of life forms: if the planet has been lifeless, there is no possibility of mass extinctions. A planet that has supported life would certainly have an abundant sedimentary record, in the form of successive layers of sediments, some of which would contain fossils. If you studied a large number of exposed sedimentary successions, correlated them by comparing their fossil content, and then determined that at a particular time a large number of species disappeared from the rock record, you could conclude that a mass extinction had occurred.

4. New data are discovered by observation, either in the natural environment or in laboratory studies. Scientists report their data mainly by:

   • having conversations with other scientists

   • giving oral presentations at scientific meetings

   • writing articles that are published in scientific journals

   • writing books

   Independent review of a scientist's data and ideas is important, because nobody (scientists included!) is perfect. Often, it takes another reader or listener to catch flaws or inadequacies in one's work.

Preparing for the Chapter Challenge

Students should now have a good idea of the possible causes of extinctions. They should have a clear sense of how organisms changed across the Mesozoic–Cenozoic boundary, and they will have collected evidence to support their ideas about these changes. Students are asked to compare the information they collected in **Activity 3**, regarding biological changes in Pleistocene. The work that students complete here can be inserted into their **Chapter Challenge** projects.

Chapter 3

Relevant criteria for assessing this section (see **Assessment Rubric for Chapter Challenge on Changing Life** on page 520) include information about:

- evidence that there has been change in the life forms and biodiversity of your community over time

- short-term and long-term factors that have influenced biological changes in your community

- the natural processes that have been responsible for the appearance and disappearance of life forms throughout geologic time

Inquiring Further

1. Hypotheses to explain the Mesozoic–Cenozoic boundary event

Students have spent the activity familiarizing themselves with the Mesozoic–Cenozoic boundary event. This mass extinction eradicated nearly 85% of all species, making it the second largest extinction event in the Earth's history. The two most commonly accepted hypotheses to explain the Mesozoic–Cenozoic extinction event are:

- impact of an asteroid-size body with the Earth

- an increase in global volcanism

Some scientists believe that the widespread distribution of iridium and the discovery of an impact site on the Yucatan Peninsula in Mexico support the impact theory. Additional evidence comes from the presence of shocked quartz crystals—tiny grains of quartz that are indicative of a high-impact event—in the sediment layer at the boundary. The high concentration of iridium in boundary deposits is also attributed by some to the impact theory. Some scientists believe that extensive lava deposits laid down at the boundary, known as the Deccan Traps of India and Pakistan, support this theory.

2. Other mass-extinction events

Some scientists have suggested that there is a cycle of mass extinctions, with organisms dying off every 26 million years. This would mean that there have been about 23 mass extinction events since the Cambrian Period. However, imperfections in the fossil record can make identification of these events difficult. Scientists have generally agreed upon six major extinction events: the late Cambrian, the late Ordovician, the late Devonian, the Permian–Triassic, and the Cretaceous–Paleogene events. Of these, the Permian–Triassic event was the largest, with as many as 96% of all marine species going extinct and more than 75% of all land vertebrate families dying out. Students can learn more about the specifics of the extinction events by visiting the *EarthComm* web site.

NOTES

Chapter 3

ACTIVITY 5 — HOW DIFFERENT IS YOUR COMMUNITY TODAY FROM THAT OF THE VERY DEEP PAST?

Background Information

This activity involves the use and interpretation of geologic maps, and research into some of the different species of organisms that have been preserved in the fossil record. (Specifically, those fossils that are relevant to the age of the bedrock in your community.) Brief explanations of the concept of species, the difficulties inherent classifying fossil species, and the reading and interpretation of geologic maps are given below.

Species and Their Evolution

The species is the basic unit in the classification of plants and animals. The concept of a species has been around much longer than the Darwinian theory of organic evolution by natural selection.

A species is defined as a population of organisms that can breed to produce fertile offspring. All of the members of a species are sufficiently closely related, in terms of their genetic makeup, to produce offspring that can in turn produce fertile offspring of their own. If two populations of generally similar organisms are not sufficiently closely related, they cannot interbreed, or, if they do so, their offspring are infertile. The mule, which is the offspring of a male donkey and a female horse, is an example of the latter case: mules can't reproduce.

The members of a species are not genetically identical: the assemblage of genes, which determine the biological nature of the organism, differs slightly from individual to individual. Mutations to the genes occur at random all the time, owing at least in part to external influences like exposure to radiation or chemical substances. The slightly changed genes are then transmitted by sexual reproduction.

According to the theory of evolution by natural selection, most mutations are harmful or fatal, but some give their carriers some advantage in life. Organisms carrying such beneficial mutations then have some advantage in reproduction, and the beneficial gene tends then to be present in a greater and greater proportion of the species. Over the enormous number of generations during the life span of the species—ordinarily, many hundreds of thousands of years, if not millions of years—the effects of beneficial mutations accumulate and lead to fundamental changes in the species. Change may even occur to the point at which a new species must be considered to have arisen.

New species also arise by isolation of one part of the population of a species from another part. Isolation may result from geographic factors, climatic factors, or some other circumstance that effectively prevents, or at least impedes, reproduction between the two groups of individuals. The different courses of genetic drift in the two groups, in response to at least slightly different environmental settings, leads eventually to two separate species by processes of natural selection.

In the modern world it is possible (at least in theory) to test, by means of breeding experiments, whether or not two groups of organisms constitute a single species. In the fossil record, however, it is obviously impossible to use breeding experiments to demonstrate whether a collection of similar fossils were actually members of a single species. Paleontologists have to sidestep this problem by assuming that if the morphology of the fossil specimens is closely similar, then very likely they did indeed belong to the same species.

The basic and very practical starting point in the initial study of a collection of fossils is to sort them by similarity in morphology. Commonly, the person doing the sorting is confronted immediately with the troublesome matter of making decisions about relative closeness of morphology. The personality of the sorter tends to rear its head in situations like this: are you a "lumper," meaning that you tend to group objects into a relatively small number of categories with a relatively large degree of variation within the categories, or are you a "splitter," meaning that you tend to group objects into a relatively large number of categories with a relatively small degree of variation within the categories? The only real guidance paleontologists have in this endeavor is comparing their findings with typical degrees of intraspecific (within-species) variation in known modern species.

Geologic Maps and Cross Sections

A geologic map shows the distribution of bedrock that is exposed at the Earth's surface or buried beneath a thin layer of surface soil or sediment. It is more than just a map of rock types: most geologic maps show the locations and relationships of rock units. Rock units are bodies or masses of rock that have a high degree of uniformity relative to adjacent rock units. Rock units are large: their minimum dimensions are almost always several meters to tens of meters, and their

maximum dimensions may be as great as many tens or even hundreds of kilometers.

Each rock unit is identified on the map by a symbol, and each unit is often given a distinctive color as well. The legend or key of a geologic map typically has one or more columns of little rectangles, with appropriate colors and symbols, identifying the various rock units shown on the map. The rectangles for the units are arranged in order of decreasing age upward. Usually, the ages of the units, in terms of the standard relative geologic time scale, are shown as well. There is often a very brief description of the units in this part of the legend.

All geologic maps convey certain other information as well. They show the symbols that are used to represent such features as folds, faults, and attitudes of planar features like stratification or foliation. They have information about latitude and longitude, and/or location relative to some standard geographic grid system. These maps always have a scale, expressed both as a labeled scale bar and as what is called the representative fraction—1:25,000 for example—whose first number is a unit of distance on the map and whose second number is the corresponding distance on the actual land surface.

All geologic maps (except perhaps very special-purpose maps that show all the details of an area that might be the size of a small room!) involve some degree of generalization. Such generalization is the responsibility of the geologist who is doing the mapping. Obviously, it is not practical to represent features as small as a few meters wide on a map that covers many square kilometers: the line depicting the feature on the map would be far finer than the finest possible ink line.

The degree of generalization necessarily increases as the area covered by the map increases. You could easily see this for yourself if you have access to a geologic map of some small area together with the corresponding geologic map of the entire state: the state map would show far less detail of the small area than the full map of that same small area.

Most geologic maps are accompanied by one or more vertical cross sections, which are views of what the geology would look like in an imaginary vertical plane downward from some line on the land surface. The geologist constructs these cross sections after the map is completed. Their locations are selected so as to best reveal the three-dimensional nature of the geology. Cross sections are constructed by projecting downward the geologic features and relationships that are observed at the surface. The degree of certainty about the geology shown on the cross section decreases downward with depth below the surface.

More Information – on the Web

Visit the *EarthComm* web site www.agiweb.org/earthcomm to access a variety of links to web sites that will help you deepen your understanding of content and prepare you to teach this activity. Many of the sites also contain images that you can download.

Goals and Assessment

Clarify that the goals indicate what students should understand and be able to do as a result of the activity. Make sure students understand that Chapter Assessments are based upon these goals.

Goal	Location in Activity	Assessment Opportunity
Understand that many different plants and animals evolved and became extinct during the Phanerozoic.	Investigate; Digging Deeper; Check Your Understanding Question 2 Understanding and Applying What You Have Learned Question 1	Responses to questions are reasonable.
Recognize that plant and animal fossil assemblages of the very deep geologic past (for example, in the Paleozoic) are unlike organisms alive today.	Investigate; Digging Deeper; Check Your Understanding Question 2 Understanding and Applying What You Have Learned Question 1	Responses to questions are reasonable.
Describe the variety of plants and animals that once lived in your community.	Investigate; Understanding and Applying What You Have Learned Questions 1 and 3	Diorama display is useful and informative. Responses to questions are reasonable.
Relate the increases and decreases in the numbers of different organisms during the Phanerozoic to geologic causes.	Digging Deeper; Understanding and Applying What You Have Learned Questions 1 – 2	Responses to questions are reasonable.
Describe several lines of fossil evidence to support the theory of evolution based on the fossil record.	Digging Deeper; Check Your Understanding Question 1 Understanding and Applying What You Have Learned Questions 3 – 4	Responses to questions are reasonable.

Chapter 3

Earth System Evolution Changing Life

Activity 5

How Different Is Your Community Today from that of the Very Deep Past?

Goals

In this activity you will:

• Understand that many different plants and animals evolved and became extinct during the Phanerozoic.

• Recognize that plant and animal fossil assemblages of the very deep geologic past (for example, in the Paleozoic) are unlike organisms alive today.

• Describe the variety of plants and animals that once lived in your community.

• Relate the increases and decreases in the numbers of different organisms during the Phanerozoic to geologic causes.

• Describe several lines of fossil evidence to support the theory of evolution based on the fossil record.

Think about It

You have been learning about how life has been changing throughout geologic time.

• If you could be transported back in time, would you recognize your community?

What do you think? Record your ideas in your *EarthComm* notebook. Be prepared to discuss your ideas with your working group and the class.

E 182

EarthComm

Activity Overview

Students begin the investigation by examining a geologic map of their community and determining which periods of geologic time the rock record in their area represents. Each group then selects a period of time; students investigate the kinds of plants and animals that lived at that time. On the basis of their research, students construct a diorama to illustrate what the community might have looked like during that period of geologic time. They visit the displays created by each of the student groups and prepare a written description of how life has changed in their community through time. The **Digging Deeper** reading section of **Activity 5** discusses the relationship between biodiversity and climate change. The reading looks at specific examples of organisms that have changed through time in response to a change in climate.

Preparation and Materials

Preparation

You will need to obtain geologic maps of your community well in advance of completing this activity. Visit the *EarthComm* web site for suggestions as to where to find these maps. Students will be using the map only to identify the time periods represented by the rocks in your area, so one copy will be sufficient.

Alternatively, you can view the map online and provide the information directly for your students.

You will also need to collect and make available resources about the organisms present in your community during the time periods that are represented in your community's rock record. Helpful resources include textbooks, encyclopedias, the Internet, and museum displays.

You will also need to provide students with resources to build their natural habitats. If resources are limited, you may wish to have students draw a poster display of the ecosystem found during the time period instead of actually building the ecosystem.

Materials
- Geologic map of your community*
- Resources like textbooks, encyclopedias, the Internet, and museum displays
- Materials to construct dioramas

Think about It

Student Conceptions

Student answers will vary. Most likely, students will say that it depends upon how far back in time they are transported. You may want to give them some parameters to

*Visit the *EarthComm* web site for suggestions regarding where to obtain this map.

Chapter 3

work with. For example, would they recognize the community if they went back a hundred years? a thousand years? ten thousand years? a million years? a hundred million years?

Answer for the Teacher Only

There is no simple, easy answer to this question. Geologic and biologic processes generally work slowly. However, catastrophic change, short by human time scales, is not uncommon in many parts of the world.

In the stable interiors of the continents, change is almost entirely very slow. The major exception to this slow rate of change is the advance and retreat of continental ice sheets during the Pleistocene. In broad areas in the interior of the southern tier of the United States, the natural setting of a community would look a little different now from the way it looked a million years ago, except for differing vegetation as a function of a different climate.

In the geologically active areas of the western United States, however, there can be great changes in the physical conditions over time spans as short as millennia. Over long geologic time scales of tens to hundreds of millions of years, the setting of an area typically becomes entirely different, and comparisons become largely irrelevant.

Assessment Tool

Think about It Evaluation Sheet
Use this evaluation sheet to help students understand and internalize the basic expectations for the warm-up activity.

NOTES

Investigate

1. Examine a geologic map of your region.

   a) Which geologic time periods are represented by the rocks in your community?

   b) Choose one of these periods as your "target" in the geologic past. Make sure that all of the time periods represented by the rock record in your community are selected by one or more of your classmates.

2. Using available resources, including textbooks, encyclopedias, the *EarthComm* web site, and museum displays (if possible), collect data on the kinds of plants and animals that have been found as fossils in your target geologic period.

3. Construct either a natural habitat (for land-dwelling animals) and aquarium (for marine-dwelling organisms), or a botanical garden (for land plants) diorama depicting your community during your selected interval of geologic time. Your display should present the biodiversity of the single period of time you chose to investigate.

4. Once you and your classmates have constructed your displays, assemble them in your classroom in order according to the geologic time scale.

   a) How has your community changed over time? Take notes on the similarities and differences between adjacent "exhibits."

    Have your teacher approve the materials you plan to use to construct your diorama.

Reflecting on the Activity and the Challenge

You may have seen that some major groups of organisms are restricted to one or two geologic periods. Other groups may have lived during several geologic periods. You may also find that some geological periods do not show evidence for one or more particular groups of organisms. This may be a result of the kinds of sedimentary rocks preserved in your region. Check the geologic map of your area, again, to see if all the geologic periods are represented by rock. Once you have finished comparing and contrasting the data, decide for yourself if you could recognize your community at any given point in geologic time on the basis of the plants and animals known from its fossil record.

Investigate

1. a) – b) The geologic map of your community or state should include a legend that lists the ages of the rocks found in the region. Work with your class to divide up the time periods represented by your community's rock record.

2. – 3. The **Preparation** section above provides suggestions for good reference materials and alternatives to the diorama.

4. a) Answers will vary.

> ## Assessment Tools
>
> *EarthComm* Notebook Entry-Checklist
> Use this checklist as a quick guide for student self-assessment and/or an opportunity to quickly score student work. Add further criteria specific to your classroom needs or to this particular investigation.
>
> Investigate Notebook Entry-Evaluation Sheet
> Point out the criteria listed on this evaluation sheet that are relevant to this particular investigation. Encourage students to internalize the criteria by making them part of your "assessment conversations" as you circulate around the classroom.

Reflecting on the Activity and the Challenge

This is a good opportunity to review the entire **Changing Life** chapter. Hold a class discussion to compare students' ideas about how life in the community has changed over time. Have the class revisit and discuss their responses to the **Think about It** question at the beginning of **Activity 5**.

Chapter 3

Earth System Evolution Changing Life

Geo Words

species: a group of organisms, either plant or animal, that may interbreed and produce fertile offspring.

taxonomy: the theory and practice of classifying plants and animals.

morphology: the (study of the) features that comprise or describe the shape, form, and structure of an object or organism.

anatomy: the (study of the) internal structure of organisms.

biodiversity curve: a graph that shows changes in the diversity of organisms as a function of geologic time.

Digging Deeper

BIODIVERSITY AND CLIMATE CHANGE

The classification of plants and animals into **species**, and the grouping of species into related groups, is called **taxonomy**. The species is the basic taxonomic unit. A species can be defined as a population of organisms that can interbreed to produce fertile offspring. Species that seem to be closely related, in terms of how they evolved, are grouped into a higher taxonomic unit called a genus (plural: genera). In turn, genera that are thought to be closely related are grouped into families, and so on; see *Figure 1*. In today's world, biologists can (in theory, at least!) determine what a species really is. Paleontologists can't do that, because individuals of the species are no longer alive. Paleontologists try to recognize features using both external appearances (**morphology**) and internal structures (**anatomy**). Keep in mind also that genera and higher taxonomic units cannot be directly verified, either for fossil organisms or for living organisms.

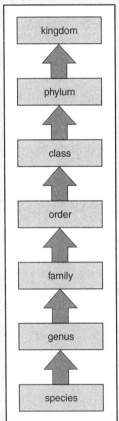

Figure 1 The hierarchy of taxonomic units.

All evolutionarily related species and genera (for example, *Homo sapiens*) are placed into the next highest larger category called a family (*H. sapiens* belongs to the family Hominidae). Related families are grouped into a higher category called an order (Hominidae belong to the order Primates). Hence, there can be many families in any order and many genera within any family. Paleontologists have tried to construct graphs, called biodiversity curves that show biodiversity as a function of geologic time. These **biodiversity curves** are based on the fossil record of higher-order taxonomic units like families or orders. That is, rather than using the number of species or genera known from an interval in geologic time, paleontologists have used the number of families or orders that have been identified.

Your community's biodiversity as reflected in the fossil record is related both to the geologic period you have examined and to the physical and chemical environment that existed there. There is

 E 184

Digging Deeper

Assign the reading for homework, along with the questions in **Check Your Understanding** if desired.

Assessment Opportunity

Reword or restructure the questions in **Check Your Understanding** for a brief quiz. Use the quiz (or a class discussion of the questions in the textbook) to assess your students' understanding of the main ideas in the reading and the activity.

Teaching Tip

Use **Blackline Master Changing Life 5.1, The Hierarchy of Taxonomic Units** of *Figure 1* on page E184 to make an overhead transparency. Incorporate the transparency into a discussion of taxonomy. Discuss with students the level at which they think most changes in biodiversity occur. They should recognize that most changes occur at the species level, with fewer changes occurring as you move up the ladder. To help them understand this, you may want to review the fact that currently there are five recognized kingdoms: Animalia, Plantae, Fungi, Protista, and Monera (bacteria). The number of named species today is over 1.5 million, and the total number of species (named and those that remain to be discovered) has been estimated to be anywhere between 3 million and 15 million.

Chapter 3

incontrovertible evidence that the number, diversity, and complexity of both marine and terrestrial organisms found in the fossil record have changed through the hundreds of millions of years of Earth's history. Many of the "big groups" that are found today in your community have fossil records that extend deep into the past. You can certainly identify vertebrates in your community, and vertebrates certainly have a fossil record that is first found in the Paleozoic. You have also discovered in this activity, however, that there are many strange and interesting creatures that once lived but are no longer found on Earth today.

Figure 2 Reconstruction of what life may have been like 540 million years ago in British Columbia, Canada, based on the fossils found within the Burgess Shale.

New fossil data are published weekly all over the globe. These data do not originate only from scientists working in North America, nor are the scientific journals in which they are found published only here. Rather, there are thousands of paleontologists (in China alone, for example) studying Earth's fossil record on all continents, including Antarctica, and there are nearly 100 different scientific journals that publish data on fossils. The results of these scientists' investigations continue to support and reinforce the general trends you have analyzed in this activity.

One principle of scientific inquiry is that a hypothesis may be falsified sometime in the future if some new, scientifically reviewed and published data are found. These new investigations may challenge certain specifics concerning an individual fossil or the group to which it belongs.

Blackline Master Changing Life 5.1
The Hierarchy of Taxonomic Units

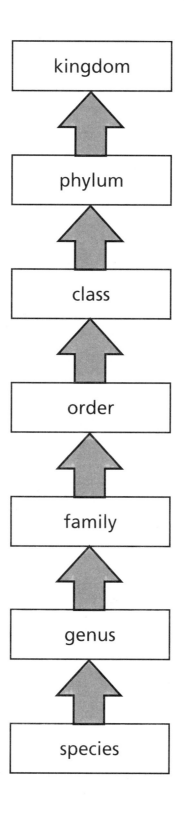

Chapter 3

Earth System Evolution Changing Life

One recent example of this is the controversy about whether or not feathers are only found in birds or if non-avian reptiles also evolved feathers. Nevertheless, the overall evolutionary pattern of life on Earth has not been dramatically altered or changed because of new scientific data.

Figure 3 Fossil of *Archaeopteryx*, which is considered to be the first ancestor to modern birds. *Archaeopteryx* is about 150 million years old.

Many different organisms have had the same or similar life strategies throughout Earth history. Some of these groups that dominated the landscape for several million years have gone extinct or have been placed into a minor role, following some change in global conditions. For example, "club moss" trees, which grew to heights greater than about 40 m (about 120 ft.), are the plants that formed the Mississippian and Pennsylvanian Period coals of North America, Europe, and China. (These two periods are sometimes jointly referred to as the Carboniferous Period because of the extensive hydrocarbon desposits of this time.)These fossil trees are known to have lasted in peat swamps for nearly 30 million years without change. Without these trees in the very deep past, there would have been no coal. Imagine how many trees must have lived over an interval of 30 million years to result in the vast resources of coal that the United States relies on now

Teaching Tip

Figure 2 on page E185 is an artist's rendition of what the organisms of the Burgess Shale might have looked like. The Burgess Shale is a Middle Cambrian deposit located in Yoho National Park in the Rocky Mountains near Field, British Columbia, Canada. This deposit is well known for its exceptional preservation of a wide variety of soft-bodied invertebrate animals. The Burgess Shale is the focus of the book *Wonderful Life* by Stephen Jay Gould. Students may be interested in reading Gould's book as an extra-credit assignment.

Teaching Tip

Figure 3 on page E186 is a photograph of a fossil of *Archaeopteryx lithographica*. This fossil was found in the Jurassic Solnhofen Limestone in southern Germany. *Archaeopteryx* is considered to be an intermediate between modern birds and predatory dinosaurs of the Mesozoic. Like modern birds, *Archaeopteryx* had feathers, wings, a "wishbone" (or furcula) and reduced fingers. Unlike modern birds, however, *Archaeopteryx* had a full set of teeth, a rather flat breastbone, a long, bony tail, "belly ribs," and three claws on the wing.

Chapter 3

Activity 5 How Different Is Your Community Today from that of the Very Deep Past?

for electricity. At one time heating, cooking, and steam engines were all powered by coal. So, why are "club moss" trees, like the ones reconstructed in the coal swamp shown in *Figure 4*, not found outside of your school today? If they were successful for 30 million years, why aren't these trees living in your community or somewhere, anywhere, on Earth?

Figure 4 Reconstruction of middle to late Carboniferous tropical coal swamp showing different plant communities.

The loss of biodiversity in the recent past and the deep geologic past appears to be related to changes in global climate. In the case of the club moss trees, their extinction is associated with climate change associated with Late Paleozoic ice ages. These trees were replaced across the Northern Hemisphere by the gymnosperm trees that reproduced by seeds. Seed plants are better adapted to dry environments, similar to the environments that are interpreted to have accompanied global deglaciation and the assembly of Pangea. Some of these same gymnosperm families are still living today, like the cycads and ginkgos, but most are extinct. Distant relatives of the club moss trees are found today as groundcover in the north temperate woodlands and elsewhere. (Change in global climate is not the only factor that contributes to the loss of biodiversity. Many species are endangered or extinct through the actions of humans.)

Figure 5 Photograph of a ginkgo tree in autumn.

Check Your Understanding

1. What factors influence biodiversity as reflected in the fossil record?

2. Why did the "club moss" tree become extinct?

3. How does the work of scientists contribute to the understanding of how life has changed through time?

Teaching Tips

(*Figure 4*) Conditions worldwide during the Pennsylvanian were conducive to lush growth of plants in many coastal environments. Sea level changed slowly and cyclically but stood relatively high, and broad areas at low and middle latitudes had a warm and humid climate. In coastal environments located well away from river mouths, lush stands of trees in swampy areas could thrive without being choked by input of sand and mud. In such areas, called "coal swamps," thick deposits of partly decayed plant matter accumulated, eventually to be buried and converted to coal.

The ginkgo tree (pictured in *Figure 5* on page E187) is part of a family of plants that extends far back into the geologic past. Ask students if they can think of any other families of organisms (like some sharks, for example) that can trace their origins back to the Mesozoic and beyond.

Check Your Understanding

1. Changes in biodiversity, as reflected in the fossil record, are influenced mainly by climate. The biodiversity reflected in the fossil record, however, depends also on the extent of fossilization; the fossil record is never a perfect reflection of the diversity of plant and animal species that existed at that time in the geologic past.

2. The club moss tree became extinct in association with climate change that occurred during the Late Paleozoic ice ages.

3. Scientists are able to use the fossil record to reconstruct changes in different species over time, and also to monitor changes in biodiversity over time.

Assessment Tool

Check Your Understanding Notebook Entry-Evaluation Sheet
Use this sheet to evaluate the extent to which students understand the key concepts explored in **Activity 5** and explained in **Digging Deeper**, and to evaluate the students' clarity of expression.

Chapter 3

Earth System Evolution Changing Life

Understanding and Applying What You Have Learned

1. Large-scale changes in biodiversity due to extinctions and evolution of new organisms thereafter have altered the aspects of your community through geologic time.

   a) How different was your community biome in the record of the deep geologic past you investigated?

   b) Are the common plants and animals you identified in your present community the same as in the fossil record? Explain.

   c) How many times during Earth's history has there been a major change in the biodiversity of your community?

2. When a group of organisms is no longer found in the fossil record, how do new organisms adapt to take its place in the ecosystem?

3. If all soft-bodied organisms and those with hard skeletons that lived in your community in the deep geologic past had been fossilized, how would this change your perspective about life on Earth?

4. What would your present community be like if one or more of the extinction events in the deep geologic past had not occurred?

Preparing for the Chapter Challenge

As a group, synthesize what you have learned about the changes in biodiversity over geologic time and the causes proposed for the mass extinction events documented in the fossil record. Review your notes on how your community has changed over time. Write an essay in which you explain whether organisms with similar life strategies have similar fossil records (i.e., do organisms with the same life strategies respond the same). Explain any differences.

Inquiring Further

1. **Extinctions**

   Is extinction common throughout Earth history, or are extinctions limited to only the "Big Ones?" One paleontologist has asked the question whether extinction is "bad timing, or bad luck?" What do you think? How do mass extinctions change the balance of Earth's ecosystems?

2. **Originations**

   What natural mechanisms are responsible for the evolution and origination of new organisms following extinction events? Is the rate of origination of new organisms slow or rapid following an extinction event? What kinds of organisms appear first in the fossil record following a mass extinction event?

Understanding and Applying What You Have Learned

1. **a)** Answers to these questions will vary with each community.

   b) Answers will vary, but only if the fossil record is as young as the Pleistocene would there be substantial overlap. Rates of extinction, even during the long periods of time between mass extinctions, are sufficiently high that fauna and flora at different times in the geologic past are largely or entirely dissimilar.

   c) Probably several times, although the perception that your community could have been a recognizable place with a continuous history that extends back into deep geologic time is not realistic for most areas. At certain times in the geologic past, a shallow ocean covered much of North America; can your community be identified as an area that was once under the sea?

2. Evolution by natural selection can be viewed as opportunistic. When an ecological niche is abandoned because of extinctions, new species evolve to fill that niche.

3. Answers will vary. Students are likely to say that they would have a much more comprehensive image of how life has changed over time.

4. Answers will vary. Most likely, the organisms found in the community would be much different.

Preparing for the Chapter Challenge

Students should examine their work from this activity and reread the **Chapter Challenge**. They should consider how they can use what they learned in **Activity 5** to explain how and why life in their community has changed through time. Students should include discussions on:

- evidence that there has been change in the life forms and biodiversity of your community over time
- short-term and long-term factors that have influenced biological changes in your community
- the natural processes that have been responsible for the appearance and disappearance of life forms throughout geologic time
- suggestions as to what could be done to reduce biodiversity loss caused by natural changes in Earth systems

Inquiring Further

1. Extinctions

Answers will vary. It should be clear to your students that extinction has been common throughout Earth's history. Background extinction is a continuing process. Everywhere paleontologists look in the record of past life, they find evidence that old species disappear and new species appear. At certain times, the rate of extinction rises

Chapter 3

sharply; these are the times that are referred to as mass extinctions. After mass extinctions, new faunal and floral groups evolve rapidly from the previous remnants to fill all of the available ecological niches. These niches are the same as those before the extinctions to some extent; however, they are also different to some extent because of climate change or sea-level change. These occurrences of very rapid evolution are called explosive radiations.

2. Originations

Rapid evolution of new species after a major extinction event is well explained by natural selection. Because of the continuous appearance of genetic mutations in any plant or animal species, the "raw material" for evolution of new species is always at hand. When ecological niches exist unfilled, species can evolve to fill those niches unhindered by competition from existing species that already occupy those niches.

NOTES

Earth Science at Work

ATMOSPHERE: *Birdwatcher*
Climate change might change the distribution of birds in North America. Birdwatchers can help scientists determine if the range of a given species of bird has changed, or if there has been a change in the time of bird migration.

BIOSPHERE: *Paleontologist*
When detectives investigate a murder, they do not just look at the dead body. They look at a great deal of other evidence. Similarly, when paleontologists find a dinosaur's bones, they can pick up extra clues from trace fossils. These clues to lifestyles help scientists conjure a much richer picture of prehistoric life.

CRYOSPHERE: *Park Ranger*
The Arctic ecosystem is in a very delicate balance and is particularly vulnerable to climate change. Climate change is likely to threaten both marine and terrestrial wildlife. From plankton to polar bears, many species could suffer or disappear entirely.

GEOSPHERE: *Urban Planner*
Urban expansion can result in deforestation of large areas. Also, pollution can make existing trees more susceptible to freeze injury and in turn to drought and insects. Urban planning needs to take into account not such social and economic factors, but environmental factors as well.

HYDROSPHERE: *Fishing Industry*
Fisheries already face the problems of overfishing, diminishing wetlands, pollution, and competition among fleets for uses of aquatic ecosystems. The industry is constantly monitoring seafood supplies for any further changes.

How is each person's work related to the Earth system, and to Changing Life?

E 189

NOTES

Changing Life and Your Community: End-of-Chapter Assessment

1. An imprint of an organism is considered
   a) to not be a fossil
   b) to be a trace fossil
   c) to be a body fossil
   d) to be the rarest kind of fossil

2. Which of the following examples is considered to be a trace fossil?
   a) a worm burrow
   b) a fragment of a tooth
   c) a rare brachiopod fossil
   d) a microscopic fossil

3. In which rock would one most likely find a fossil?
   a) a limestone
   b) a sandstone
   c) a conglomerate
   d) a granite

4. The oldest fossils on Earth came from
   a) mammals
   b) dinosaurs
   c) shell fish
   d) algae and bacteria

5. In an ecosystem, the lowest trophic level is occupied by _____, which are considered to be _____.
   a) herbivores; primary producers
   b) green plants, primary producers
   c) herbivores; primary consumers
   d) insects; primary consumers

6. Which of the following organisms would be at the highest trophic level?
   a) grass
   b) eagle
   c) mouse
   d) cow

7. Identify the most appropriate biome to fit the following description:

   A biome common at high northern latitudes where winters are long and cold and summers are short and cool. Cold-tolerant evergreen trees are common and include spruce and fir. Common animals include a range of mammals and birds including caribou, moose, bear, bald eagles, woodpeckers and warblers.
   a) taiga
   b) chaparral
   c) tundra
   d) temperate deciduous forest

8. Biomes are classified primarily on the basis of
   a) annual precipitation
   b) geographic location
   c) the main type of vegetation
   d) average temperature

9. The distribution of oak and spruce pollen in North America has changed over the past 20,000 years in response to
   a) the extinction of many of the Pleistocene plant species
   b) changes in climate caused by receding glaciers
   c) the movement of the North American continent
   d) changing regional wind patterns since that time

10. During the height of the last glacial period, glaciers expanded to cover
   a) everything north and south of 20° latitude
   b) much of North America and Eurasia
   c) Antarctica to a thickness of 20 km
   d) the United States as far south as Florida

11. Which of the following are commonly associated with the time period that follows a mass extinction event?
   a) slow rates of species recovery and evolution
   b) the evolution of marine organisms onto land
   c) rapid evolution of new species with similar adaptations
   d) an extensive period of forest growth

12. The most accepted hypothesis for the mass extinction that occurred at the end of the Mesozoic Era involves
   a) a prolonged ice age
   b) the impact of an asteroid
   c) a long period of drought
   d) a huge volcanic explosion

13. Which of the following criteria can scientists use to determine the trophic level of an animal (or fossil of an animal) within an ecosystem?
   a) the animal's teeth
   b) the animal's size
   c) the animal's limbs
   d) the animal's senses

Chapter 3

14. The end of the Cretaceous Period is associated with
    a) a rapid cooling of global climate
    b) a rapid warming of global climate
    c) depletion of oxygen in the atmosphere
    d) an extended ice age and growth of glaciers

15. Many of the major changes in biodiversity observed in the fossil record can be related to
    a) the biases that are inherent to the fossilization process
    b) significant changes in the global climate at that time
    c) plate tectonic activity and volcanic eruptions
    d) catastrophic flooding and rapid rise in sea level

Answer Key

1. c
2. a
3. a
4. d
5. b
6. b
7. a
8. c
9. b
10. b
11. c
12. b
13. a
14. a
15. b

Teacher Review

Use this section to reflect on and review the investigation. Keep in mind that your notes here are likely to be especially helpful when you teach this investigation again. Questions listed here are examples only.

Student Achievement

What evidence do you have that all students have met the science content objectives?

_____

Are there any students who need more help in reaching these objectives? If so, how can you provide this?_____

_____

What evidence do you have that all students have demonstrated their understanding of the inquiry processes?_____

_____

Which of these inquiry objectives do your students need to improve upon in future investigations? _____

_____

What evidence do the journal entries contain about what your students learned from this investigation? _____

_____

Planning

How well did this investigation fit into your class time?_____

_____

What changes can you make to improve your planning next time? _____

_____

Guiding and Facilitating Learning

How well did you focus and support inquiry while interacting with students?

_____

What changes can you make to improve classroom management for the next investigation or the next time you teach this investigation? _____

_____

How successful were you in encouraging all students to participate fully in science learning? _____

_____

How did you encourage and model the skills values, and attitudes of scientific inquiry? _____

_____

How did you nurture collaboration among students? _____

_____

Materials and Resources

What challenges did you encounter obtaining or using materials and/or resources needed for the activity? _____

_____

What changes can you make to better obtain and better manage materials and resources next time? _____

_____

Student Evaluation

Describe how you evaluated student progress. What worked well? What needs to be improved? _____

_____

How will you adapt your evaluation methods for next time? _____

_____

Describe how you guided students in self-assessment. _____

_____

Self Evaluation

How would you rate your teaching of this investigation? _____

_____

What advice would you give to a colleague who is planning to teach this investigation? _____

_____

Chapter 3

NOTES

EarthComm Assessments

Assessing the *EarthComm* Notebook

- *EarthComm* Notebook Entry-Evaluation Sheet
- *EarthComm* Notebook Entry-Checklist
- Think About It Evaluation Sheet
- Investigate Notebook Entry-Evaluation Sheet
- Check Your Understanding Notebook Entry-Evaluation Sheet

Assisting Students with Self Evaluation

- Student Evaluation of Group Participation
- Student Ratings and Self Evaluation

Assessing Student Presentations

Student Presentation Evaluation Forms

- Student Presentation Evaluation Form
- Student Ratings and Self Evaluation

References

- Doran, R., Chan, F., and Tamir, P. (1998). *Science Educator's Guide to Assessment.*
- Leonard, W.H., and Penick, J.E. (1998). *Biology – A Community Context.* South-Western Educational Publishing. Cincinnati, Ohio.

Assessment Blackline Master I-Journal Entry

Name: _____ Date: _____ Module: _____

Explanation: The *EarthComm* notebook is an important component of each *EarthComm* module. In using the notebook as you investigate Earth science questions, you are mirroring what scientists do. The criteria, along with others that your teacher may add, will be used to evaluate the quality of your Notebook entries. Use these criteria, along with instructions within investigations, as a guide.

Criteria

1. Entry Made
       1   2   3   4   5   6   7   8   9   10   _____
   Blank   Nominal   Above average Thorough

2. Detail
       1   2   3   4   5   6   7   8   9   10   _____
   Few dates   Half the time   Most days   Daily
   Little detail Some detail   Good detail   Excellent detail

3. Clarity,
       1   2   3   4   5   6   7   8   9   10   _____
   Vague   Becoming clearer   Clearly expressed
   Disorganized   well organized

4. Data Collection/Analysis
       1   2   3   4   5   6   7   8   9   10   _____
   Data collected   Data collected,   Data collected
   Not analyzed   some analyzed   and analyzed

5. Originality
       1   2   3   4   5   6   7   8   9   10   _____
   Little evidence   Some evidence   Strong evidence,
   of originality   of originality   of originality

6. Reasoning/Higher-Order Thinking
       1   2   3   4   5   6   7   8   9   10   _____
   Little evidence   Some evidence   Strong evidence
   of thoughtfulness of thoughtfulness   of thoughtfulness

7. Other
       1   2   3   4   5   6   7   8   9   10   _____

8. Other
       1   2   3   4   5   6   7   8   9   10   _____

EarthComm Notebook Entry-Checklist

Name: _____ Date: _____ Module: _____

Explanation: The *EarthComm* Notebook is an important component of each *EarthComm* module. In using the notebook as you investigate Earth science questions, you are mirroring what scientists do. The criteria, along with others that your teacher may add, will be used to evaluate the quality of your Notebook entries. Use these criteria, along with instructions within investigations, as a guide.

Criteria:

1. Makes entries   _____

2. Provides dates and details   _____

3. Entry is clear and organized   _____

4. Shows data collected   _____

5. Analyzes data collected   _____

6. Shows originality in presentation   _____

7. Shows evidence of higher-order thinking   _____

8. Other   _____

9. Other   _____

                                                _____

Comments:

Think about It Evaluation Sheet

Name: _____ Date: _____ Module: _____

	Strong		Fair		No Entry
Shows evidence of prior knowledge	4	3	2	1	0
	Strong		Fair		No Entry
Reflects discussion with classmates	4	3	2	1	0

Think about It Evaluation Sheet

Name: _____ Date: _____ Module: _____

	Strong		Fair		No Entry
Shows evidence of prior knowledge	4	3	2	1	0
	Strong		Fair		No Entry
Reflects discussion with classmates	4	3	2	1	0

Think about It Evaluation Sheet

Name: _____ Date: _____ Module: _____

	Strong		Fair		No Entry
Shows evidence of prior knowledge	4	3	2	1	0
	Strong		Fair		No Entry
Reflects discussion with classmates	4	3	2	1	0

Investigate Notebook Entry-Evaluation Sheet

Name: _____ Date: _____ Module: _____

Criteria

1. Completeness of written investigation
    1   2   3   4   5   6   7   8   9   10   _____
Blank   Incomplete   Thorough

2. Participation in investigations
    1   2   3   4   5   6   7   8   9   10   _____
None or little;   Needs minimal guidance, Leads, is inquisitive
unable to guide   sometimes helping others persistent, focused.
self

3. Skills attained
    1   2   3   4   5   6   7   8   9   10   _____
Few skills   Tends to use some   High degree of
evident   appropriate skills   appropriate skills used

4. Investigation Design
    1   2   3   4   5   6   7   8   9   10   _____
Variables not   Sometimes   Considers variables.
considered   considers variables,   Sound rationale for
techniques   uses logical techniques techniques
illogical

5. Conceptual understanding of content
    1   2   3   4   5   6   7   8   9   10   _____
No evidence   Approaches understanding Exceeds expectations
of understanding of most concepts   for content attainment

6. Ability to explain/discuss inquiry
    1   2   3   4   5   6   7   8   9   10   _____
Unable to   Some ability to   Uses scientific reasoning
articulate   explain/discuss   to explain any
scientific thought the inquiry   aspect of the inquiry

7. Other
    1   2   3   4   5   6   7   8   9   10   _____

8. Other
    1   2   3   4   5   6   7   8   9   10   _____

Check Your Understanding Notebook Entry-Evaluation Sheet

Name: _____ Date: _____ Module: _____

	Strong		Fair		No Entry
Shows evidence of prior knowledge	4	3	2	1	0
Reflects discussion with classmates	4	3	2	1	0

Check Your Understanding Notebook Entry-Evaluation Sheet

Name: _____ Date: _____ Module: _____

	Strong		Fair		No Entry
Shows evidence of prior knowledge	4	3	2	1	0
Reflects discussion with classmates	4	3	2	1	0

Check Your Understanding Notebook Entry-Evaluation Sheet

Name: _____ Date: _____ Module: _____

	Strong		Fair		No Entry
Shows evidence of prior knowledge	4	3	2	1	0
Reflects discussion with classmates	4	3	2	1	0

Student Presentation Evaluation Form

Student Name_____ Date_____

Topic_____

	Excellent		Fair		Poor
Quality of ideas	4	3	2	1
Ability to answer questions	4	3	2	1
Overall comprehension	4	3	2	1

COMMENTS:

Student Presentation Evaluation Form

Student Name_____ Date_____

Topic_____

	Excellent		Fair		Poor
Quality of ideas	4	3	2	1
Ability to answer questions	4	3	2	1
Overall comprehension	4	3	2	1

COMMENTS:

Student Evaluation of Group Participation

Key:
4 = Worked on his/her part and assisted others
3 = Worked on his/her part
2 = Worked on part less than half the time
1 = Interfered with the work of others
0 = No work

My name is _____ .   I give myself a _____

The other people in my group are: I give each person:

A. _____   _____

B. _____   _____

C. _____   _____

D. _____   _____

Key:
4 = Worked on his/her part and assisted others
3 = Worked on his/her part
2 = Worked on part less than half the time
1 = Interfered with the work of others
0 = No work

My name is _____ .   I give myself a _____

The other people in my group are: I give each person:

A. _____   _   _____

B. _____   _____

C. _____   _   _____

D. _____   _____

Student Ratings and Self Evaluation

Name: _____ Date: _____ Module: _____

Key:
Highest rating _____
Lowest rating _____

1. In the chart, rate each person in your group, including yourself.

	Names of Group Members				
Quality of Work					
Quantity of Work					
Cooperativeness					
Other Comments					

2. What went well in your investigation?

3. If you could repeat the investigation, how would you change it?

NOTES

NOTES

NOTES

NOTES